Technology and the Future of Europe

TECHNOLOGY AND THE FUTURE OF EUROPE:
Global Competition and the Environment in the 1990s

Edited by
Christopher Freeman, Margaret Sharp and William Walker

Pinter Publishers
London and New York

First published in Great Britain in 1991 by
Pinter Publishers Limited
25 Floral Street, London WC2E 9DS

British Library Cataloguing in Publication Data

CIP cataloguing record for this book is available
from the British Library
ISBN 0-86187 075 1

Library of Congress Cataloging-in-Publication Data

A CIP catalog record for this book is available
from the Library of Congress

For enquiries in North America please contact
PO Box 197, Irvington, NY 10533

Typeset by Koinonia Ltd, Bury, Lancashire
Printed and bound in Great Britain by Biddles Ltd.

Contents

IV. Energy, Environment and Regulation

V. The International Context and Policy Agenda

Acknowledgements

The production of this book was truly a collective effort involving most members of the Science Policy Research Unit and many people outside it.

In particular, we would like to thank six people who commented in detail on an earlier draft of the book: Giovanni Dosi, Detleff Eckert, Ove Granstrand, Aubrey Jones, Arthur Knight and Christopher Saunders. Their forthright comments and criticisms were invaluable in the process of editorial revision, although of course they are not responsible for the views and conclusions expressed.

The book went through many successive drafts and revisions and involved more than 30 different authors. It would have been quite impossible to organise this complex process without the extraordinary contribution made by Susan Lees. Her work went far beyond the purely secretarial and we are deeply grateful to her for her many-sided and extremely efficient efforts. She was assisted by many others but particularly we would like to thank Chris Gaterell and Janet French.

The authors of the chapters were a long-suffering group. Not only did they have to put up with an almost endless series of requests for changes, additions and deletions from several different editors and external referees, they were also faced with editorial policy changes as we gradually attempted to achieve greater integration between the chapters and as we ourselves went through a learning process. In some cases this involved elimination or total re-write of earlier chapter drafts. We were only too well aware that our colleagues had to meet our requests as one more extra demand over and above the normal work pressures, which are in any case intense. Our heartfelt thanks to them all and especially to those who suffered most from our editorial process.

It is impossible to write a book of this kind without resorting to a certain amount of technical jargon. We ourselves do not like jargon and we are only too well aware that it may sometimes be a barrier to the understanding of our readers, who we hope will come from all walks of life. We tried to cope with this problem in two ways: first by editorial simplification in the text and second (in order to avoid cumbersome detail), by providing a glossary of technical terms and acronyms at the front of the book. We are very grateful to Graham Thomas and to Richard Walker, who were responsible for the glossary.

Many SPRU colleagues assisted the editors in this work by commenting helpfully on one or more chapters. In particular, we would like to acknowledge the help of Graham Thomas in commenting on Part II of the book. There were two colleagues who contributed substantially to the editing of the book as a whole, even though they chose not to be named as editors: Marie Jahoda and Keith Pavitt. They know better than anyone how much we valued their consistent contribution. They took a full part in all our meetings and were editors in all but name.

This book was not the result of a specific research project or grant. It represents an attempt to bring together the results of many different projects on information technology, on biotechnology, on the world vehicle industry, on construction, on mechatronics and on more general topics such as technological competitiveness, energy, defence, environmental policy, industrial policy and technical innovation. This research has been supported by a variety of different sponsors and agencies but we would like especially to acknowledge the support of the Economic and Social Research Council (ESRC) – their support for two major research activities has been particularly valuable: The Designated Research Centre on Science, Technology and Energy Policy in British Economic Development and the Centre on Information and Communication Technology.

Finally, we would like to make a somewhat unusual acknowledgement: to our publishers. The relationship between editors, authors and publishers can often be conflict ridden. But we have been fortunate in finding a publisher who is uniquely qualified to handle our material and who has been consistently patient, constructive and efficient in coping with our difficulties and failings. Frances Pinter is herself a model of a business innovator and we are deeply grateful to her and to her colleagues for their help in the production of this book.

C.F., M.L.S. and W.B.W.

Glossary

Technical terms

Germany has been unified since the main bulk of this book was written and West Germany in the text relates to the FRG before unification. All statistics referring to FRG prior to October 1990 thus relate to former West Germany.

4GL	Fourth Generation Language. A computer programming language which automates many of the details of coding, thus allowing programmers to concentrate on purposes and structure of a program.
ACTV	Advanced Compatible Television. A European version of IDTV (see below) supplied by Thomson.
AD/Cycle	Applications Development Cycle (IBM). A standardised set of procedures, from design through construction to testing, for the development of computer application programs, using software engineering principles.
AI	Artificial Intelligence. Techniques for the development of machines which can exhibit behaviour (for example, learning) characteristic of intelligent beings.
ALARA	As Low As Reasonably Achievable.
ALI	Annual Limit of Intake.
AMT	Advanced Manufacturing Technology.
APSE	Ada programming Support Environment. Software which provides programming and project management tools to facilitate the production of (usually large) programs written in the Ada language.
ASIC	Application Specific Integrated Circuit. A design-intensive microchip with logic tailored for a specific purpose (and often for a specific user). To be contrasted with general-purpose ICs such as microprocessors or memory devices.
ATM	Automated Teller Machine. A local computer-controlled machine, connected to a central database, which allows customers to carry out common banking functions such as withdrawing cash or enquiring about account balances.
BATNEEC	Best Available Technology Not Entailing Excessive Cost.
BDAT	Best Demonstrated Available Technology.
BPEO	Best Practicable Environmental Option.
BPM	Best Practicable Means.
BSE	Bovine Spongiform Encephalopathy. A disease affecting the brains of cows and other animals. Commonly known as 'mad cow disease'.

CAD	Computer-Aided Design. Design work which is partially automated using a computer program.
CAE	Computer-Aided Engineering. The use of computers to assist in engineering processes.
CAIS	Common Ada Interface Standard. A military standard regulating the ways in which devices interact with computers via programs written in the Ada language.
CAM	Computer-Aided Manufacturing. Manufacturing processes at least partly controlled by computer programs.
CD	Compact Disc. A small disc containing information stored digitally by burning minute pits in the disc's surface with a laser beam.
CD-I	Interactive Compact Disc. A compact disc containing sound, image and textual material which can be retrieved in combinations which allow a user to explore a multimedia database.
CD-ROM	Compact Disc – Read-Only Memory. A compact disc containing data which can be retrieved using a computer.
CE	Consumer Electronics.
CFC	Chloro-Fluoro-Carbon. A chemical once widely used in refrigerators and aerosols but now being withdrawn by international agreement because of its effects on the ozone layer.
CHD	Coronary Heart Disease.
CIM	Computer-Integrated Manufacturing. The use of computers to automate and control an integrated set of manufacturing processes.
CMOS	Complementary Metal Oxide Semiconductor. A kind of microchip (to be contrasted with bipolar) suitable for high levels of circuit integration and simple structures, with low power consumption and low cost.
CNC	Computer Numerically Controlled Machine Tool.
CO_2	Carbon Dioxide.
CPU	Central Processing Unit. The 'calculating engine' of a computer. Now being made less central by the advent of parallel computing which uses multiple processing units to carry out different tasks at the same time.
CS	Commercial (net) Sales.
CTOS	Convergent Technologies Operating System. The proprietary operating system of computers made by the firm Convergent Technologies
CTV	Colour Television.
DAT	Digital Audio Tape. Cassette tape containing sounds stored digitally which provides accurate reproduction free of the 'tape-hiss' endemic to analogue-recorded tape.
DBS	Direct Broadcasting by Satellite. The transmission of television signals from satellites to end-user receiving dishes.
DNC	Direct/Distributed Numeric Control.
DRAM	Dynamic Random Access Memory. Microchips which allow information to be read and written in non-linear fashion. In

contrast to static RAM, the contents of these chips need constant refreshing from the computer's CPU and are lost when power is switched off.

ECL Emitter Coupled Logic. A transistor logic circuit, characterised by fast action and high power dissipation and used in high-speed computers. cf TTL.

EDI Electronic Data Interchange. Computer-to-computer transfer of documentation relating to transactions. Sometimes known as 'paperless trading'.

EDP Electronic Data Processing. Originally used to differentiate 'business' computing from that involving scientific and technical calculations.

EEPROM Electronically Erasable Programmable Read-Only Memory. A microchip which can have data or instructions programmed into it (and removed from it) electronically.

EFTPOS Electronic Funds Transfer at Point Of Sale. A means of debiting customers' bank accounts at (or near) the time when they buy goods or services, using computer communications from terminals located at the sales point.

EISA Extended Industry Standard Architecture. A format for connecting slotted-in circuit boards to the CPU of an IBM personal computer – promoted by makers of IBM-compatible PCs to counter IBM's adoption of its proprietary 'Micro Channel Architecture' for the PS/2 range of PCs introduced in 1986.

EP Environmental Protection.

EPROM Erasable Programmable Read-Only Memory. A shorthand for EEPROM (see above).

EPT Environmental Protection Technology.

ES Expert System. Computer program which aims to encode the knowledge of experts in order subsequently to guide non-experts through a decision-making process.

ESI Electricity Supply Industry.

FGD Flue Gas Desulphuration.

FMC Flexible Manufacturing Cell. A self-contained Flexible Manufacturing system built around a central machining centre and sold as a customised package, often specified by the manufacturer in response to the application described by the customer.

FMS Flexible Manufacturing System. An arrangement of computer-controlled machine tools and transport and handling systems which operates as an integrated system under the supervisory control of a larger computer. Such systems began to emerge in the 1980s and offer considerable improvements in the speed with which products can be made.

FMU Flexible Manufacturing Unit.

FTL Flexible Transfer Line. A section of a flexible manufacturing system.

GDP Gross Domestic Product. Gross National Product (see below) minus net investment income from foreign nations.

GNP	Gross National Product. The total value of all final goods and services produced by a nation in a given year.
GWe	Gigawatt electric. A thousand million watts.
HDTV	High-Definition Television. Signal and display format allowing an image to be represented by more dots within a given area than in the systems used in the 1980s, giving a sharper picture with more accurate colour representation.
HLW	High-Level Waste. Highly radioactive rubbish from nuclear reactors.
IBN	Integrated Broadband Network. High-capacity telecommunications network allowing transmission of sound, data and both still and moving images.
IC	Integrated Circuit. A complete electronic circuit with more than one functioning element manufactured on a single base. Most commonly incarnated as a microchip.
ICT	Information and Communication Technologies. Term used for the whole range of industries based on technologies which allow the digital storage, transmission and manipulation of information. Sometimes IT is used as a shorter form.
IDTV	Improved Definition Television. A set of technologies which aim to enhance the quality of existing television signals, for instance by using digital processing techniques.
IKBS	Intelligent Knowledge-Based Systems. Another term for 'expert systems' (see above).
ILW	Intermediate-Level Waste. Radioactive rubbish from nuclear reactors.
IPC	Integrated Pollution Control.
IPSS	International Packet Switched System.
IPSE	Integrated Project Support Environment. Software which provides programming and project management tools to facilitate the production of (usually large) programs.
IS	Information Systems.
ISDN	Integrated Services Digital Network. Telecommunications network which allows the transmission of both voice and data in digital form, even over ordinary copper telephone wires.
ISIC	International Standard Industrial Classification.
IT	Information Technology. See ICT. Sometimes used to distinguish the non-communications areas of ICT.
KJ	Kilo-Joule. A unit of work or energy.
LAN	Local Area Network. A data communications network within a single site.
LCD	Liquid Crystal Display. One method of displaying electronic signals. Uses cells which change their reflectivity when an electric field is applied to them.
LCP	Large Combustion Plant.
LLW	Low-Level Waste. Moderately radioactive rubbish from nuclear reactors.
LRTAP	Long-Range Transport of Air Pollutants.
MAC	Multiplexed Analogue Components. A family of formats for High-Definition Television, including D-MAC, HD-MAC,

	D2-MAC etc.
MCA	Micro Channel Architecture. See EISA above.
MDNS	Managed Data Network Service. A telecommunications service providing transport and management functions for the wide-area data network of another organisation.
Mips	Millions of Instructions Processed per Second. A measurement of the speed of a computer processor.
MNC	Multinational Company.
Modem	Modulator-Demodulator. A device which converts digital information from a computer into signals which can be sent down an analogue telephone line, and vice-versa.
MOS	Metal Oxide Semiconductor. A group of processes for the manufacture of semiconductors, in contrast to bipolar. Includes CMOS and NMOS.
MPC	Maximum Possible Concentration.
MPU	Multi-Processing Unit. The 'calculating engine' of a parallel computer which can harness several individual processors to perform simultaneous elements of a common task.
MS-DOS	Microsoft Disc Operating System. Operating system for IBM personal computers and compatibles.
NC	Numeric Control. Digital computer control (generally of a machine tool).
NFC	Nuclear Fuel Cycle.
NGO	Non-Governmental Organisation.
NIC	Newly-Industrialising Country.
NMOS	N-channel Metal Oxide Semiconductor. See MOS, CMOS.
NO_x	Oxides of Nitrogen.
NTS	New Telematic Services. Services which combine substantial elements of telecommunications and computing, for example, remote database access, electronic mail.
NTSC	National Television System Committee. A format for 525-line colour television used in the United States, Japan and elsewhere.
OOP	Object-Orientated Programming. Programming style (and associated languages) in which functions of a program are separated from each other and communicate using standard 'messages', thus allowing individual program modules to be recombined to form new programs.
ONP	Open Network Provision. European Commission programme designed to set the technical and commercial groundrules which will allow organisations to offer telecommunications-based services across the networks of public telecoms operators.
OS	Operating System. Computer software which controls the resources of a computer (for example, processors, memory, input/output devices) and manages their interaction.
OSI	Open Systems Interconnection. A set of rules and specifications which aims to allow computer-based application programs to communicate with similar applications across telecommunications networks, regardless of which firms supply the computers hosting the applications or the intervening networks.

PAL	Phase Alternation Line. A format for 625-line colour television, used in Britain, Germany and elsewhere.
PC	Personal Computer.
PCM	Plutonium-Coated Material.
PLC	Programmable Logic Controller.
PSTN	Public Switched Telephone Network. The 'ordinary' telephone network supplied by public operators, as distinct from an organisation's private network.
PTO	Public Telecommunications Operator. An organisation licensed by a government to provide a wide range of telecommunications services to individuals and organisations within a specific geographical area.
PTT	Post, Telegraph and Telephone. An organisation, which used to be common in Europe, with monopoly provision of both postal and telecommunications services within a country.
PVC	Polyvinyl Chloride. A type of plastic.
QUID	Quantitative Ingredient Declaration.
RAM	Random Access Memory. See DRAM
RCA	Revealed Comparative Advantage.
R & D	Research and Development.
RISC	Reduced Instruction Set Computer. Computer with a simple processor containing 'hardwiring' for only the most basic and commonly-used instructions (which can then be executed very quickly). Other instructions are carried out in software by combining the simpler ones.
RPI	Retail Price Index.
RTA	Revealed Technological Advantage.
SC	Semiconductor. A device, such as a transistor, based on a substance, such as silicon, that has an electrical conductivity which increases with temperature and is intermediate between a metal and an insulator.
SE	Software Engineering. The attempt to apply the principles and disciplines of engineering to the design, construction and maintenance of computer programs.
SECAM	Sequential Couleur à Memoire. A format for 625-line colour television, used in France, the Soviet Union and elsewhere.
SEM	Single European Market.
SITC	Standard International Trade Classification.
SME/SMF	Small or Medium-Sized Enterprise/Firm.
SO_2	Sulphur Dioxide.
SQL	Structured Query Language. A standardised language for the interrogation of computerised databases.
SRAM	Static Random Access Memory. See DRAM.
SST	State of Science and Technology.
T&D	Treatment and Disposal.
TIP	Technology Integration Project.
TORS	Tour Operator Reservation System. System for booking holidays, generally using computers and telecommunications and carried out from the premises of a travel agent.
TTL	Transistor-Transistor Logic. A transistor logic circuit using

	direct (bipolar) transistor to transistor coupling. Often used in small-to-medium-scale integrated circuits. Cf ECL.
TWh	Tera-Watt Hours. A measurement of electricity production and consumption. A million million watts produced/used in an hour.
VADS	Value-Added and Data Services. A term introduced by the United Kingdom Department of Trade and Industry in 1987 to amalgamate VANS and MDNS.
VANS	Value-Added Network Services. A British term, introduced in 1982 and abandoned in 1987, to denote telecommunications-based services offering more than just 'basic' telecommunication (and therefore subject to different regulation).
VCR	Video Cassette Recorder.
VDU	Visual Display Unit. A screen, generally using a cathode ray tube, to display output from a computer.
VER	Voluntary Export Restraint.
VLSI	Very Large-Scale Integration. The packing and integration of a very large number of circuit elements (typically at least hundreds of thousands) on to a single semiconductor.
WAN	Wide Area Network. A data communications network extending across several sites which are not necessarily adjacent.
X.25	A set of standards, set by the CCITT (see above), defining the interface between terminals and (generally public) data networks operating in packet-switched mode.
X.75	A set of standards for the interface between (generally national) packet-switched data networks on international circuits.
X.400	A set of message-handling standards for text communication (including electronic mail) conforming to OSI (see above) rules.

Institutions and programmes

ACARD	Advisory Council for Applied Research and Development (United Kingdom)
ACF	Advisory Committee on Foodstuffs (Commission of the European Communities)
AEG	Allgemeine Elektrizitäts Gesellschaft
ANSI	American National Standards Institute
ASEAN	Association of South-East Asian Nations
AT&T/ATT	American Telegraph and Telephone
BAP	Biotechnology Action Programme
BP	British Petroleum
BRIDGE	Biotechnology Research for Innovative Development and Growth in Europe
BRIE	Berkeley Round Table on International Economics
BRITE	Basic Research in Industrial Technologies
BT	British Telecom
CASE	Computer-Aided Software Engineering
CBI	Confederation of British Industries
CCC	Consumers' Consultative Committee

CCITT	International Telegraph and Telephone Consultative Committee
CDC	Control Data Corporation
CDEP	Central Directorate on Environmental Pollution (United Kingdom, Department of the Environment)
CDU	Central Democratic Union (Germany)
CEC	Commission of the European Communities
CECG	Consumers in the European Community Group
CEGB	Central Electricity Generating Board
CEPT	Conference of European Postal and Telecommunication Administrations
CERN	Conseil Européen pour la Recherche Nucléaire
CFE	Conventional Forces in Europe
CFP	Compagnie Française de Petrole
CGE	Compagnie Générale d'Electricité
CGS	Cap Gemini Sogeti
CIA	Central Intelligence Agency (United States)
COMMET	Community Programme in Education and Training for Technology
COST	European Cooperation in the field of Scientific and Technological Research
CSCE	Conference on Security and Cooperation in Europe
CSU	Christian Socialist Union (Germany)
DANW	Directorate of Air, Noise and Waste (United Kingdom, Department of the Environment)
DBP	Deutsche Bundespost
DEC	Digital Equipment Corporation
DG	CEC Directorate General
DIN	Deutsche Institut für Normung
DoE	United Kingdom, Department of the Environment
DSM	Dutch State Mining
DTI	United Kingdom, Department of Trade and Industry
EC	European Community
ECSA	European Computer Services Association
ECSC	European Coal and Steel Community
EdF	Electricité de France
EFTA	European Free Trade Association
EMI	Electrical and Musical Industries
ESA	European Space Agency
ESPRIT	European Strategic Programme for Research in Information Technology
ESRO	European Space Research Organisation
ESSI	European Systems and Software Initiative
ETSI	European Telecommunication Standards Institute
EUCLID	European Cooperation for the Longer Term in Defence
EUREKA	European Research Coordinating Agency
EUSIDIC	European Association of Information Services
FDP	Free Democratic Party (Germany)
FRG	Federal Republic of Germany
FT	Financial Times

GATT	General Agreement on Tariffs and Trade
GDR	German Democratic Republic
GEC	General Electric Company
GFAVo	Grossfeuerungsanlagenverordnung
GM	General Motors
GPO	General Post Office
HMIP	Her Majesty's Inspectorate of Pollution (United Kingdom)
IAPI	Industrial Air Pollution Inspectorate (United Kingdom)
IBM	International Business Machines
ICI	Imperial Chemical Industries
ICL	International Computers Limited
ICRP	International Commission on Radiation Protection
IDC	International Data Corporation
IEA	International Energy Agency
IEPG	Independent European Programme Group
IIASA	International Institute for Applied Systems Analysis
IMO	Information Market Observatory
INS	ICL-GEIS Co. International Network Service
INTUG	International Telecommunication User Group
IPCC	Intergovernmental Panel for Climate Change
ITU	International Telecommunications Union
JET	Joint European Torus
JRC	Joint Research Centre
KDD	Kakusai Denshin Denwa
LSE	London School of Economics
MAP	General Motors Manufacturing Automation Protocol
MBB	Messerschmitt Bolkow Blohm
MCC	Microelectronics and Computer Technology Corporation
MEP	Member of the European Parliament
MIT	Massachusetts Institute of Technology
MITI	Ministry of International Trade and Industry (Japan)
MMC	Monopolies and Mergers Commission (United Kingdom)
MTU	Motoren Turbo Union
NATO	North Atlantic Treaty Organisation
NCR	National Cash Registers
NEC	Nippon Electric Company
NRW	North Rhine Westfalia
OECD	Organisation for Economic Cooperation and Development
OPEC	Organisation of Petroleum-Exporting Countries
OSF	Open Software Foundation
OTA(F)	Office of Technology Assessment (and Forecast)
PACG	P.A. Consultancy Group
PACT	P.A. Computers and Telecommunications
RACE	Research in Advanced Communications for Europe
RTZ	Rio Tinto Zinc
SCF	Scientific Committee for Food
SCS	Scientific Control Systems
SEA	Single European Act
SEI	Software Engineering Institute
SEM	Single European Market

SERM	Software Engineering Research Centre
SPAG	Standards Promotion Action Group
SPRU	Science Policy Research Unit
STC	Standard Telephone and Cable
STP	Software Technology Programme
SWIFT	Society for Worldwide Interbank Financial Telecommunications
TOPIC	London Stock Exchange Price Information System
UI	Unix International
UK	United Kingdom
UBA	Umweltbundesamt (German Environmental Office)
UN	United Nations
UNCTAD	United Nations Commission on Trade and Development
UNECE	United Nations Economic Commission for Europe
UNEP	United Nations Environmental Programme
UNESCO	United Nations Education and Science Cooperation Organisation
UNIDO	United Nations Industrial Development Organisation
WEU	Western European Union

Preface

The authors of this book come from a variety of disciplines in the natural and social sciences and engineering, but they are united by a common interest in technical change. In the 25 years since the Science Policy Research Unit (SPRU) was set up in 1966 at the University of Sussex, it has tried to improve understanding of policies for science and technology through research on the R & D system, on innovations and their diffusion and on the complex interaction between technical, institutional and social change.

The European dimension of analysis has become increasingly important in this work. We have always interpreted our mission as an international one, in the sense of both its field of interest and involvement with a world-wide research community. Research centres similar to our own have become increasingly active in the 1980s in other countries in Europe, in North America, in Japan and in the Third World. This includes also our colleagues in the Centre for Business Research at Brighton Polytechnic. This book embodies some of their findings as well as our own and we hope that they will find it useful in their own work.

However, it is not primarily addressed to them but to a much wider public as a small contribution to the development of an informed European public opinion in relation to policies for science and technology. We hope it will be read by some of those who carry direct political responsibility for those policies both in institutions of the European Community (EC) and the various member countries. But more still, we hope it will be read by a wider public in Europe and elsewhere which is concerned with the long-term trends in the growth of the world economy, with the future of the global environment and with the role of technology policy in contributing to an improved quality of life in the 1990s and beyond.

The book is organised in five Parts. Part I presents an overview of the global context in which the EC has been evolving in the 1980s, especially in relation to the structure of industry and trade, science and technology and growth of the economy. An introductory chapter warns against some dangers for the EC in the 1990s: underestimating the strength of technological competition, complacency about the effects of trade liberalisation and the Single European Market; underestimation of the scale and intensity of environmental problems; reversion to protectionism in trade and technology.

Parts II and III of the book consider in detail the performance of the EC in a selection of industries and services. It was not possible to cover every sector although we recognise of course the vital importance of other industries, such as transport and steel, and of social services, such as health and education, which are not included. The sectors which were chosen were those with the fastest rate of growth in the world economy in the last quarter century and in which we had some research competence.

Part II deals entirely with information and communication technology (ICT) since this is the most pervasive and important area of technical change for the economy as a whole, as well as the focus for several very fast-growing industries. These include computers, electronic components, software, telecommunications, telematic services and electronic consumer goods, all of which are the subject of separate chapters. Part II shows that in none of these sectors does the performance of European firms match up to their Japanese and American competitors. A final chapter discusses the thorny problems of harmonisation of standards in information and communication technology (ICT).

Part III deals with a group of industries in which European firms have performed much better: chemicals, pharmaceuticals (including biotechnology), construction, machinery and automobiles. But in all of these there are also serious problems of pollution, regulation and industrial standards, as well as intensifying world competition.

Part IV brings together some of the key regulatory problems and policy issues which will increasingly confront the EC in the 1990s, since they cannot be simply or easily resolved by the magic wand of '1992'. They include industrial wastes and their disposal, the regulation of the energy industries, the associated problems of environmental policy and the quality of food.

Finally, in Part V, we review the international context in which the EC will evolve in the 1990s, especially the relationship between Eastern and Western Europe and the problems of the defence sector.

Within a diverse and large group of authors there are necessarily some differences of emphasis and opinion and we have not attempted to iron them out. Nor are there any simple solutions to all the complex social, economic and environmental problems which arise from our analysis. But in the concluding chapter we outline some policies for the Community which might help in tackling them, while enhancing European technological capacity and competitiveness.

C. Freeman, M. L. Sharp and W. B. Walker
SPRU, University of Sussex
January 1991

Part I
The background of European industry and technology

1 Introduction: beyond the single market

Christopher Freeman and Geoffrey Oldham

I Introduction

At the close of the 1980s, there were good reasons for some euphoria in Europe. Ideals of civil liberty and of parliamentary government, which were born and nurtured here appeared to be triumphant throughout the continent for the first time in history. The end of the Cold War offered the prospect of a substantial reduction in the burden of armaments and of closer cooperation between the two halves of what had for so long been a divided continent. Economic and political cooperation within Europe achieved a new impetus in the 1980s with the Single European Act and the removal of many remaining barriers to trade. As 1992 approached, the European Community (EC) achieved a new lease of life and its institutions visibly grew in self-confidence and in stature within the world community of nations. The economic achievements of the EC in the second half of the 1980s were good by world standards and expanding flows of trade with Eastern Europe seemed to offer exceptionally favourable prospects for the 1990s.

We share these feelings of satisfaction and hope. If in this book we raise some thorny problems, it is not because we wish this progress to stop but because we wish it to continue, and for the prosperity which many citizens of Europe now enjoy to be spread more widely throughout our Continent and beyond. Already in 1990 the acute economic crisis in Eastern Europe and the slowdown in the West European economies pointed to some of the difficulties ahead. There are four points in particular which we shall stress. The first is the problem of the continuing technological competitiveness of the EC. As we shall attempt to show in detail in Parts II and III of the book, the competitive strength of European industry is by no means assured and the technological challenge in the 1990s will be more severe than hitherto, particularly from Japan and other countries of eastern Asia.

Most people would now agree that foremost among the key technologies of the 1990s is ICT. As Part II unequivocally demonstrates, European firms were behind the world leaders in the 1980s in almost every sector of the ICT industries. But with ICT it is not just a question of computers or software. This is a technology which affects almost every function in every industry. Part III shows that even in areas of great European comparative strength, such as machinery, automobiles and chemicals, continued pre-eminence is by no means assured. The diffusion of information technology is transforming all kinds of industries and services and the rules of the game are no longer what they used to be. Moreover, successful adoption of ICT whether in products, processes, design, research, marketing or services is not merely or even mainly a question of introducing competitive markets. It involves many institutional changes in management organisation and behaviour, in industrial relations, in training, in quality control, in standards and in public policies.

The second point relates to the limits of what free markets and scale economies can deliver. The market is an extremely useful social mechanism, as the East Europeans have now re-discovered. But from Adam Smith onwards economists have pointed out that it cannot solve all economic problems and that the phenomenon of 'market failure' is by no means a rarity. As Keynes insisted:

> The World is *not* so governed from above that private and social interests always coincide. It is *not* so managed here below that in practice they coincide. It is *not* a correct deduction from the Principles of Economics that enlightened self-interest generally *is* enlightened; more often individuals seeking separately to promote their ends are too ignorant or too weak to attain even these. [Quoted in Eatwell, 1982]

The most rational investors and consumers in the 1890s or even the 1930s could never have anticipated the social and environmental consequences of the mass ownership of passenger cars in the 1990s. Nor today can we anticipate all the consequences of computerisation, biotechnology or nuclear power.

Since there is no way that the market can assess or anticipate many of the future social costs and benefits of technical change, some form of political debate and decision-making is unavoidable. The conflict of interest groups will always remain and so too will some degree of uncertainty. The best that can be hoped for is to disclose some of the possible and probable consequences of technical change and to stimulate a widespread public debate among those who may be affected for good or ill. Expert opinions and judgements can and should be sought, but there may often be disagreement among experts and in the end many judgements are ethical and political in nature.

Since they are themselves social institutions, markets function well only within a wider institutional regulatory framework. Among the many problems which cannot be simply left to the market to resolve are those of equity in income distribution, of preventing monopoly or regulating monopolies where they cannot be avoided, of adequate infra-structural investment and regional development, of education and training, of the business cycle and the money supply, of industrial relations and of social welfare. All of these are legitimate concerns of the institutions of the European Community and all affect technological competitiveness directly or indirectly. However, in this book we do not have the space or the competence to address them all. Our focus on technology is a narrower one but even within this perspective there are many issues which necessitate social and political initiatives which go beyond the market. These are the subject of Parts IV and V of the book.

The third point relates to the biggest single example of market failure – the environment. The scale and intensity of some environmental hazards are now so great that they endanger the very existence of the human and other animal species. The East European neglect of environmental problems did enormous damage and will take many years to repair. But they were not the only types of economy to cause environmental damage and if they are now replaced by rampant individualistic materialism and short-term market pricing, the dangers to the global and European environment will remain.

We share the view that the economic, social and political problems of the day must be approached in the context of a world-wide danger to the planet, even though their time-perspectives transcend those of billions of people who have more urgent immediate needs. Pollution on an unprecedented scale and the risks of the greenhouse effect follow from the energy-intensive and materials-intensive pattern of growth characteristic of the 20th century mass-production system, whether in East or West, North or South. While fertility in the EC is, with the exception of Ireland, below replacement rate, the populations of many poor countries are still increasing very rapidly. If they were simply to repeat the pattern of the rich world in their justified demand for higher living standards, the ecological consequences could be catastrophic. A new pattern of ecomonic development and sustainable growth is essential for all parts of the world economy.

Although the main part of this book is concerned with the technological competitiveness of the EC, in Part IV we also attempt to address this wider question, which we recognise as pre-eminent. The two goals are not necessarily in conflict. Technologies which improve the quality of life, which protect the environment and which are energy- or material-saving, may well become the cutting edge of world technological competition. But this will not occur automatically. It will require active regulatory policies for the environment and new policies for R & D. Now and still more from 1992 onwards, the EC will have a major voice in world affairs. For the EC to develop satisfactory policies for technology and the environment is therefore of paramount importance.

The final point concerns the international relationships of the EC. In Eastern Europe frustration with all the problems of an inefficient and totalitarian command economy has given rise to exaggerated expectations of what a market economy can achieve and how quickly. We argue in Chapter 21 that simply to open up Eastern Europe to the international market will not solve the problems of those countries but is more likely to relegate them to the Third World. The shortages which exist there, the disparity in living standards and the scale of environmental damage and indebtedness have created a situation similar to that at the end of World War II. A comparably imaginative and ambitious programme for recovery in Eastern Europe is needed now. As in the case of the United States and the Marshall Plan, the EC needs to show a comparable long-term patience and readiness to provide technical as well as financial assistance on a large scale.

As in our relationships with Eastern Europe, so in our relationships with the Third World the EC has a major role to play in the promotion of international cooperation in science and technology and of international technology transfer. The dangers of virulent nationalism are everywhere apparent; they are likely to intensify and degenerate into racism and fundamentalism of the worst kind if the growth of the rich nations widens the gaps between rich and poor. These gaps have indeed widened in the 1980s both internally *within* the EC, the United States and Japan and between these richer countries and many parts of the Third World in Africa, Asia and Latin

America. Only in East Asia and some of the oil-exporting countries has the growth of per capita incomes in the 1980s begun to close the North-South gap. This is a danger signal which must not be ignored, both for reasons of social equity and because, having at last got rid of the Cold War between East and West, it would be tragic to replace it by an even more dangerous and prolonged North-South conflict.

This introductory chapter now considers each of these four points and introduces those sections of the book in which they are more thoroughly discussed.

II Information and communication technology

There are, of course, many new technologies being developed and applied in the world economy. In almost every industry and service activity there is a continuous process of technical and organisational improvement and innovation. It is not easy to reduce this complex variety of discoveries, inventions and innovations to a coherent pattern. There are many innovations which are specific to one particular industry or even to one particular firm. Biotechnology is obviously of great and growing importance and is discussed in particular in Chapter 13. But in our own work in Sussex on technical change in European industry we have found, like other researchers, that one set of interrelated innovations is having particularly pervasive and far-reaching effects in every branch of the economy. This is the constellation of innovations associated with computer, telecommunication and micro-electronic technologies. Some people (Perez 1983) have used the expression 'change of techno-economic paradigm' to describe this wave of new technology and the related social changes. Other people prefer Schumpeter's terminology – in their view the computer revolution is the latest in the series of 'creative waves of destruction' or 'successive industrial revolutions' which he identified.

Whatever the terminology, and whatever the relationship to earlier waves of technical change, there is now rather widespread agreement that 'Information and Communication Technology' does indeed constitute an exceptionally important cluster of innovations, which are profoundly affecting technological competitiveness in virtually every branch of the world economy. The main characteristics of this contemporary wave of technical change and their relationship with organisational change may be summarised as follows with a reference to the relevant chapters of our book:

(i) A continuing dramatic improvement in large-scale integration of electronic circuits (VLSI) and the reduction in costs which this permits (see Chapter 5).

(ii) These successive generations of micro-electronics permit equally dramatic improvements in computer performance in storing, processing and communicating vast quantities of information (Chapter 6). New applications of computers lead to the development of new generations of software (Chapter 7).

(iii) The availability of cheap and reliable micro-processors and increasingly powerful small computers and super-computers facilitate the design and introduction of new generations of a very wide range of capital and consumer goods, ranging from word processors to VCRs, robots and machine tools (Chapters 10, 15 and 16).

(iv) Parallel developments in optical fibres and the convergence of computer technology with telecommunication technology have vastly reduced the cost of communicating data to all parts of the world. This permits the development of a new range of data banks, computer-based networks and information services, and new forms of cooperation between firms in different locations (Chapters 8 and 9).

(v) Similarly within the enterprise these developments in computer and communication technology facilitate the introduction of Computer-Aided Design (CAD), Flexible Manufacturing Systems (FMS) etc. (Chapter 15) and the integration of design, production, administration and marketing functions. They also facilitate new systems of production planning and inventory control. Information technology affects not only every industry and service but also every *function* within each industry and their interaction.

(vi) As a result of this combination of technical and organisational changes, there is scope both for a more varied product mix and for more rapid changes in both products and processes. This strengthens the requirement for new services and skills in design, software engineering and development, affecting both in-house activities and consultancy (see for example the case of the automobile industry in Chapter 16).

(vii) While the information technology revolution affects all functions in all industries, the requirement for new micro-electronic components, computers, electronic instruments, telecommunication equipment and electronic consumer goods has led to very rapid growth of the industries supplying these products. These new sectors have displaced the older industries such as automobiles, petrochemicals and metal products as the most dynamic fast-growing sectors of the economy. (See Figure 1.1, Chapter 2 and Part II).

(viii) Companies all over the world are confronted with the need to gain access to new sources of information and communication technology through a variety of cooperative arrangements combined with their own R & D. The pressures of world competition have driven banks to cooperate with computer companies and software companies and with each other; they have driven semiconductor companies to organise joint research and production, machine tool companies to collaborate with electronic companies, governments to organise joint research programmes involving industry and universities (see, for example, Arnold and Guy, 1986) and so on. Networking has become the name of the game.

It is because of these characteristics that we devote the whole of Part II to an examination of the relative performance of the European information technology industries and why, in Part III, when we discuss a wide variety of industrial sectors, we are particularly concerned with the diffusion of computer-based organisational and technical change (see especially Chapter 15).

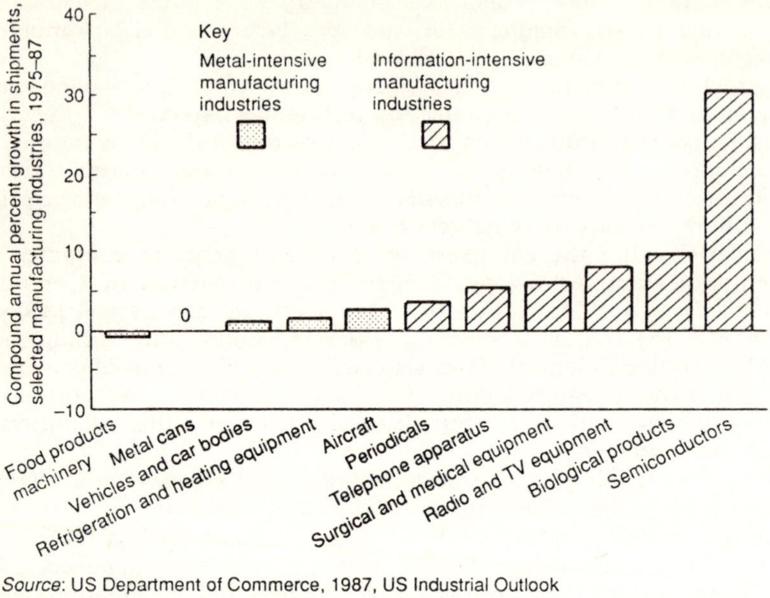

Figure 1.1 Variable growth rates among American manufacturing industries

III Institutional change and the diffusion of ICT

Some large Japanese firms have been particularly innovative in reorganising production systems, in developing new types of sub-contractor networks, in their approach to training and in the integration of design, development and marketing (see for example, Chapter 16). But new patterns of industrial organisation are not, of course, confined to Japan. There are certainly many examples of European companies which have embarked on revolutionary changes in their structures and procedures as well as in their equipment and products.

Nor should the Japanese pattern be regarded as an ideal model of social organisation. It has many unattractive features for Europeans. The political corruption at the highest levels and the social strains of conformism have become increasingly apparent in the 1980s. It is neither possible nor desirable to transpose this model simplistically to the different circumstances of Europe. In several respects German and Swedish practice offer a more attractive way of conducting industrial relations for most European countries. Neither the workaholic lifestyle of Japanese business executives nor the American 'hire and fire' philosophy of employment are congenial to most Europeans. The European iconoclastic tradition in basic science has proved a fertile source of original and creative ideas. We can and should develop our own distinctive ways of managing the assimilation, shaping and diffusion of new technologies appropriate to European circumstances and social goals. But we also have a great deal to learn from Japanese and American experience; technological rivalry with Japanese and American industry can bring Europe great benefits.

European firms in ICT are significantly weaker than their Japanese and American competitors (See Table 1.1), and we cannot simply assume that the enlargement of the European market and the removal of remaining barriers to trade and competition will in themselves be sufficient to ensure improvements in the electronics, telecommunications, automobile and every other industry. As the MIT Commission on Industrial Productivity cogently points out in relation to the United States (which hitherto has had the largest 'single market'):

> The organizational patterns and attitudes that we believe are at the root of the productivity problem are notoriously hard to change, even though the need for change is recognised ... Change if it is to occur will have to take place on a broad front involving firms, government, educational institutions and organised labor. There is consequently every reason to open public discussion of these systemic rigidities and of ways to remedy them. [Dertouzos *et al.*, 1989:45].

Scale economies are, of course, important, and firms all over the world have benefited from the reduction of trade barriers since World War II. But in many cases it is essential to operate on the global market and sometimes Japanese and American firms will benefit more from '1992' than European firms. In any case, with the new technologies it is not the static economies of plant size which are the most important. In some industries, such as steel, there is actually a tendency to reduction in optimal size of plant. The scale economies which are of the greatest significance now are of two sorts:

(i) Economies of scale and economies of scope in research, design, development and world-wide marketing networks. In many branches of manufacturing and services these are now more important than plant production scale economies. The costs of designing, developing and testing new generations of telephone exchange, of VLSI, of super-computers, of aircraft, of drugs, of automobiles and of many other products and services are often such that commercial viability is only possible if they are spread over very large markets.

(ii) Dynamic economies of scale, which are associated with the accumulation of knowledge within a firm from R & D, design, production and marketing. Much research at SPRU has demonstrated the fundamental importance of these dynamic learning processes (Pavitt, 1984; 1986; Bell *et al.*, 1982). They are now increasingly associated with the types of organisational change identified in the MIT report (Dertouzos *et al.*, 1989).

The benefits which can be derived from these scale economies are certainly not the automatic and inevitable result of operating in a larger home market, but are rather the product of superior *organisation* of research, design, development, production and marketing in the firm and supportive infra-structures and policies.

The changes which are needed go far beyond a simple change of technology. They are summed up in Table 1.2, which contrasts the old 'Fordist' mass production approach with the new demands of ICT. Consequently '1992' needs to be seen not simply as a larger market, still less as an opportunity to keep others out of this market, but primarily as a process of social, structural and organisational change within the European countries, enabling them to realise the enormous potential of the new techno-economic para-

Table 1.1 How European electronic companies rank in the World, 1989, by total sales (US$bn) and sector.

Computer Company	Country	Sales $bn	Telecommunications (1988) Company	Country	Sales $bn	Semiconductors* Company	Country	Sales $bn
1 IBM	USA	62.71	AT&T	USA	10.24	NEC	Japan	5.01
2 DEC	USA	12.74	Alcatel	France	9.41	Toshiba	Japan	4.93
3 Fujitsu	Japan	11.88	Siemens	FRG	6.81	Hitachi	Japan	3.97
4 Unisys	USA	10.10	NEC	Japan	5.82	Motorola	USA	3.32
5 NEC	Japan	10.02	Northern Telecom	Canada	5.40	Fujitsu	Japan	2.96
6 Hitachi	Japan	9.84	Ericsson	Sweden	5.04	Texas Instruments	USA	2.79
7 Hewlett-Packard	USA	8.10	Motorola	USA	3.02	Mitsubishi	Japan	2.58
8 Olivetti	Italy	7.26	Philips	Neths.	2.80	Intel	USA	2.43
9 Bull	France	6.47	Fujitsu	Japan	2.49	Matsushita	Japan	1.88
10 NCR	USA	5.96	Bosch	W.Germany	2.16	Philips	Neths.	1.72

* This table excludes intra-company captive market sales. IBM is certainly a major manufacturer of integrated circuits but does not sell on the open market. See Part II for more recent changes in company rankings following mergers such as the Siemens takeover of Nixdorf (computers) and of Plessey (telecomms).

Source: G. de Jonquieres, 'Shadows over the Sunrise Sector', *Financial Times*, 25 July 1990

Table 1.2 Change of techno-economic paradigm

'Fordist' Old	ICT New
Energy-intensive	Information-intensive
Standardised	Customised
Rather stable product mix	Rapid changes in product mix
Dedicated plant and equipment	Flexible production systems
Automation	Systemation
Single firm	Networks
Hierarchical management structures	Flat horizontal management structures
Departmental	Integrated
Product with service	Service with products
Centralisation	Distributed intelligence
Specialized skills	Multi-skilling
Minimal training requirements	Continuous training and re-training
Adversarial industrial relations; Collective agreements codify provisional armistices	Moves towards long-term consultative and participative industrial relations
Government control and planning and sometimes ownership	Government information, facilitating, co-ordination, regulation and 'Vision'
'Full employment' (Emphasis on full-time employment for adult (16-65) male workers)	'Active Society' (More flexible hours and involvement of part-time workers and post-retirement people)

Source: Based on Perez, 1990; Boyer, 1989

digm. Active policies for R & D, for education and training, for consultancy, for standards and for industry are all essential as well as social and macro-economic policies. We return to these policy issues in Chapter 23.

IV Economic growth and the global environment

The competitive advantage of the EC is, however, far from being the only consideration in technology policy. Progress in disarmament, overcoming Third World poverty and overcoming global environmental hazards must take priority over competitive market pressures for obvious reasons. The common interest in survival is more important than the sectional advantage of firms, nations or groups of nations. Technological competition can only be justified as a means of achieving wider and more important social, cultural and environmental objectives.

In the 1970s SPRU was involved in publishing two earlier contributions to the world-wide debate which was triggered by the MIT model of *Limits to Growth* (Meadows *et al.,* 1972). On the basis of a computer simulation model the MIT report suggested that unless economic growth slowed down to zero the world economy would collapse in the middle of the 21st Century, either because of materials shortage, world-wide pollution or a classic Malthusian population problem leading to food shortage and famine.

In our books (Cole *et al.*, 1973; Freeman and Jahoda, 1978), we criticised some of the basic assumptions and methods used in the MIT models. In any computer modelling, assumptions determine the outcome ('Malthus in, Malthus out'). We argued in particular that technical and social changes were either neglected or under-represented in the MIT equations and models and that if they were included, then a pattern of sustainable growth or at least of a more prolonged period of world economic growth could be envisaged in the 2lst century, during which the Third World countries could reasonably hope to catch up in living standards.

For example, we argued that the materials-intensive pattern of growth characteristic of the American economy in the l9th and 20th centuries would not necessarily be the standard for other countries in the next century. If materials became scarce and expensive, or if their use involved unacceptable pollution hazards, we argued that technical changes could (and very likely would) lead to economising in and substitution between materials, to re-cycling, to innovation in processes and anti-pollution technologies which would lead to a different pattern of growth. That this was not too far removed from the realms of the possible is suggested by Figures 1.2 and 1.3, which indicate that in the leading industrial countries there was indeed a significant break in the trend of materials (particularly metals) consumption during the l970s. It is of course true that a significant part of that reduction was due to substitution with new materials, some of which are equally or more polluting. But a major part of the reduction is related to the introduction of ICT, which can vastly reduce the number and size of components in many products and facilitate designs and process control systems which reduce rejects and waste.

Similar considerations apply to food production and population growth. We argued that the main problems were those of diffusion of known technologies and that agricultural production was constrained more by social policies and attitudes than by technical possibilities. Hunger was and is a terrible social problem but it is more the result of unequal income distribution and of political conflicts than of technical limits to food production. Moreover, higher living standards, associated with economic growth, are more conducive to a slowdown in population growth than the persistence of mass poverty in the Third World.

We did not use the expressions 'sustainable growth' or 'sustainable development' in our books in the l970s. These expressions entered the debate somewhat later, but they convey the sense of what we were suggesting in our comments on the *'composition'* of growth and the distribution of the fruits of growth (Cole *et al.,* 1973:l0).

Does this mean then that we do not take seriously the argument about global environmental hazards and in particular the greenhouse effect? Nothing could be further from the truth. As we said at the time:

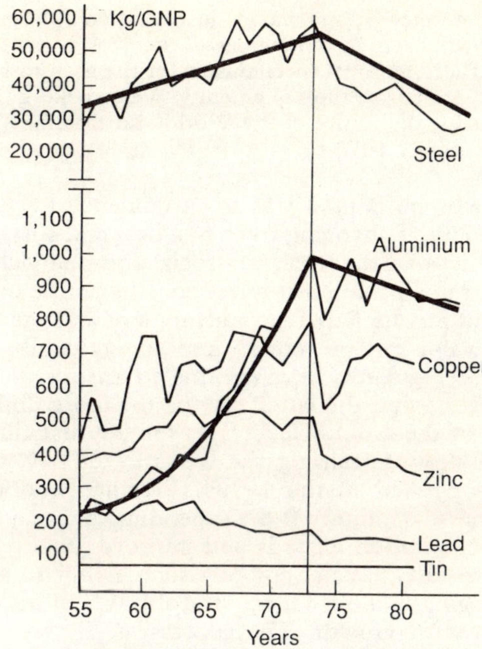

Source: Cohendet et al, 1988

Figure 1.2 Metals consumption per unit of GNP in Japan, 1955–1984

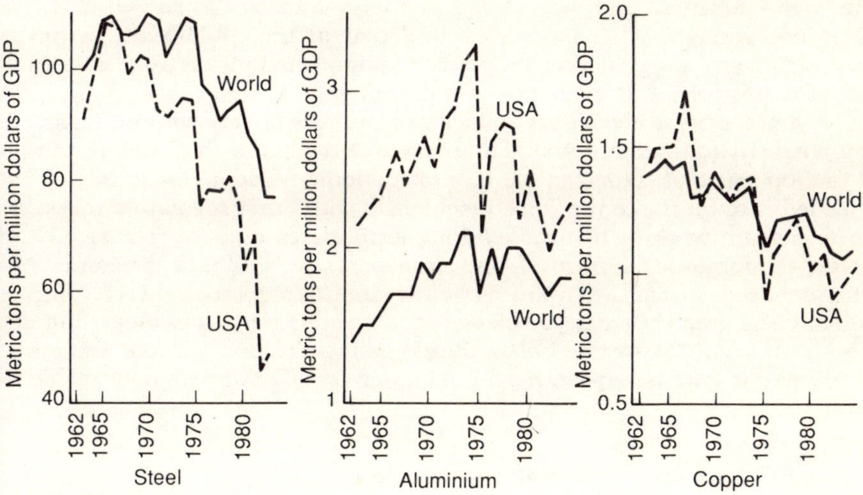

Source: Sousa, 1988

Figure 1.3 Trends in the intensity of metal usage

Our critique must not be taken in any way as justifying complacency about such issues as population growth rates in some areas of the World, the development of satisfactory national and international mechanisms for the monitoring and prevention of pollution hazards or conservation of amenity. We *do* believe that these issues are extremely important for the future of the World and that they are urgent and of global concern [Cole *et al.*, 1973:6].

Where we could perhaps be faulted was for a tendency towards over-optimism about the possibility of changing track to a new and sustainable pattern of growth or in our belief that the necessary social and political changes were feasible. This was and remains an open question. It is both a question of new priorities for public and private R & D to nurture a new range of possibilities in such areas as renewable energy sources and energy conservation devices *and* a question of new regulatory mechanisms to ensure their world-wide diffusion. Part IV suggests that the latter may be the more difficult problem. However, the ending of the Cold War creates for the first time for a generation a serious possibility of realising the scenario for new public R & D priorities which we suggested in the 1970s (Freeman and Jahoda, 1978: Ch.7) – a shift from massive military R & D spending to a major public effort to combat a variety of pollution hazards and improve quality of life. The R & D programmes of the EC already embody such a shift to a small extent.

Even more encouraging is the evidence in Table 1.3 of a shift away from energy-wasteful patterns of growth. The successive oil crises in 1973 and 1979 did have some lasting effect, and Saddam Hussein may unwittingly have given the OECD countries a further push in the direction which they must all travel. But the changes will have to be far more rapid in the next 20 years than those which have already occurred. A global scenario of zero growth in energy production and consumption over the next 20 years would be possible only if there is a significant *absolute reduction* in consumption in the EC and other rich countries – the *relative* fall of energy-intensity as shown in Table 1.3 is not enough. The experience of Japan in the 1973-1988 period is encouraging and suggests that the shift to an information-intensive pattern of growth is possible. But even this is not enough.

The market-price mechanism allocates resources largely without regard to long-term environmental hazards and resource depletion. It therefore cannot do the job unaided. Nor can we rely on periodic Middle East crises. What is needed is to reinforce the price mechanism with other regulatory measures and to make it work by a sustained long-term increase in the relative cost of energy to domestic, industrial and transport users. Transport and construction need special attention as shown in Chapters 14 and 16. This is probably the greatest single challenge to EC industrial, technology and environmental policies in the 1990s. But it cannot be met without the cooperation of our East European neighbours, our OECD partners and the Third World.

V The EC and international cooperation

Although we are concerned about the technological competitiveness of the EC, this book is not written from the standpoint of European nationalism, still less of British nationalism. Neither 'Fortress Europe' nor 'Fortress

Table 1.3 Final energy intensity* of GDP 1973–1988 (Tonnes per 1,000 ECU)

Country	1973	1985	1988
Belgium	0.641	0.495	0.486
Denmark	0.456	0.343	0.312
FRGermany	0.520	0.429	0.403
Greece	0.512	0.567	0.601
Spain	NA	0.430	0.416
France	0.438	0.376	0.362
Ireland	0.591	0.567	0.567
Italy	0.457	0.371	0.364
Luxemburg	1.511	0.829	0.744
Netherlands	0.594	0.478	0.474
Portugal	NA	0.543	0.598
United Kingdom	0.617	0.482	0.447
EC	0.522	0.426	0.407

*Energy Intensity – Gross Inland Consumption divided by GDP at 1980 prices and 1980 exchange rates

Source: Eurostat (CEC, Panorama of Industry 1990)

America' nor 'Fortress Japan' would be comfortable places to live in. Above all in Europe, where nationalistic rivalries gave rise to two world wars in this century, we should be aware of the dangers of aggravating international rivalry whether in commerce or in weapons technology. International cooperation both within Europe and outside Europe is the only sane way forward in a world of multi-polarity. In the 1990s world prosperity and world security will depend on a continuing process of international negotiation in which neither the EC, nor the United States, nor Japan, nor the Soviet Union, nor China can be a 'hegemonic' power and in which Third World countries must increasingly take part. Similarly, as Mary Kaldor argues in her chapter on Eastern Europe, the problems of East-West relations in Europe can only be satisfactorily handled within this wider framework of international cooperation.

Consequently, when we discuss the relative strengths and weaknesses of the EC in technological competition with other countries, this is always within a context which assumes a common interest in strengthening a global international framework of regulation – of 'rules of the game' – for the environment, for trade, investment and technology flows. As Europeans know from the experience of '1992' and many earlier negotiations from the inception of the Community, it is by no means easy to reach agreements for such common rules and the machinery to ensure their observance, even among countries which are geographically and culturally fairly close. The examples discussed in Chapters 11, 18, 19 and 20 show how complex the problems can be. It is even more difficult, but even more important, at the level of the United Nations and other global international institutions.

World leadership in basic science has shifted to the United States since World War II, in part because of the migration of European scientists to

North America. More recently, leadership in some areas of technology has given a competitive edge to many Japanese products in world trade. Other Asian countries, such as South Korea, are also now serious competitors in some areas of advanced technology.

None of these developments is unwelcome. On the contrary, Europe has everything to gain from the greater participation of other countries in advancing science and technology, including those at present (or until recently) classified as 'underdeveloped' or 'Third World'. Neither the problems of poverty, nor the global environmental problems which threaten us all, can be resolved without such active and increasing involvement. Science and technology are quintessentially international activities from which all can benefit and in which all can participate. In so far as there is competition, it is a race in which all can have prizes.

In debating policies for science and technology it is often argued that the nation-state is no longer the most appropriate level of analysis (Ohmae, 1990). The results of basic science are universally available through the world scientific literature and international conferences; multinational corporations increasingly conduct their activities on a global basis including their R & D; international institutions take increasing responsibility and national 'sovereignty' is diminished everywhere. The development of the EC itself demonstrates the limitations of the nation-state.

These are powerful arguments, especially in the context of the transition marked by '1992'. However, there is still a justification for examining national systems of innovation both within Europe and outside Europe. It is still largely within the national arena that the most vital decisions are taken about the allocation of resources to scientific and technological activities. National and regional networks are still important even alongside the growth of international networks. The research of Lundvall (1988) and his colleagues has shown the critical importance of continuing user-producer relationships in the domestic market for the evolution of new and improved products and processes and their argument has been powerfully reinforced by the statistical analysis of Patel and Pavitt (1990 and Chapter 3).

The concept of a 'national system of innovation' is much wider than a network of R & D and other scientific and technical institutions; it also involves the production system and a continuous process of learning by doing. For a whole variety of cultural, linguistic, historical, economic, geographical and institutional reasons, a great deal of this learning by doing and by using is conditioned by national and local circumstances. Much research on innovation networks has conclusively shown that regional (sub-national) factors such as geographical and cultural proximity are particularly important in advanced technological development. Despite the growth of multinational corporations and the international character of science and of markets, there are still considerable differences between the behaviour of Japanese corporations, American corporations and European corporations with respect to the management of innovations and the educational background of their labour force in many of their operations. Indeed, it is precisely because of these differences that comparisons between countries and regions can be such a fruitful method of analysis and of learning.

The differences *within* Europe are also still significant. It is obvious that there are major differences between British and German institutions, or between Greek and Danish institutions, which have substantial effects on the way in which technical change occurs. Some of these differences may diminish or even disappear as a result of the Single Market and further steps towards European integration. As Patel and Pavitt (1990) show, there are already some signs of this in the case of Belgium and the Netherlands. But even the most ardent advocate of European Union would accept that this will be a fairly prolonged process and that much diversity will remain at the national level. Indeed, many argue that some sorts of diversity and pluralism could be an important source of strength for Europe.

Even if EC institutions come to predominate over national and regional governments in wide areas, including some policies for science and technology in the course of the 1990s, it is very unlikely that nation-states will cease to be important. Again and again the continuing emotional and political strength of nationalism has been demonstrated in the 1980s in every corner of the globe, not least in such federal states as the Soviet Union and Yugoslavia, but also in the Middle East, in Latin America, in Africa, and in such unhappy divided states as Cyprus, Sri Lanka and Ireland. Even in North America, both in the United States and Canada, a conspicuous feature of the 1980s was the growing role of state and provincial governments in policies for innovation and for education.

It is, therefore, a reasonable proposition that the national and regional levels of decision-making and analysis will continue to be important for a long time to come. In the transition to global and supra-national forms of decision-making for some activities, it is essential to preserve national, regional and local autonomy in many other areas. Many EC initiatives in R & D and innovation take the form of coordination and networking, rather than total control from the centre.

There have been other attempts at integration, such as the Andean Pact and the Association of South-East Asian Nations (ASEAN), which put considerable stress on joint scientific and technical activities but which did not have great success or lead to supra-national integration. However, despite these reasons for caution and without underestimating the continued strength of divisive forms of nationalism, we do believe that the EC experiment is likely to prove more enduring and successful and that the European dimension of scientific, technical and industrial policies is already of great and growing importance. Indeed, as Chapter 4 by Margaret Sharp so clearly demonstrates, the recognition of the need for some common policies for industry and technology was one of the main sources for the renewed impetus towards European integration in 1992. Therefore, although we make use of many intra-EC comparisons, we also make many comparisons between the EC as a whole and the principal industrial competitors of the EC: the United States and Japan. This wider background of global developments is the main focus of the three chapters which follow in Part I.

2 Structural change and patterns of production and trade

Christopher Saunders, Mark Matthews and Pari Patel

I Introduction

The first chapter suggested that the European economies, alongside those of their main industrial competitors, are currently in the throes of major change, with the advent of information technology and the shift from Fordist-based industries of mass production to new information-based production systems. We shall be looking in more detail later in this book at how these changes are coming into effect and at their impact on product, process and market in specific sectors. In this chapter we explore the impact to date of these changes on the overall structure of the economy and how far there are discernible trends affecting both the overall pattern of production and trade and the distribution of activities between different members of the Community and the non-EC European countries. The following chapter focuses specifically on changing patterns and capabilities among technological activities and highlights inter-country and inter-sectoral differences.

II The EC within the global context

The contribution of manufacturing industry to West European GDP was 28 per cent in 1990, much the same level as in 1965. Faced with the severe global environmental problems associated with industrial pollution and above all with the 'greenhouse effect', some ecologists have advocated wholesale de-industrialisation as the solution for the rich countries and a non-industrial path of development for the poor. Even if this extreme path is not pursued it is not a wholly implausible scenario for the world economy that over a long period there could be a shift of manufacturing employment and output towards the Third World, especially Asia and Latin America.

Certainly the present international division of labour is not sacrosanct or permanent, but in our view it is not realistic to envisage a European withdrawal from manufacturing over the next two or three decades. It is more likely (and far more desirable) that some of the newly industrialising economies will catch up with the OECD countries and will become efficient producers in all three sectors – agriculture, manufacturing and services – while the OECD countries will also continue as major producers in all three sectors. Experience suggests that such developments, far from limiting opportunities for trade, lead to a rapid growth of intra-sectoral trade between

Table 2.1(a) Inflows and Outflows of Foreign Direct Investment 1986-88

	Average annual outflow		Average annual inflow	
	$bn	%	$bn	%
USA	27.2	22	39.4	32
EC	42.6	35	26.3	21
Japan	22.7	18	0.3	0.2
Rest of world	30.9	25	57.4	46
Total	123.4	100	123.4	100

Table 2.1(b) Projected 1995 matrix of Foreign Direct Investment Flows ($bn)

To/from	USA	EC	Japan	Rest of world	Inward Total
USA	—	40.0	15.0	4.2	59.2
EC	13.0	25.0	14.0	5.2	57.2
Japan	1.5	0.8	—	0.7	3.0
Rest of world	23.8	24.2	31.4	30.0	109.4
Outward total	38.3	90.0	60.4	40.1	228.8

Source: Julius, 1990: Tables 2.5, 3.8 and 3.9

countries with broadly similar industrial structures and a highly specific pattern of industrial specialisation *within* sectors, often indeed between subsidiaries of the same multinational company.

Indeed one of the features of the post-war world has been a resurgence in capital flows between nations, much of it in terms of investment by multinational companies and partly, it must be admitted, stimulated by trade barriers or fear of the imposition of trade barriers of one sort or another. Much Japanese investment in the EC in the 1980s has come because voluntary export restraints (VERs) inhibit direct exports (see for example Chapter 10). The first wave of this overseas investment was from the United States to Europe during the 1950s and 1960s. In the 1970s it spread more widely around the globe, largely in search of oil reserves. In the 1980s, British and Japanese firms expanded strongly overseas, mostly into the United States; German and French firms have been less active but more diversified in their targets. The indications are that this more diverse pattern of flows will continue through the 1990s. Tables 2.1(a) and 2.1(b) illustrate this global pattern of investment flows; Table 2.1(a) showing the importance to date of the United States and EC both as recipients and sources of foreign direct investment, and Tables 2.1(b) projecting the flows of investment between the main trading blocs in 1995. Such projections are of course only 'guesstimates' based on recent trends.

Nevertheless, these projections are an ample illustration of the likely growth of industrialisation on a global scale in the next decade. The term 'globalisation' has two interpretations: first, it can mean the expansion of industries and markets world wide as an increasing number of countries join the ranks of the industrialised; second, the more specific expansion of marketing and production activities into new markets around the world by

Table 2.2 European market economies: shares of value added and employment by sector (%)

	Past values			'Base-line' scenario
	1965	1973	1985	2000
Value added				
Agriculture	6.6	5.4	5.2	4.5
Mining and quarrying	2.0	1.4	1.9	1.5
Manufacturing	28.0	30.6	28.2	28.8
Electricity, gas and water	2.2	2.8	2.9	3.0
Construction	9.0	8.0	6.1	5.9
Trade	13.5	13.5	13.2	13.8
Transport and communication	6.1	6.4	6.9	7.8
Finance	12.5	13.4	15.6	15.6
Collective services	20.1	18.6	19.9	19.1
Employment				
Agriculture	16.2	10.9	7.6	5.2
Mining and quarrying	1.5	0.9	0.8	0.6
Manufacturing	30.1	29.0	26.3	22.1
Electricity, gas and water	0.9	0.9	0.9	0.8
Construction	8.7	8.4	7.2	6.5
Trade	14.3	15.5	17.0	20.2
Transport and communication	5.7	6.0	5.8	5.1
Finance	3.5	5.2	7.0	8.4
Collective services	19.0	23.3	27.3	31.2

Source: ECE Secretariat data base and estimates, Geneva, ECE, 1989

established large firms who are then said to be 'globalising' their activities. Both interpretations impinge on our theme. How far is technology changing industrial structures and forcing previously national firms into world wide activities? If so, what implications does this have for industrial policy? And how far does the rapid industrialisation of previously less developed countries and the pollution problems this will create, force the pace of technology change towards energy- and materials-effiecient technologies in production, transport and distribution? We return to these issues in Parts IV and V of the book.

III Manufacturing and its changing role in production

Table 2.2 shows European value added (GDP) broken down by the main sectors of activity with figures covering the period 1965-1985 and a projection through to the year 2000. There are two notable features about this table: first, the relative stability in output shares between the main sectors of the economy. Looking back over the last two decades, there have been only minor changes in output shares – some increases in the case of energy, transport and communications, a somewhat bigger increase in the finance sector, and decreases in agriculture, construction and manufacturing; second, the fact that these changes are generally more marked in terms of employment than in output. In employment terms, the share of manufacturing dropped between 1965 and 1985 from 30 per cent to 26.3 per cent, that of agriculture from 16.2 per cent to 7.6 per cent, while that of finance rose from 3.5 per cent to 7 per cent of the workforce. In the United States, often seen as a precursor to trends in Europe, the share of manufacturing in employment is now only 17 per cent, topped by trade (mainly retailing) at 24 per cent, while finance stands at nearly 12 per cent.

It is sometimes suggested that services can substitute for manufacturing and that a decline of manufacturing in some or all European countries can be looked upon with equanimity so long as services take up the slack in employment, output and trade. While it does seem probable from all the work which has been done on technical change and employment in the OECD countries in the 1980s that the slow secular decline of manufacturing employment will continue, it is a fallacy to jump from an analysis of this declining share of employment to the conclusion that it no longer matters to a national economy whether there is a manufacturing sector or not, how competitive it is or how rapidly it declines.

As already noted, the share of manufacturing in value added has declined less than the share in employment (Table 2.2). There has also been an increased blurring of boundaries between manufacturing and service activities. Many firms, for example, sub-contract the maintenance and servicing of machines to specialist (service sector) firms whereas in the past the maintenance engineers would have been part of the (manufacturing) workforce. The increased use of computers in manufacturing processes exacerbates such trends, but it also illustrates how closely knit together manufacturing and services have become. Many service activities are now ancillary to manufacturing and depend for their livelihood on a thriving manufacturing sector. It is often hard to say with firms in the information technology area whether they should be classified as 'manufacturing' or 'services'.

For some time we have seen very small economies – such as Monte Carlo – concentrating almost entirely on service activities. Given the interdependence of member states within the EC, logically it would be feasible for any one economy within the EC to specialise in service activities – say software – which were ancillary to manufacturing activities elsewhere in the EC. We do not think, however, that it is sensible or feasible for the EC as a whole to neglect manufacturing activities. On the contrary, for several reasons, a healthy manufacturing sector is essential for the EC and its major member countries over the next few decades at least, even though individual regions within the EC may indeed derive great benefits from specialisation in tourism, financial services or other service industries.

First, as already mentioned, there is an obvious and direct dependence of many, if not most, service activities such as wholesale and retail distribution, transport and financial services on the flow of manufacturing production and vice-versa. Those services which are provided for other productive firms rather than services direct to consumers are sometimes called 'producer services'. If they are added to manufacturing, construction and energy then it is clear that their combined contribution to European GDP and to total employment is nearer one half than one quarter (Table 2.2). Manufacturing thus acts as a dynamo within the core economy, and we depend upon this dynamo to create the wealth to support other services such as health, education and administration on which our quality of life depends.

Second, and most important of all from the standpoint of the analysis in this book, the interdependence of manufacturing and services is crucial in relation to technical change. Numerous studies of innovation have shown that the design and development of new products, processes and services depend heavily on the availability of a supporting network of suppliers and information sources (Saxenian, 1990; Lundvall, 1988).

Finally, manufacturing is crucial, still, to competitiveness in international trade. While traded services (the so-called 'invisibles' such as insurance, consultancy, software or tourism) are certainly of growing importance, goods still account for about two-thirds of combined world trade in goods and services.

IV Changing patterns of production and trade – the approach adopted

This chapter now attempts to provide a broad-brush picture of the changing patterns of world production and trade within which European integration is taking place; in this respect it provides the background for the more detailed examination of the various technologies and industries made in Parts II and III. Before outlining the salient features of these changes and attempting to draw some basic conclusions as regards their impact upon, and implications for, European technological activities, it is worth making explicit the basic theoretical focus underlying this investigation.

The fundamental feature of the approach adopted here is that we assume patterns of trade, and therefore revealed patterns of comparative advantage, reflect underlying technological advantage which, in turn, reflects *ex post facto* social and economic choices made in earlier time periods. For their part, these choices are often embodied in institutional and corporate structures and contribute to distinctive national and corporate 'cultures'. For example, the apprenticeship system established in Germany in the 19th century underlies the present-day skill training system in the former West Germany and helps to explain that country's strength in the engineering industries.

The importance of these factors should not be underestimated. Many economists tend to see competition mainly in price terms. What they neglect is that fairly small but persistent differences in product characteristics – even something as seemingly unimportant as the quality of the lubrication system – can lead to long-run and important differences in trade performance. Non-price factors are often as important as – or even more important than – price factors in competition, and explanations of differential performance need to home in, in part, on the social and institutional influences that underlie the apparently small differences in technical detail. To adopt, if in a rather unusual sense, a phrase increasingly used by those who apply non-linear dynamics to basic technological issues (Arthur, 1988), these obscure 'small events' are not mere noise; they can become determinants of the evolutionary behaviour of the whole system.

One reasonably self-evident implication of the above position is that we should not necessarily expect, at the outset, to explain very much about trade flows from big differences in national structures of production. It is conceivable that virtually identical structures of production as measured by aggregative two-digit industrial classifications could lead to large differences in trade performance, both because of product differentation based on technical and other attributes and because of product innovation conferring temporary monopoly advantages for specific products within a broad industrial category (that is, pure 'technology gap' trade). In either case, process innovation may lead to competitive cost advantages.

V Changing patterns of production

It is now commonplace to view the 1980s as a period of substantial structural change. We consider in this section the nature of these changes and how far there are significant differences between Europe and its main competitors or between the major countries in Europe.

Table 2.3 echoes Table 2.2 in providing the broad breakdown between sectors, but concentrates only on the 1980s and gives some detail for each of the main countries. The broad picture is similar for all the countries listed in the table – in terms of contribution to GDP – agriculture continues to decline and in none of the countries listed (except Spain) now contributes more than 3 or 4 per cent to GDP; it is at its lowest in the United Kingdom (1.4 per cent) and Federal Republic of Germany (1.5 per cent). In most countries, the contribution of industry to GDP also fell, although it increased very slightly in Spain. The fall was less marked in Germany (42.7 to 40.1 per cent), the Netherlands (32.8 to 30.7 per cent), Belgium (33.0 to 31.3 per cent), Sweden (30.8 to 30.3 per cent) and Japan (41.9 to 40.6 per cent) than the United Kingdom (36.7 to 31.8 per cent) or Italy (39.0 to 33.4 per cent), which topped the 'deindustrialisation' league. The United Kingdom, France, Belgium, Sweden and the United States all have service sectors which now contribute more than two-thirds of their GDP.

In employment terms there was an interesting variation in experience from country to country. In all countries except the Netherlands employment in agriculture declined during the 1980s; the Netherlands, the United States and Japan all experienced positive growth in industrial employment, and the Federal Republic of Germany saw only a 0.4 per cent decline, compared with a 21.2 per cent decline in industry employment in the United Kingdom and a 21.8 per cent decline in Belgium. All countries saw a positive increase in service employment, the fastest rise being in Italy (31.2 per cent) and the United States (28.9 per cent). Interestingly, the increase in service employment in the United Kingdom (20.9 per cent) was less than that of both Italy and Spain.

In general, however, the message is one of similarity – of convergence, not divergence – in industrial structures. The same message emerges if we investigate changes within the industrial sectors. Here we take, as measures of manufacturing structure, the percentage shares of 28 industries classified according to the ISIC classification. As a convenient approach for summary analysis we have divided the 28 branches into three groups according to the rank order for growth rates in total OECD production over the whole period 1963 through 1988 (see Table 2.4).

Table 2.3 The structural composition of output, 1980 and 1988 (%)

Country	Contribution to output						Change in employment 1978–1988		
	Agriculture		Industry*		Services		Agriculture	Industry	Services
	1980	1988	1980	1988	1980	1988			
United Kingdom	1.8	1.4[a]	36.7	31.8[a]	61.5	66.8[a]	-14.6	-21.2	20.9
Federal Republic of Germany	2.1	1.5[a]	42.7	40.1[a]	55.2	58.4[a]	-3.5	-0.4	1.9
France	4.2	3.3	33.7	29.3	62.0	67.3	-26.5	-18.0	16.0
Italy	5.8	3.7	39.0	33.4	55.2	62.7	-33.0	-10.4	31.2
Netherlands	3.5	4.0	32.8	30.7	63.7	65.3	1.1	1.2	3.5
Belgium	2.3	2.1[a]	35.9	30.9[a]	61.8	67.1[a]	-15.3	-21.8	14.2
Spain	7.1	5.7[b]	38.6	39.3[b]	54.3	55.0[b]	-31.8	-14.6	23.3
Sweden	3.4	2.7	30.8	30.3	65.8	67.0	-33.1	-4.6	17.2
United States	2.6	2.0[a]	33.6	29.2[a]	63.8	68.8[a]	-6.3	3.6	28.9
Japan	3.7	2.8[a]	41.9	40.6[a]	54.4	56.7[a]	-25.1	8.4	20.9

* Including utilities, construction, mining.

Source: OECD, statistics on member countries; OECD in figures, *Supplement* to *OECD Observer*, June/July 1990: Tables on pp. 28–9.

[a] 1987
[b] 1986

Table 2.4 The structure and growth rates of OECD output ('world demand,) by growth sectors and branches

ISIC		Composition % 1980-88 (average)	Growth rates 1963-88 % p.a.	Rank	Growth rates 1979/80-88 ('1980s') % p.a.	Rank
	Fast Growth					
356	Plastic products	2.5	7.9	1	4.1	2
383	Electrical machinery	11.0	6.1	2	7.2	1
351	Industrial chemicals	4.9	5.0	3	2.1	9
352	Other chemicals	5.0	4.7	4	3.6	5
385	Scientific instruments	2.4	4.6	5	3.1	6
382	Non-electrical machinery	11.8	4.2	6	4.0	3
	Medium Growth					
341	Paper products	3.5	3.4	7	3.0	7
342	Printing and publishing	5.6	3.4	8	3.8	4
313	Beverages	2.0	3.2	9	1.3	14
362	Glass products	0.9	3.1	10	0.8	17
332	Furniture	1.4	2.8	11	1.2	15
384	Transport equipment	11.0	2.8	12	2.6	8
355	Rubber products	1.2	2.8	13	2.0	10
372	Non-ferrous metals	1.5	2.6	14	1.5	13
311	Food products	8.8	2.5	15	1.7	12
381	Fabricated metal products	6.4	2.5	16	0.9	16
	Slow Growth					
390	Other manufacturing	1.3	2.2	17	-0.1	20
369	Other non-metallic mineral products	2.7	2.1	18	0.6	18
331	Wood products, excluding furniture	1.8	2.1	19	1.8	11
361	Pottery, china etc.	0.3	1.8	20	0.4	19
353	Petroleum refining	2.4	1.8	21	-2.2	27
314	Tobacco	1.1	1.3	22	-0.3	23
321	Textiles	3.4	1.2	23	-0.2	21
322	Apparel	2.1	0.8	24	-0.8	25
371	Iron and steel	4.0	0.8	25	-0.3	22
354	Misc. petroleum and coal products	0.3	0.4	26	-0.8	24
323	Leather products	0.3	-1.0	27	-2.1	26
324	Footwear	0.4	-2.4	28	-3.9	28
300	Total manufacturing	100.0	3.1		2.9	

Notes: 1. For definitions of fast, medium and slow growth, see text.
2. Composition refers to percentage of total manufacturing value-added.

Source: UNIDO Industrial Statistics Database

The two-digit ISIC classification has the advantage of international comparability over a very extended period. But the use of such a broad classification has the disadvantage that quite substantial shifts *within* product groups which do involve significant structural change will not be picked up. This disadvantage affects especially ISIC 383 (electrical machinery) and ISIC 384 (transport equipment). ISIC 383 includes both the very fast-growing electronic industries and some relatively slow-growing older sectors, while ISIC 384 includes such very diverse industries as aircraft, automobiles, ships and railway equipment. Some disaggregation would be particularly important for comparisons with Japan since until recently that country had only a very small aircraft industry but a very large electronics industry. It is also important to examine electronics separately for our overall objective of assessing comparative performance of the EC in information and communication techno-logy (ICT). We have therefore given very detailed attention to the electronic industries in Part II.

The rank order of branches changed little between the three periods, despite the substantial decline in total OECD growth (4.6 per cent a year in the 1960s, falling to 1.6 per cent in the depressed 1970s and recovering to around 3.0 per cent in the 1980s). The relative stability by groups of industries in the rank order reflects primarily the much higher growth rates of capital goods and slow growth or decline in the consumer goods sectors (with the exception of consumer durables). But some important changes can be noted: in particular, electrical machinery (383) is exceptional in that the growth rate in the 1980s exceeded that in the longer period 1963-88 and took first place in the 1980s. This is, of course, primarily due to the very rapid growth of electronics. The rise of printing and publishing to fourth place also reflects the progress of the information technology revolution. Finally, ISIC 352, other chemicals (that is, fine chemicals, pharmaceuticals etc.), completely outstripped industrial chemicals (ISIC 351 comprising mainly the older basic 'commodity' chemicals). This change is analysed in Chapters 12 and 13.

From Table 2.4 we may further note:

(a) the heavy weight of the machinery industries – electrical (383), non-electrical (382) and the small scientific instruments branch (385), all in the fast-growth group, plus transport equipment (384). Together, ISIC 382, 383, 384 and 385 accounted in 1980-88 for 36 per cent of total OECD production;
(b) a lesser but still heavy weight for the two chemical branches – industrial (351) and other (352), making up 10 per cent of the total; food products (311) adds another 9 per cent, plastic products (356) and beverages (313) each add another 2 per cent;
(c) hence OECD manufacturing production is dominated by the eight branches of machinery, chemicals and food processing – together nearly 60 per cent of the total. None of the remaining 20 branches accounts for more than 5 per cent.

The fast-growth branches include the main carriers of new technologies (even if they also include some older products). Together with aircraft, they are the most R & D-intensive sectors of industry, accounting in most countries for over 70 per cent of all industrial R & D (Chapter 3). Because they are the main engines of growth and largely influence the products and processes of the rest of the economy, all of these fast-growth industries are discussed in some detail in Parts II and III of the book.

Table 2.5 Manufacturing output growth rates by periods: structural composition and structural change summary by three growth sectors

	OECD	EC (12)	JAPAN	USA
A *Growth rates (% p.a.) of total manufacturing value added*				
'The 1960s' 1963–73/74	4.6	3.5	9.2	4.8
'The 1970s' 1973/4–79/80	1.6	0.8	2.4	2.2
'The 1980s' 1979/80–88	2.9	1.4	4.7	2.9
B *Composition of manufacturing in 1980–88 (%)*				
Fast Growth	37.6	37.0	41.0	38.9
Medium Growth	42.3	40.6	38.6	44.2
Slow Growth	20.1	22.4	20.4	16.4
	100.0	100.0	100.0	100.0
C *'Adjustment' changes in compositions between 1963–73 and 1980–88 (%)*				
Fast Growth	5.0	6.0	5.5	4.5
Medium Growth	0.0	2.0	0.8	0.4
Slow Growth	-5.4	-7.3	-6.2	-5.2
Sum of changes	10.4	15.3	12.5	10.1

Source: UNIDO Industrial Statistics Database

How far is this pattern common to the individual countries and areas? Tables 2.5 and 2.6 give country-based summaries of output growth rates, structural composition and structural change in terms of the same fast-, medium- and slow-growth sectors. The broad impression is, indeed, one of similarity of national patterns – but with significant exceptions.

The proportion of output accounted by the fast-growth branches (which might be considered a major determinant of overall growth rates) is in fact fairly similar among the countries – despite the large differences between national overall growth rates. But the Federal Republic of Germany (43 per cent), Japan (41 per cent) and the United States (39 per cent) are significantly above the OECD (Europe) average of 36 per cent (Tables 2.5 and 2.6). Spain (26 per cent) falls far below average and some of the other smaller EC member countries (not separately shown in the tables) such as Greece and Portugal also resemble the Spanish pattern. Only Japan clearly displays the expected association with faster than average overall growth.

Considerable differences appear between the weights of individual branches in the various countries. Such differences denote branch specialisations (or partial retirement from particular branches).

Examples of specialisations in the 1980s may be noted:

(a) in the fast-growth group, Japan's weighting is largely the result of its now well-established specialisation in electronics (part of ISIC 383) and to some extent in the much less important plastic products (ISIC 356). The Netherlands and the Federal Republic of Germany are also more specialised than elsewhere in electrical machinery (including electronics);
(b) in chemicals, on the other hand, despite the rapid growth of the Japanese chemical industry, the EC is much more specialised than Japan, especially the Federal Republic of Germany, United Kingdom and the Netherlands;
(c) in the slow-growth group there are few specialisations. Conspicuous exceptions are Italy and Spain with heavy weights in some 'mature' industries: Italy

Table 2.6 Manufacturing output growth rates by periods, structural composition and structural change by country and OECD (Europe average)

	UK	FRG	France	Italy	NL	Belgium	Spain	Sweden	OECD(E)
A *Growth rate (% p.a.) of total manufacturing*									
1963–1973/74	2.6	4.1	5.0	4.9	3.6	4.1	9.7	4.2	3.6
1973/74–79/80	-1.6	1.4	0.7	2.2	1.3	0.8	1.7	0.8	0.7
1979/80–1988	1.6	1.5	0.3	0.6	0.7	0.1	1.6	2.0	1.5
B *Composition of manufacturing in 1980–88 (%) by growth rates in 1963–88*									
Fast	36.5	43.4	31.3	35.3	36.4	32.9	25.9	31.8	36.2
Medium	44.2	36.1	43.6	35.7	47.2	45.1	45.3	50.6	41.1
Slow	19.3	20.5	25.4	29.0	16.4	22.0	28.8	17.6	22.7
	100	100	100	100	100	100	100	100	100
C *'Adjustment' changes in composition between 1963–73 and 1980–88 (%)*									
Fast	3.5	9.1	4.0	2.4	2.2	6.5	3.4	3.5	5.8
Medium	0.5	1.1	5.5	-0.2	3.7	0.1	2.1	3.7	1.5
Slow	-4.0	-11.2	-9.1	-2.3	-6.1	-6.4	-5.2	-7.2	-7.0
Sum of Changes	8.0	21.4	18.6	4.9	12.0	13.0	10.7	14.4	14.3

Source: UNIDO Industrial Statistics Database

Table 2.7 The magnitude of structural change

Country/area	1963/73–1974/79	1974/79–1980/88
OECD	5.9	11.1
OECD-Europe	8.7	8.3
EC (12)	8.8	9.3
Japan	9.4	13.9
United States	8.2	7.0
Federal Republic of Germany	13.4	19.2
France	15.5	7.5
United Kingdom	7.0	16.0
Italy	11.5	10.6
Netherlands	13.8	9.8
Spain	15.7	19.6
Belgium	14.3	12.3
Sweden	13.6	10.4

Notes: These magnitudes are given by the sum of absolute differences in industry shares of total value added between given time periods.

Source: UNIDO Industrial Statistics Database

in textiles, apparel and footwear; Spain in leather goods; Belgium and Japan in iron and steel (where Japan in the 1980s accounts for a quarter of OECD production); and Sweden in wood products, paper and furniture (one of the few examples of natural resource specialisation). In some cases, notably Spanish leather and footwear or Italian apparel, these specialisations are very clearly design- and export-driven.

Such exceptions to the general similarity of national structures show that structural development need not be regarded as a set or deterministic pattern nor, therefore, as a reliable basis for forecasting. A degree of specialisation in older industries, if superficially appearing as a failure to adapt to world demand, can still be successful in international trade, particularly if it is based on process innovations (paper and wood products in Sweden) and/or on design innovations (furniture in Sweden, clothing in Italy). Correspondingly, retreat from slow-growth industries, marked in most OECD countries, can be associated with successful performance of the economy as a whole as long as it is offset by competitiveness in more promising lines – as has happened in Japan, but not always elsewhere.

Turning to the issue of the *degree* of structural change between industries, Table 2.7 uses a simple summary measure of the degree of change, namely the sum of absolute difference in industry shares of total manufacturing, irrespective of size. A larger measure therefore provides a rough indication of greater overall structural change and vice versa. A bigger change could of course be explained by either a more rapid expansion of the high-growth industries or the elimination of parts of the slower ones. Looking at the OECD as a whole, it would seem that there has been a significant acceleration in structural change in the last two decades. But this quickening of change is by no means universal. Countries exhibiting an increase in structural change were Japan, Federal Republic of Germany, the United Kingdom and Spain; but France, Italy, the Netherlands, Belgium (all members of the

Table 2.8 Overall patterns of trade

Country	Change volume, 1978–88 (%)		Export market shares, 1988		Trade coverage ratios, 1988		
	Imports	Exports	Manuf.	High RDI*	High RDI*	Med RDI†	Low RDI‡
			(OECD=100)		(exports/imports)		
United Kingdom	5.2	2.8	7.25	8.47	0.93	0.93	0.67
Federal Republic of Germany	3.3	4.0	18.18	14.95	1.26	2.30	0.93
France	3.9	3.4	8.43	7.51	0.96	1.05	0.85
Italy	4.6	2.9	7.43	4.34	0.77	1.18	1.33
Netherlands	3.2	4.2	5.16	3.74	0.83	1.04	1.19
Belgium	3.3	4.2	4.76	2.13	0.83	1.06	1.18
Spain	6.8	5.9	1.97	0.90	0.36	0.73	1.29
Sweden	4.1	4.2	2.80	2.06	0.89	1.18	1.41
United States	5.5	6.0	12.81	21.76	0.86	0.52	0.35
Japan	4.4	6.6	14.71	22.66	5.15	4.23	0.92

*RDI = R & D intensity

* High RDI = aerospace, office machinery, computers, electronic components, drugs and medicines, instruments, electrical machinery.

† Medium RDI = motor vehicles, chemicals, non-electrical machinery, rubber, plastics, non-ferrous metals, other transport equipment.

‡ Low RDI = all other industries.

Source: OECD, *OECD in Figures*, Statistics on Member countries, Supplement to *OECD Observer*. June/July 1990, Tables 60 and 61; 56 & 57

original EC6), the United States and Sweden all show a slowing down of structural change by this measure. In general, this accords with earlier analyses which suggested that the original EC6 (the Federal Republic of Germany, France, Italy and the Benelux countries) experienced major changes in the first 15 years of integration (1958-73), whereas the newcomers in 1973 (Britain, Ireland, Denmark), and particularly Britain, had retained old 'vertical' patterns of trade and production. All these countries had rapidly to adjust within the EC to a new trading pattern involving extensive exchange (intra-trade) within manufacturing sectors (Sharp and Shepherd, 1985). The most marked change in Table 2.7 is in fact in the position of the United Kingdom, which came bottom of the earlier league table and, next to Spain, top of the second.

VI The changing patterns of trade

As indicated in the introduction to this section, the existence of significant intra-industry trade flows testifies to the importance of small differences in product characteristics. Table 2.8 provides a broad overview of trade flows similar to Table 2.3 in production. The first two columns give the percentage growth of imports and exports over the decade 1978 to 1988 and show clearly the endemic tendencies of the United Kingdom, and Italy to import more than they export, while both the Federal Republic of Germany and Japan tend to export more than they import. Columns 3 and 4 show clearly the dominance of the Federal Republic of Germany in manufacturing exports and Japan in high-technology exports, Japan having overtaken the United States as the leading exporter of high-technology products only in the 1980s. The final three columns indicate clearly *where* the Federal Republic of Germany and Japan gain their advantage – both are well represented in the high- and medium-R & D intensity range of products and score poorly in low-R & D intensity areas, in direct contrast to the United Kingdom and the United States' capabilities which are spread more evenly across the three areas.

Table 2.9 analyses disaggregated trade performance by the same fast-, medium- and slow-growth categories used earlier (Tables 2.4, 2.5 and 2.6) to analyse production performance – not quite, but nearly, the same categories as the OECD high, medium and low R & D intensity used in Table 2.8. In this regard, the concept of revealed comparative advantage (RCA) is useful. RCA is the measure, for any given product, of market share for that product compared with the country's average market for all products. Hence the United Kingdom, whose average market for manufacturing share for the period 1984-7 was 7.35 per cent, is said to have comparative advantage in such sectors as scientific instruments, where market share is above the average. RCA is not, of course, a pure measure of competitiveness – other factors beside competition underlie trade shares – but it gives a rough indication of comparative strengths between sectors and between countries.

Comparing the EC countries identified in Table 2.9, it is notable that the Federal Republic of Germany's overall market share – at nearly 18 per cent – is almost double that of any other EC country – France, with 8.64 per cent being the next highest. Germany's strong position in market share is mirrored by its strong presence in fast- and in medium-fast-growth sectors, particularly

Table 2.9 Shares of total OECD export by product groups, 1984–87 (%)

	UK	FRG	France	Italy	NL	EC	Japan	USA
Fast Growth								
3560 Plastic products	6.73	23.71	9.49	14.72	6.52	68.09	7.34	11.49
3830 Electrical machinery	6.23	15.12	6.78	4.42	3.65	39.09	35.10	15.93
3510 Industrial chemicals	8.99	21.82	10.37	5.60	10.28	60.93	7.96	17.06
3520 Other chemicals	12.92	18.72	14.75	4.87	6.38	63.45	6.89	14.85
3850 Scientific instruments	8.58	16.64	6.21	3.32	4.46	42.19	23.51	19.49
3820 Non-electrical machinery	8.32	19.97	6.29	8.72	3.00	50.32	17.04	19.76
Medium Growth								
3410 Paper products	2.95	10.40	5.21	2.85	3.65	28.35	2.67	12.19
3130 Beverages	18.36	8.06	36.77	9.68	5.92	88.88	0.99	3.16
3420 Printing and publishing	13.99	20.99	9.83	6.52	6.08	64.27	4.47	16.37
3620 Glass products	6.42	21.46	17.54	11.15	3.94	66.13	9.88	10.84
3320 Furniture	4.41	22.27	7.25	24.98	4.15	72.71	2.68	7.30
3840 Transport equipment	5.66	19.57	7.83	3.75	1.37	40.88	24.98	17.51
3112 Food	4.97	13.48	11.38	5.06	16.29	65.74	1.63	14.28
3550 Rubber products	7.44	17.86	15.78	9.14	3.71	58.82	17.77	9.33
Slow Growth								
3720 Non-ferrous metals	8.99	17.08	8.43	3.50	4.89	46.84	6.20	9.28
3810 Fabricated metal products	7.18	23.31	8.95	11.91	5.12	61.89	11.93	9.08
3900 Other manufacturing	14.96	11.42	7.05	17.52	2.11	56.40	16.03	9.72
3690 Cement, clay etc.	4.95	17.86	8.64	22.96	3.61	69.92	9.90	7.72
3310 Wood	1.05	7.60	4.06	3.10	1.99	24.17	0.87	12.30
3530 Petroleum refining	10.36	6.35	7.21	9.66	27.87	71.37	1.23	11.77
3610 Pottery and china	11.13	18.08	6.69	15.85	2.11	61.40	26.41	6.10
3140 Tobacco	15.31	14.37	1.93	0.12	20.80	55.67	0.28	6.91
3210 Textiles	7.02	19.07	10.00	15.10	5.85	64.02	13.85	6.91
3220 Apparel	7.36	15.85	9.93	28.03	4.17	78.69	2.98	3.90
3540 Coal and petroleum	7.43	28.21	5.98	4.63	6.25	55.11	10.59	19.13
3710 Iron and steel	5.39	20.07	11.73	7.49	3.92	54.32	26.18	2.65
3230 Leather products	8.28	14.76	10.19	29.68	3.12	76.24	5.92	7.95
3240 Footwear	2.76	7.74	6.90	51.14	1.98	87.78	0.49	3.01
3000 Manufacturing	7.35	17.82	8.64	7.57	5.44	51.97	16.43	14.80

the plastics and chemical sectors, transport equipment, furniture, printing and publishing. By contrast, the United Kingdom's average export share of manufactures is only 7.35 per cent, and its strongest growing sectors are beverages and tobacco, both medium- to slow-growth industries. Among high- technology industries the United Kingdom has tended, in the 1980s, to specialise increasingly in those with major defence linkages, particularly aerospace (see Chapter 22). It has also been increasing its presence in the pharmaceutical industry, reflected in Table 2.9 in the 'Other Chemicals' category (see Chapter 13).

Both the United Kingdom and the Federal Republic of Germany present a fairly even spread of competitiveness, with a few cases of outstanding comparative advantage and obvious specialisations. The difference in overall trade performance between them cannot be explained by any great difference in their export structures but rather by factors which affect almost all industries – such as skills, investment in new equipment, marketing and commitment to R & D. We shall be exploring some of these factors in the next chapter.

Italy presents a strong contrast. Italy's overall market share – just under 8 per cent – is now slightly greater than that of the United Kingdom. But Italy's product pattern is much more varied. The strong points – market shares of 12 per cent or more – are in plastics (fast growth), in furniture (where Italy, with 25 per cent, is by far the largest exporter) and in five slow-growth products including cement, clothing, leather goods and footwear (in footwear, Italy, with 51 per cent, dominates the market). The weakest points of the Italian structure (5 per cent or lower market shares) are in three fast-growth groups (electrical machinery, other chemicals, scientific instruments); and also in eight medium- or slow-growth products (including transport equipment and paper products). Yet despite a pattern of comparative advantages apparently out of line with trends in world demand, Italy has succeeded in increasing its overall market share, in part through retaining its strong position in some slow-growth products.

The French pattern of trade shows strength in the fast-growing product areas only in chemical products, but considerable strength in medium-growth areas: beverages (mainly wine – a natural resource advantage), glass (a strong historical tradition) and rubber products (tyres, sports goods). France also has modest comparative advantage in iron and steel and displays few cases of very marked comparative disadvantage – tobacco being an outstanding exception.

The Netherlands' export pattern is notable for its high degree of specialisation in a handful of products: industrial chemicals (10 per cent market share against the average of 5.5 per cent), food products, petroleum products (28 per cent market share – far ahead of the other countries) and tobacco products (20 per cent). But the earlier Dutch specialisation in electrical machinery (Philips) is no longer conspicuous.

Outside the EC, the United States and Japan each have overall market shares of 15 and 16.5 per cent respectively. Superficially, the United States seems well placed. It has above-average market shares in five of the six fast-growing areas and in aircraft. However, this first impression needs to be seen in the light of more recent trends. The United States was the world leader in the introduction of computers, TV, integrated circuits, many electronic

Table 2.10 Export/import ratios (x/m) in ICT products, 1965–87

	EC–4*			USA			Japan		
	1965	1979	1987	1965	1979	1987	1965	1979	1987
Typewriters and office machines	1.08	1.33	1.27	2.82	0.48	0.36	0.52	4.54	19.65
Computers and peripherals	1.39	0.83	0.77	4.30	3.48	1.11	0.14	2.20	6.24
Telecommunications equipment	2.15	1.78	1.33	2.84	0.88	0.47	10.18	11.95	15.57
Consumer electronics	1.93	0.59	0.45	0.36	0.17	0.06	86.01	42.01	28.63
Scientific instruments, photographic supplies, watches and clocks	1.55	1.04	0.96	1.94	1.04	0.59	4.38	4.85	6.47
Electrical instruments (measurement and control) including medical	1.24	1.21	1.32	6.42	4.54	1.75	1.04	1.06	2.08

*EC–4 = Former West Germany, France, United Kingdom, Italy. Figures *exclude* intra-trade

Source: IKE Aalborg Institute of Production, 1990

instruments and other innovations and at one time dominated world trade in most electronic products. Its market share in such products has, however, slipped markedly in the 1980s. Table 2.10 details export to import (X/M) ratios in the main electronics sectors and illustrates well how American dominance in the 1960s had by 1987 given way to the Japanese. Only in computers had the ratio remained above unity (that is, did export values exceed imports) and even here the ratio had deteriorated from 4.3 to 1.1.

Table 2.10 illustrates the strength of the Japanese advance in all sectors of electronics and the broadening of its base from consumer electronics (where its X/M ratio was 86 in 1965) to other categories. Even in instrumentation, long regarded as an area of strength for European and American firms, Japan's X/M ratio has doubled to 2.08, exceeding both American and European ratios. In all other categories, the Japanese X/M ratios range from 6.24 (for computers) to 28.63 (for consumer electronics) – extraordinarily high by any standard. In Table 2.9 these are reflected in Japan's 35 per cent share for electrical machinery, and 23 per cent share for scientific instruments. Japan also has particularly high shares in transport equipment (25 per cent, but more in autos alone), pottery (26 per cent) and iron and steel (26 per cent). On the other hand, the Japanese position in three fast-growth products (plastics and both chemical groups) is comparatively weak (7-8 per cent) and little better than average in non-electric machinery; and in a wide range of medium- and slow-growth products, Japan is a relatively weak competitor. Indeed, Japan presents the most conspicuous case of success by a limited range of specialisations – at the cost of accepting relative weaknesses elsewhere. However, its specialisation is in some of the largest and fastest-growing product groups in world trade.

VII Conclusions: implications for European technological activities

The earlier sections of this chapter have examined the changing patterns of industrial structures and trade within the Community and between the Community and its major trading partners. Two conclusions stand out. In the first place, one is struck by the similarity of economic and industrial structures. While differences of course remain, the trends are the same – the fast-growing industries of the United States and Japan are the fast-growing industries of the EC; and within the EC, all the major economies have seen their fastest growth in broadly the same sectors. However, the extraordinarily rapid growth of the ICT industries and of the automobile industry in Japan is notable.

The second conclusion follows directly from the first. If all the main economies of the world are similar in structure, then increasingly trade is taking place not *between* sectors, as in the 19th century, but *within* sectors, with firms trading increasingly specialised products among themselves. Again, the Japanese specialisation in electronics is something of an exception. The specialisation of the United States and United Kingdom in aircraft has political rather than economic origins.

The 'performance' of an economy is usually measured in comparative terms, above all in terms of market share – how far it is gaining or losing market share overall or in particular market segments. The analysis in this chapter suggests that industrial structures, as measured here, are only loosely related to performance or competitiveness, which depend more on the varying efficiency of firms and the particular characteristics of their products than on the average performance of industries. Only in the case of electronics, among the very fast-growing industries, is there a clear case of structural specialisation in Japan and a relative lag in Europe.

What has still to be explained is why some national systems of innovation are more effective than others in improving efficiency across many sectors of the economy, as well as the comparatively enduring strength of such national comparative advantages as clothing and footwear (Italy), machinery and chemicals (Germany), drugs (Switzerland and the United Kingdom). The next chapter relates the discussion of comparative advantages by sector and by country to more direct measures of technological performance, and Parts II and III discuss in detail the relative performance of the EC in each major fast-growth sector.

3 Europe's technological performance

Pari Patel and Keith Pavitt

I Introduction

The central importance of technology for power and welfare has long been obvious to practicing politicians and to scholars in such disciplines as archaeology and history. Over the past 20 years, students and practitioners of management and economics have once again been reminded of its importance for what they study and what they do. Until the early 1970s, it was commonly assumed that the open trading system was sufficient to allow the rapid international diffusion of technology (embodied in machinery and blueprints) and as a consequence enable the smooth catching up of Western Europe and Japan to the levels of technology and efficiency of the world's leading country, the United States.

Technology-related events since then have turned out to be anything but smooth. There has been uneven development among sectors, in part as a consequence of the micro-electronics revolution. There has also been uneven development among countries. Some (for example, the United Kingdom) have caught up only very partially, while others (for example, the Federal Republic of Germany and Japan) have actually overtaken the world's technological leading country in certain important sectors (Nelson, 1990).

This is because the international diffusion of technology is neither automatic nor easy. In addition to – and just as important as – the transfer of machines and blueprints, it requires the conscious accumulation of complex technological skills through experience and analysis. Differences among countries in the resources devoted to such 'technological accumulation' have led to international technological gaps which, in turn, have led to international differences in economic performance. The 'neo-technology' theories of trade and growth pioneered by Posner (1961) and Vernon (1966) in the 1960s have been amply confirmed by the events of the 1970s and 1980s, and by related econometric analyses of Soete (1981) and Fagerberg (1987; 1988).

While international differences in technology are therefore clearly important, commentary and analysis leave much to be desired. During the Cold War, descriptions and predictions about competitors' achievements in military technology often turned out to be inaccurate because they were either ill-informed, self-serving or both. The same often holds in debates about capitalist countries' civilian technology. J-J Servan-Screiber's 'American Challenge' of the 1960s quickly evaporated. Japan's growing technology-based competitiveness in the 1970s was often ignored or described as nothing more than a cunning mixture of copying, meek trade unions and cheating.

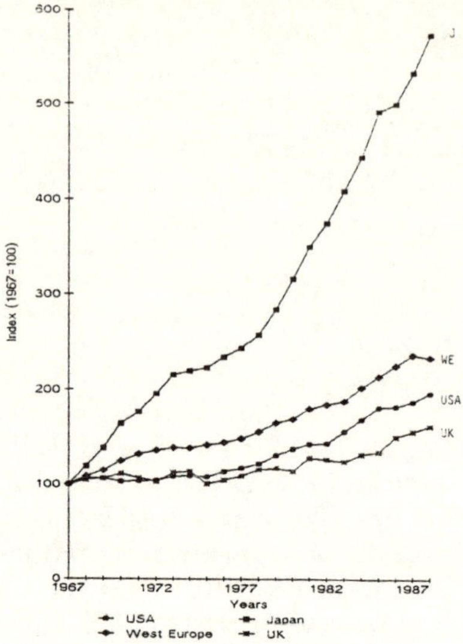

Note: Europe (WE) comprises Belgium, France, FR Germany, Italy, Netherlands, Sweden, Switzerland, UK

Source: OECD Science and Technology Indicators Unit

Figure 3.1 Trends in real expenditure on civilian R & D, 1967–88

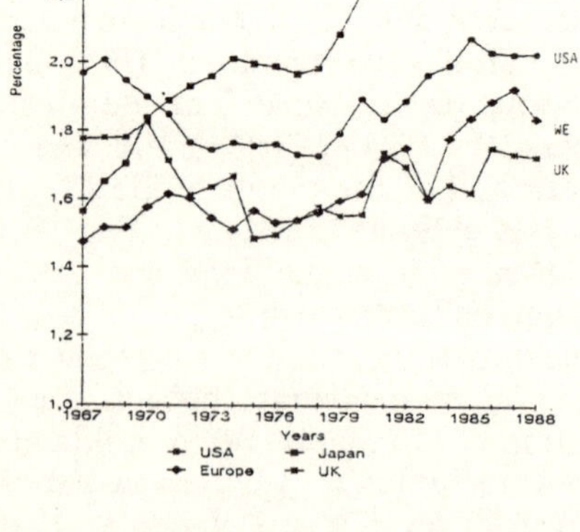

Figure 3.1 Trends in civilian R & D as a percentage of GDP, 1967–88

And the West German-based 'Eurosclerosis' of the early 1980s has turned out to be only very partial as recent events in Central and Eastern Europe testify.

One reason for these shortcomings is the very real difficulty of measuring and comparing technological accumulation across firms, sectors and countries. The activities contributing to technological accumulation are complex and varied, encompassing fundamental research by qualified scientists at one end of the spectrum and routine tinkering on the production line at the other. If only for this reason, all measures of technological activities are bound to be imperfect.

However, as a result of the growing demands from public and private policy-makers for better data, progress has been made in measurement and conceptualisation, where academic analysis and advances in information technology have both played an important role. The advantages and drawbacks of the various measures have been extensively reviewed elsewhere (Freeman, 1987; Pavitt, 1988a; Patel and Pavitt, 1988). In particular, we have shown in our earlier work that the combined use of data on R & D activities, and on patenting in the United States by country of origin, gives a plausible and consistent picture of technological activities at the world's technological frontier.

R & D is a better measure of rates of change in real resources over time, but it measures technological activities in small firms only very imperfectly. American patenting is a better measure of technological activities in small firms and can be broken down quite finely to specific firms and specific technical fields. Neither measure is satisfactory for software technology, but no alternative yet exists. And neither measure captures all the activities that lead to product and process innovations such as design, management, production engineering, marketing and learning by doing.

Using these two measures, we shall show that the main features of the past 20 years have been: (i) the emergence of Japan as a major technological force, especially in electronics and automobiles; (ii) the relative decline of the United States, with its increasing relative strength in natural resource- and defence-based technologies; (iii) uneven development in Europe; across country with the increasing strength of the Federal Republic of Germany and decline of the United Kingdom; and across sector, with increasing strength in chemicals and decline in electronics.

II The volume of technological activities in Japan, the United States and Western Europe

Civilian R & D activities

Figures 3.1 and 3.2 show long-term national trends in expenditures on civilian R & D, in real terms and as a percentage of Gross Domestic Product (GDP). With the exception of the United Kingdom, we see a continuous upward trend in all regions, with an acceleration in the 1980s after a slowdown in the 1970s. They show that by the early 1970s Japan was already spending a higher proportion of GDP on R & D than both Western Europe (as defined in Figures 3.1 and 3.2) and the United States. In spite of rapid growth in real terms, the Western European proportion remains below that

Table 3.1 Trends in industry-financed R & D as a percentage of GDP in 12 OECD countries, 1967-88

	1967	1975	1985	1988
United States	1.01	1.01	1.35	1.38
Japan	0.83	1.12	1.84	1.95
Belgium	0.59	0.75	1.06	1.13
Denmark	0.34	0.41	0.60	0.68
France	0.61	0.69	0.94	0.96
Federal Republic of Germany	0.94	1.12	1.58	1.78
Ireland	0.17	0.22	0.33	0.49
Italy	0.35	0.47	0.58	0.54
Netherlands	1.12	1.02	0.96	1.14
Spain	0.08	0.19	0.26	0.32
Sweden	0.72	0.96	1.71	1.74
Switzerland	1.78	1.67	1.59	2.20
United Kingdom	1.00	0.80	0.96	1.06
Western Europe	0.79	0.83	1.07	1.17

Source: OECD

Table 3.2 Per capita inputs and outputs of basic research: Japan, the United States and four Western European countries

(West Europe* = 100)

Measure	Japan	USA	West Europe*
Government Funding**			
1987	52	104	100
1980	55	97	100
Publications***			
1986	51	118	100
1981	46	125	100

* France, the Federal Republic of Germany, the Netherlands and the United Kingdom

** At purchasing power parities exchange rates. See Irvine *et al.*, 1990: Table 8.2

*** World publication output using the CHI/NSF Science Literature Indicators Database, 1981 journal selection. See Martin et al., 1990: Table 1.

of the United States, although – as we shall see later – the spread within Western Europe remains considerable.

Business-funded R & D

The largest component of civilian R & D is expenditure by business firms. Table 3.1 shows a marked change in national rankings in the levels of such expenditures as a percentage of GDP between 1967 and 1988. Again, Japan had already overtaken the United States by 1975, while Western Europe still remains below the United States. However, Table 3.1 also shows a remark-

Table 3.3 Trends in defence R & D as a percentage of GDP in 12 OECD countries, 1967-88

	1967	1975	1985	1988
United States	1.01	0.63	0.85	0.83
Japan	0.01	0.01	0.02	n.a.
Belgium	n.a.	n.a.	n.a.	n.a.
Denmark	n.a.	n.a.	n.a.	n.a.
France	0.54	0.35	0.46	0.52
Federal Republic of Germany	0.21	0.14	0.14	0.13
Ireland	n.a.	n.a.	n.a.	n.a.
Italy	0.02	0.01	0.08	0.08
Netherlands	0.04	0.03	0.03	0.03
Sweden	0.43	0.26	0.30	0.29
Switzerland	0.05	0.05	0.03	n.a.
United Kingdom	0.60	0.64	0.67	0.51
Western Europe	0.28	0.23	0.28	0.26

Source: OECD

Table 3.4 Shares of Western European patenting in the United States: 1963-88

	1963–8	1969–73	1974–8	1979–83	1984–8
Fed. Republic of Germany	33.74	35.62	37.16	40.10	40.91
United Kingdom	24.80	21.56	18.16	15.91	14.67
France	13.42	13.99	14.45	14.29	14.60
Netherlands	4.67	4.36	4.32	4.49	4.62
Italy	4.27	4.77	4.72	5.37	5.83
Denmark	0.92	1.08	1.03	0.93	1.03
Belgium	1.63	1.88	1.84	1.64	1.55
Ireland	0.06	0.14	0.11	0.12	0.20
Spain	0.42	0.47	0.59	0.40	0.57
Greece	0.06	0.08	0.07	0.04	0.05
Portugal	0.03	0.03	0.03	0.02	0.02
Switzerland	8.66	8.41	8.81	8.33	7.31
Sweden	5.24	5.08	5.65	5.09	4.89
Austria	1.33	1.57	1.76	1.85	1.89
Norway	0.50	0.53	0.62	0.56	0.60
Finland	0.25	0.42	0.68	0.85	1.27
Total	100.00	100.00	100.00	100.00	100.00

Source: United States Department of Commerce, Patent and Trademark Office.

able diversity of performance within Western Europe, most notably the relative decline of the Netherlands and the United Kingdom, and the improved rankings of the Federal Republic of Germany, Sweden and (from more modest beginnings) Belgium and Denmark. Four countries are now well ahead of the United States in the proportion of GDP spent by business firms on R & D: Japan, Sweden, Switzerland and the Federal Republic of Germany. As George Ray (1989) has recently shown, these are also the countries

Table 3.5 Sectoral patterns of relative technological advantage 1963–88

	Fine chem.	Indus. chem.	Mate-rials	Mech eng.	Motor vehic.	Elect. mach.	Elec-tron cap.	Tele-comm.	Elect tron cons.	Raw mat. based	Def-ence
United States											
1963–68	0.89	0.93	1.04	1.01	0.89	1.00	1.02	1.03	0.94	1.09	1.00
1984–88	0.91	1.00	0.95	0.98	0.56	1.01	1.01	1.06	0.71	1.29	1.16
Japan											
1963–68	2.95	1.61	1.01	0.77	0.83	1.17	1.46	1.06	1.99	0.44	0.35
1984–88	0.81	1.91	1.41	0.86	2.41	1.13	1.59	0.86	2.53	0.35	0.10
West Europe											
1963–68	1.24	1.28	0.94	0.97	1.60	1.01	0.98	0.91	1.38	0.62	1.04
1984–88	1.38	1.15	0.90	1.10	1.09	0.90	0.67	1.03	0.66	0.74	1.27
Federal Republic of Germany											
1963–68	1.13	1.48	0.67	0.96	1.43	0.77	0.93	0.73	1.88	0.54	1.03
1984–88	1.13	1.29	0.91	1.16	1.35	0.81	0.60	0.80	0.55	0.61	1.36
United Kingdom											
1963–68	0.87	1.03	1.09	1.02	1.99	1.22	1.03	1.02	0.92	0.68	1.28
1984–88	1.92	1.05	1.07	1.03	1.01	0.94	0.68	1.00	0.68	0.87	1.05
France											
1963–68	1.87	1.01	1.06	1.02	2.10	1.22	0.83	1.15	0.80	0.49	1.10
1984–88	1.40	0.95	0.77	1.06	0.70	1.15	0.84	1.69	0.53	0.77	1.77
Netherlands											
1963–68	1.71	1.45	1.23	0.75	0.15	1.34	1.90	1.09	1.93	1.15	0.15
1984–88	0.53	1.02	0.73	0.84	0.42	1.21	1.24	1.48	2.03	1.21	0.35
Italy											
1963–68	1.29	1.93	0.52	0.93	1.25	0.67	0.87	0.69	0.53	0.71	0.78
1984–88	2.11	1.00	0.71	1.14	1.24	0.65	0.58	1.06	0.39	0.76	0.86
Denmark											
1963–68	3.07	0.77	0.98	1.11	0.39	0.87	0.47	0.55	1.17	0.92	0.12
1984–88	2.10	0.86	0.79	1.22	0.18	0.74	0.27	0.48	0.41	1.02	0.64
Belgium											
1963–68	1.23	1.37	4.03	0.71	0.44	0.91	0.66	1.02	3.96	0.42	0.78
1984–88	1.82	1.55	2.07	0.75	0.19	0.80	0.14	0.53	1.55	1.14	1.53

Source: Data supplied to SPRU from the United States Department of Commerce, Patent and Trademark Office

where new technologies find rapid and wide acceptance: business-funded R & D is correlated with performance in the diffusion of technology.

Basic research

Basic research – undertaken mainly in academic and related laboratories, and to a significant extent by large firms – provides the knowledge and skills on which national systems of business R & D can build. It is difficult to measure and compare basic research, given inevitably arbitrary judgements about boundaries with related activities such as teaching in universities and applied research in government and business laboratories.

Table 3.2 is based on the most thorough attempts made so far to compare per capita inputs and outputs of basic research in Japan, the United States

	Fine chem.	Indus. chem.	Mate-rials	Mech eng.	Motor vehic.	Elect. mach.	Elec-tron cap.	Tele-comm.	Elect tron cons.	Raw mat. based	Def-ence
Ireland											
1963–68	0.00	0.39	0.00	1.28	2.99	0.00	0.00	0.65	0.00	0.55	0.00
1984–88	1.54	1.68	0.00	0.69	0.00	1.06	0.90	0.00	0.76	1.27	0.00
Spain											
1963–68	0.85	0.56	0.48	1.18	2.99	0.85	0.13	1.02	0.47	0.47	1.93
1984–88	1.46	0.74	0.15	1.25	2.64	0.54	0.08	0.34	0.07	0.64	0.90
Portugal											
1963–68	10.61	1.40	0.00	0.98	0.00	0.67	0.00	0.00	0.00	0.98	3.45
1984–88	0.00	1.99	0.00	1.42	0.00	1.88	0.00	0.00	0.00	0.00	0.00
Greece											
1963–68	0.00	1.01	0.00	1.00	3.08	0.38	0.93	0.67	1.69	0.56	1.98
1984–88	0.83	2.83	0.00	0.55	3.41	0.98	0.00	0.00	0.00	1.64	4.08
Switzerland											
1963–68	2.61	2.00	0.23	0.89	0.56	0.84	0.57	0.73	0.63	0.47	1.44
1984–88	1.86	1.44	0.65	1.09	0.50	0.88	0.45	0.71	0.52	0.61	1.04
Sweden											
1963–68	0.94	0.44	0.44	1.22	1.12	1.13	0.71	1.37	0.38	0.67	2.34
1984–88	0.53	0.50	0.69	1.46	0.79	0.90	0.34	0.72	0.27	0.96	2.17
Austria											
1963–68	1.42	0.80	1.06	1.25	1.21	0.62	0.28	0.38	1.69	0.39	0.17
1984–88	0.38	0.59	0.64	1.49	1.43	0.94	0.30	0.41	0.48	0.79	0.59
Norway											
1963–68	0.94	0.61	0.00	1.24	0.36	1.15	0.65	0.46	0.29	0.91	0.46
1984–88	0.69	0.70	0.44	1.09	0.53	1.07	0.23	0.64	0.51	2.06	2.14
Finland											
1963–68	0.00	0.62	0.00	1.30	2.18	0.72	0.00	0.32	0.20	1.33	0.47
1984–88	1.04	0.62	0.72	1.47	0.64	0.80	0.17	0.33	0.14	1.35	1.72

Notes
1. See text for a definition of 'revealed technology advantage'.
2. See Table 3.7 for the full names of the technology classes.

and four Western European countries: France, the Federal Republic of Germany, the Netherlands and the United Kingdom. It shows that Western Europe, thus defined, compares very favourably with Japan, with nearly double the per capita effort, and is at about the same level of per capita inputs as the United States; the 20 per cent higher per capita American output can probably be attributed to the favourable bias towards the English language in the selection of journals used to count publications.

However, there is no obvious explanation for the paradoxical combinations of improving Japanese publication performance, compared to Western Europe, coupled with declining inputs; and with the opposite happening for the United States. In addition to inevitable measurement errors, we have no means of assessing trends in expenditures on basic research from non-government sources, especially foundations, charities and companies. Nor can we assess the degree to which Japanese basic research is simply becoming better known and more widely diffused in the English language.

Defence R & D

Table 3.3 shows the very uneven distribution of government-funded defence R & D among regions and countries. It remains highest in the United States, and is still very small in Japan. Within Europe it is heavily concentrated in France, the Federal Republic of Germany, Sweden and the United Kingdom. A high proportion of defence R & D is performed in firms in the aerospace and electronics industries. In aerospace, it is the dominant source of funds in all countries performing aerospace R & D, and (as we shall see in section III) is a source of national technological advantage. In electronics, it is a major source only in France, the United Kingdom and the United States, and (again as we shall see in section III) is *not* systematically associated with national technological advantage.

The uneven distribution of technological activities within Western Europe

Table 3.4 shows West European countries' shares of American patenting and confirms the diversity of technological performance and the impossibility of treating 'Europe' as a technologically homogeneous entity. In particular, we see the growing dominance of the Federal Republic of Germany (40 per cent more than the combined total of France, Italy and the United Kingdom in 1984-8), the decline of the United Kingdom, and the growing shares of Italy, Ireland, Spain, Austria and Finland. It also shows the concentration of technological activities (96 per cent of the EC total in the top five countries), and the importance of the EFTA countries (equivalent to 19 per cent of the EC total).

In principle, we should include the Central European countries (Czechoslovakia, Hungary and Poland) in our comparison. All may be associated with European economic integration in future. All have histories of significant technological achievement going back at least to the beginning of the century. Unfortunately, an accurate assessment of their aggregate technological capa-city is for the moment impossible. On the one hand, their very high expenditures on R & D as a percentage of GDP include activities often uncoupled from business and production; moreover, their definitions of R & D include some activities which are excluded from the 'Frascati' (OECD) definitions. On the other hand, their very small share of American patenting reflects in part their previously hostile political relations with the United States. However, as we shall see in section III below, their sectoral specialisations, as revealed by American patenting, are similar to those of their Western Euro-pean neighbours, and they clearly do have considerable scientific and technological potential.

III Sectoral patterns of national technological specialisation

So far, we have compared countries' aggregate technological performance. Table 3.5 summarises the sectoral patterns of technological advantage of the countries included in our analysis. Based on the American patent classifications, it divides technologies into 11 specified fields. The content of most of them will be clear from their titles: technologies for extracting and processing raw materials are related mainly to food, oil and gas; defence-related technologies are defined as aerospace and munitions.

Table 3.5 shows each country's or region's index of 'Revealed Technology Advantage' (RTA) in each of the technological fields for the periods 1963-8 and 1984-8. RTA is defined as a country's or region's share of all American patenting in a technological field, divided by its share of all American patenting in all fields. An RTA of more than one therefore shows a country's or region's relative strength in a technology and less than one its relative weakness. These measures correspond broadly to the measures of comparative advantage in world trade as shown in Chapter 2. In both cases, all countries by definition have sectors of relative advantage and of relative disadvantage. However, *relative* advantage does not necessarily denote *absolute* advantage. For example, Table 3.5 shows that, from 1984 to 1988, the United Kingdom had a more favourable RTA than the Federal Republic of Germany in telecommunications (1.00 *versus* 0.80). However, the Federal Republic of Germany's *absolute* share of American patenting in telecommunications remained greater than that of the United Kingdom, since Table 3.4 shows that Germany's aggregate share of American patenting in all sectors was much greater (share of Germany in European total in telecommunications = 0.80 x 40.91 = 32.72 per cent; United Kingdom share = 1.00 x 14.67 = 14.67 per cent). The very small numbers of patents in some sectors for the smaller countries (for example, Portugal and Ireland) should also be taken into account in interpreting the data in Table 3.5.

Readers may find useful the conclusions of some of our earlier papers, where we try to explain the patterns of sectoral technological advantage observable in each country (Patel and Pavitt, 1988; Pavitt, 1988b). Briefly stated, we conclude that a country's sectoral technological advantage can come from two sources. First, there are country-specific inducement mechanisms to which local firms have preferential access: chief among these will be access to defence and other public sector markets, to abundant raw materials and to idiosyncratic consumer tastes. Second, there are firm- or region-specific skills that enable firms to respond to a variety of technological opportunities and market needs: chief among these will be their own in-house mastery of one or more of the three pervasive 'technological families': electrical-electronic, chemical and mechanical (and related automobile technology).

At any one time, both these sets of factors are likely to be at work, at least in the industrially advanced countries with which we are concerned here. However, over time, we would expect countries to evolve – through firms' and countries' investments in R & D and related activities – from reliance on country-specific inducement mechanisms to mastery of one or more of the pervasive technologies. Examples would include Sweden's evolution from mining, through metals and machinery, to mastery of mechanical technology; Switzerland's evolution from textiles, through dyestuffs and textile machinery, to its present mastery of fine chemicals and mechanical technology; and even Japan's evolution from the requirements of the Imperial Navy, through the internal combustion engine and communications, to present mastery of automobiles and electronics. Needless to say, none of these evolutions are automatic or pre-ordained but require conscious entrepreneurial investment, associated with supportive public policies.

Table 3.6 Large firms in national technological activities, 1981–86

Country	National Sources of Patenting in USA (3 columns add up to 100 per cent)			Patenting in USA by national firms from outside home country (per cent of national total)
	Large Firms		Other	
	National	Foreign		
Belgium	8.8	39.7	51.5	14.7
France	36.8	10.0	53.2	3.4
Federal Republic of Germany	44.8	10.5	44.2	6.9
Italy	24.1	11.6	64.3	2.2
Netherlands	51.9	8.7	39.4	82.0
Sweden	27.5	3.9	68.6	11.3
Switzerland	40.1	6.0	53.9	28.0
United Kingdom	32.0	19.1	49.0	16.7
Western Europe	44.1	6.2	49.7	8.1
Canada	11.0	16.9	72.1	8.0
Japan	62.5	1.2	36.3	0.6
USA	42.8	3.1	54.1	3.2

Note: All columns as percentage of total national patenting in USA, 1981–6
Shell is assumed to be Dutch and Unilever British

Source: Patel and Pavitt, forthcoming

Table 3.5 suggests that countries with relatively small increases in total business-funded R & D over the past 20 years have increased their specialisation in technologies where country-specific inducement mechanisms count, while countries with relatively large increases have become more specialised in one or more of the pervasive technological families. In particular, the United States has shifted its specialisation markedly towards defence- and raw materials-related technologies, and away from electronics and automobiles, although there has also been a slightly increased specialisation in chemicals. The Japanese pattern has been the reverse: increasing specialisation in electronics and automobiles, decline in chemicals and continuing weakness in raw materials- and defence-related technologies.

The European patterns of specialisation are not so clearcut. In aggregate, they reflect German strengths in chemicals, mechanical engineering, automobiles and defence, coupled with decline in electronics (with the exception of telecommunications) and continuing weakness in raw materials. This pattern is reproduced – in whole or in part – in some of Germany's neighbours: Austria, Switzerland, and even in Czechoslovakia, Hungary and Poland. Italy and Spain share relative strengths in mechanical engineering, automobiles, fine chemicals and (as shown by more detailed data) textiles, while the earlier Swedish model is being reproduced elsewhere in Scandinavia, based on the extraction and processing of raw materials. France shows relative strength in fields closely linked to state markets (military and civilian), while the United Kingdom has become increasingly strong in defence and in a sector dependent on a strong base in fundamental science (fine chemicals), while showing particularly rapid decline in electronic capital goods and components. Only the Netherlands in Europe shows consistent specialisation over the whole range of electrical and electronic technologies, related of course to the exceptional position of Philips.

IV Structures and strategies of large firms

The strategies of large firms, and the market structures within which they operate, are major factors in the level, trends and sectoral composition of the technological activities in the countries that we are studying. This conclusion is based on data that we have compiled on the world's 660 largest, technologically active firms, as measured by their patenting activity in the United States in the period 1981-86.

Structures

Table 3.6 uses our data on patenting in the United States to compare the structure of technological activities in the early 1980s of the 11 countries that account for more than 95 per cent of the OECD's total. The first two columns show the shares of total national patenting in the United States granted to nationally-controlled large firms, and to foreign-controlled large firms, while the third column gives the combined share for other national sources (that is, other firms, government agencies and individuals). Thus, assuming that American patenting reflects national technological activities, Table 3.6 shows that 32.0 per cent of British technological activities were performed by British-controlled large firms, 19 per cent by foreign-controlled large firms and 49 per cent by other firms and organisations. The fourth column shows American patenting by nationally-controlled firms from outside their home country, expressed – like the other three columns – as a percentage of total national patenting in the United States. Thus, again by way of illustration, the technological activities of British-controlled large firms undertaken outside the United Kingdom amounted to 16.7 per cent of total British technological activities, while the equivalent proportion varied from a massive 82.0 per cent for Dutch-controlled large firms, to a miniscule 0.6 per cent for Japanese-controlled large firms. Two major conclusions emerge from Table 3.6.

First, there is considerable variety among countries in the structure of technological activities, whether measured in terms of the relative importance of national large firms, foreign large firms, other organisations, or of the importance of nationally-controlled firms' technological activities abroad. We have shown elsewhere that there is no statistically significant relationship between any of these structural characteristics and countries' performance measured in terms of either levels or rates of increase in technological activities (Patel and Pavitt, 1991).

Table 3.7 Shares of American patenting for top 20 firms in 11 technological areas (sorted according to the share in 1984-88

Technology class: 1 fine chemicals		*Years*		
	69–73	*74–78*	*79–83*	*84–88*
1 Bayer (FRG)	2.68	4.05	3.70	3.55
2 Merek (USA)	2.66	2.50	3.27	2.60
3 Ciba-Geigy (Switz.)	4.39	3.17	2.86	2.45
4 Hoechst (FRG)	1.57	2.55	2.51	2.40
5 E I du Pont de Nemours (USA)	1.79	1.27	1.27	2.05
6 Eli Lilly (USA)	1.39	2.01	1.98	1.61
7 Warner-Lambert (USA)	0.76	0.79	0.51	1.52
8 BASF (FRG)	0.55	0.86	1.06	1.26
9 Pfizer (USA)	1.34	1.13	1.38	1.22
10 Dow Chemical (USA)	1.27	1.29	1.40	1.21
11 Squibb (USA)	1.09	1.05	1.21	1.21
12 Imperial Chemical Industries (UK)	0.84	1.31	1.08	1.18
13 Johnson & Johnson (USA)	0.33	0.67	1.15	1.14
14 Hoffmann La Roche (Switz.)	1.71	1.41	1.37	1.10
15 Smith Kline & Beckman (USA)	1.26	2.22	1.44	1.10
16 American Cyanamid (USA)	2.49	1.89	1.34	1.07
17 Boehringer Mannheim (FRG)	0.99	1.62	1.50	0.97
18 Stauffer Chemical (USA)	1.14	1.57	1.11	0.94
19 Beecham Group (UK)	0.15	0.75	1.56	0.93
20 Fujisawa Pharmaceutical (Japan)	0.07	0.08	0.67	0.90
Share of the top 20	28.47	32.20	32.39	30.41
Share of all large firms	66.61	66.05	62.72	55.04
Technology class: 2 industrial chemicals				
	69–73	*74–78*	*79–83*	*84–88*
1 Bayer (FRG)	2.52	3.07	3.30	2.88
2 Dow Chemical (USA)	2.44	1.91	1.81	2.75
3 BASF (FRG)	1.40	1.70	2.16	2.18
4 General Electric (USA)	1.17	1.36	1.88	1.90
5 Ciba-Geigy (Switz.)	2.51	2.76	2.25	1.89
6 Hoechst (FRG)	1.69	2.14	2.08	1.86
7 E I du Pont de Nemours (USA)	3.44	2.49	1.78	1.66
8 Union Carbide (USA)	1.14	0.96	1.24	1.11
9 Phillips Petroleum (USA)	1.50	1.24	1.09	0.90
10 Exxon (USA)	0.84	0.66	0.87	0.87
11 Shell (NL)	0.93	0.60	0.68	0.83
12 Amoco (USA)	0.34	0.60	0.75	0.81
13 Imperial Chemical Industries (UK)	1.51	1.47	1.04	0.78
14 Texaco (USA)	0.40	0.69	0.66	0.74
15 Eastman Kodak (USA)	1.05	0.68	0.81	0.73
16 Henkel (FRG)	0.36	0.37	0.55	0.68
17 Sumitomo Chemical (Japan)	0.52	0.80	0.60	0.62
18 Rhone-Poulenc (France)	0.59	0.77	0.60	0.61
19 Mobil Oil (USA)	0.46	0.36	0.56	0.61
20 Allied (USA)	1.17	0.69	0.69	0.60
Share of the top 20	25.97	25.31	25.41	25.02
Share of all large firms	70.84	70.51	68.26	64.85

Technology class: 3 materials		*Years*		
	69–73	*74–78*	*79–83*	*84–88*
1 Fuji Photo Film (Japan)	0.24	0.59	1.06	2.61
2 PPG Industries (USA)	2.51	4.21	3.33	2.06
3 Minnesota Mining and Manufacturing (USA)	1.39	1.58	1.43	2.01
4 Corning Glass Works (USA)	2.98	2.79	2.83	1.68
5 General Electric (USA)	2.08	1.80	2.27	1.59
6 E I du Pont de Nemours (USA)	3.72	2.29	1.59	1.51
7 Hitachi (Japan)	0.18	0.33	0.71	1.33
8 Owens-Corning Fiberglass (USA)	1.82	1.89	2.27	1.33
9 Emhart (USA)	0.58	0.41	0.36	1.16
10 TDK (Japan)	0.12	0.15	0.94	1.13
11 Dow Chemical (USA)	1.25	0.84	0.53	0.92
12 Matsushita Electric Industrial (Japan)	0.30	0.19	0.43	0.92
13 Hoechst (FRG)	0.53	0.46	0.62	0.83
14 Bayer (FRG)	0.58	0.73	0.71	0.82
15 Allied (USA)	0.63	0.62	0.52	0.80
16 Saint-Gobain Industries (France)	0.94	0.76	0.76	0.77
17 GTE (USA)	0.26	0.39	0.42	0.74
18 United Technologies (USA)	0.57	0.86	0.73	0.74
19 Toshiba (Japan)	0.42	0.21	0.42	0.72
20 IBM (USA)	0.49	0.68	0.39	0.71
Share of the top 20	21.59	21.79	22.31	24.41
Share of all large firms	65.99	63.97	60.44	60.49

Technology class: 4 mechanical engineering				
	69–73	*74–78*	*79–83*	*84–88*
1 Hitachi (Japan)	0.15	0.37	0.59	0.86
2 Toshiba (Japan)	0.09	0.09	0.30	0.71
3 General Electric (USA)	1.04	0.92	0.91	0.68
4 General Motors (USA)	1.14	0.85	0.65	0.67
5 Siemens (FRG)	0.27	0.55	0.67	0.66
6 Canon Camera (Japan)	0.03	0.18	0.33	0.65
7 Toyota (Japan)	0.10	0.24	0.44	0.58
8 Aisin Seiki (Japan)	0.19	0.29	0.43	0.58
9 Nissan (Japan)	0.10	0.27	0.50	0.55
10 United Technologies (USA)	0.52	0.47	0.52	0.54
11 Philips (Neth.)	0.38	0.45	0.42	0.52
12 Westinghouse Electric (USA)	0.56	0.49	0.43	0.49
13 Allied (USA)	0.72	0.51	0.61	0.47
14 Honda (Japan)	0.03	0.06	0.15	0.46
15 ITT (USA)	0.39	0.33	0.53	0.45
16 IBM (USA)	0.48	0.40	0.40	0.44
17 Robert Bosch (FRG)	0.22	0.32	0.36	0.39
18 AT&T (USA)	0.59	0.38	0.44	0.35
19 Xerox (USA)	0.31	0.60	0.30	0.34
20 Mitsubishi Electric (Japan)	0.03	0.05	0.07	0.32
Share of the top 20	7.33	7.80	9.06	10.69
Share of all large firms	34.76	35.20	36.93	36.61

Technology class: 5 motor vehicles

	Years			
	69–73	*74–78*	*79–83*	*84–88*
1 Honda (Japan)	0.74	2.11	2.99	9.48
2 Nissan (Japan)	1.43	4.92	7.18	6.12
3 Toyota (Japan)	0.47	4.98	6.79	5.79
4 Robert Bosch (FRG)	3.67	4.80	7.18	5.14
5 Nippondenso (Japan)	1.13	2.55	3.76	3.87
6 General Motors (USA)	4.80	4.89	3.45	2.83
7 Mazda (Japan)	0.77	0.73	0.86	2.76
8 Aisin Seiki (Japan)	0.55	0.77	1.10	2.28
9 Yamaha (Japan)	0.33	0.40	1.72	1.91
10 Hitachi (Japan)	0.36	1.00	1.09	1.83
11 Mitsubishi Electric (Japan)	0.17	0.14	0.30	1.67
12 Fuji Heavy Industries (Japan)	0.08	0.33	0.74	1.61
13 Ford (USA)	2.40	1.82	1.38	1.51
14 Daimler-Benz (FRG)	2.92	1.59	1.16	1.12
15 Lucas Industries (UK)	0.99	1.48	1.68	0.98
16 Outboard Marine (USA)	0.72	0.59	0.82	0.84
17 Allied (USA)	1.24	1.33	1.93	0.78
18 Fiat (Italy)	0.63	0.49	0.89	0.77
19 Brunswick (USA)	0.33	0.52	0.61	0.73
20 Porsche (FRG)	0.44	0.24	0.21	0.72
Share of the top 20	24.19	35.68	45.84	52.76
Share of all large firms	45.62	57.12	64.77	68.14

Technology class: 6 electrical machinery

	69–73	*74–78*	*79–83*	*84–88*
1 General Electric (USA)	5.08	4.51	4.23	3.15
2 Westinghouse Electric (USA)	3.26	2.97	2.50	2.71
3 Hitachi (Japan)	0.45	1.15	1.71	2.08
4 Philips Corporation (Netherlands)	1.37	1.87	1.55	1.97
5 Siemens (FRG)	1.38	2.21	2.09	1.91
6 Mitsubishi Electric (Japan)	0.19	0.33	0.67	1.87
7 AMP (USA)	1.09	1.26	1.34	1.82
8 Toshiba (Japan)	0.45	0.39	1.19	1.70
9 GTE (USA)	1.17	1.47	1.50	1.40
10 Motorola (USA)	0.56	0.53	0.66	1.01
11 NEC (Japan)	0.39	0.50	0.45	0.99
12 Matsushita Electric (Japan)	0.88	1.14	0.96	0.98
13 General Motors (USA)	1.50	0.87	0.74	0.81
14 Allied (USA)	1.10	0.95	1.07	0.75
15 United Technologies (USA)	0.98	0.76	0.83	0.69
16 Brown Boveri (Switz.)	0.40	0.48	0.60	0.65
17 AT&T (USA)	1.39	0.92	1.00	0.62
18 RCA (USA)	0.74	1.04	0.94	0.60
19 Robert Bosch (FRG)	0.48	0.55	0.81	0.59
20 ITT (USA)	0.92	0.90	0.61	0.59
Share of the top 20	23.78	24.81	25.44	26.88
Share of all large firms	61.12	61.62	53.04	51.80

Technology class: 7 electronic capital goods and components *Years*

	69–73	*74–78*	*79–83*	*84–88*
1 IBM (USA)	8.84	8.08	7.72	5.84
2 Toshiba (Japan)	0.48	1.30	3.30	5.31
3 NEC (Japan)	2.03	1.87	2.52	4.42
4 Hitachi (Japan)	1.49	3.25	3.40	4.15
5 Motorola (USA)	2.11	2.72	2.63	3.04
6 Fujitsu (Japan)	0.38	0.42	1.90	2.98
7 Texas Instruments (USA)	2.03	2.58	2.61	2.77
8 Philips (Neth.)	2.81	3.37	2.60	2.63
9 AT&T (USA)	6.25	3.72	3.28	2.57
10 Siemens (FRG)	1.76	2.71	2.67	2.06
11 Mitsubishi Electric (Japan)	0.13	0.42	0.47	1.75
12 RCA (USA)	3.61	3.94	3.52	1.70
13 General Electric (USA)	3.40	2.38	1.81	1.55
14 Burroughs Machines (USA)	3.66	2.47	2.77	1.44
15 Honeywell (USA)	2.16	2.78	2.49	1.39
16 ITT (USA)	1.68	1.07	1.03	1.06
17 Sharp (Japan)	0.04	0.26	0.63	1.01
18 Sony (Japan)	0.35	1.35	0.66	0.94
19 Thomson-Brandt (France)	0.34	0.67	1.06	0.92
20 Schlumberger (USA)	0.79	0.79	0.92	0.85
Share of the top 20	44.34	46.15	48.00	48.37
Share of all large firms	70.08	71.15	74.31	73.34

Technology class: 8 telecommunications

	69–73	*74–78*	*79–83*	*84–88*
1 AT&T (USA)	6.51	4.06	3.83	4.29
2 NEC (Japan)	1.17	1.27	1.86	4.16
3 Siemens (FRG)	1.97	3.60	4.06	3.50
4 Philips (Neth.)	1.45	1.84	2.19	2.64
5 Motorola (USA)	1.12	1.25	1.89	2.51
6 General Electric (USA)	3.27	2.72	2.33	2.13
7 Westinghouse Electric (USA)	3.17	3.36	2.71	1.94
8 GTE (USA)	1.18	2.21	1.69	1.85
9 ITT (USA)	3.73	3.13	2.28	1.69
10 Thomson-Brandt (France)	0.76	1.05	1.84	1.50
11 Toshiba (Japan)	0.27	0.33	0.93	1.50
12 Mitsubishi Electric (Japan)	0.23	0.29	0.70	1.49
13 Hitachi (Japan)	0.41	0.85	1.38	1.39
14 RCA (USA)	1.13	1.62	1.81	1.29
15 Allegheny (USA)	0.52	0.59	0.74	1.17
16 Northern Telecom (Canada)	0.43	0.65	0.85	1.15
17 IBM (USA)	1.25	0.86	0.68	1.00
18 Rockwell International (USA)	0.52	0.64	1.30	0.88
19 Burroughs Machines (USA)	0.70	0.82	0.96	0.81
20 Compagnie Générale d'Electricité (France)	0.62	0.63	0.88	0.76
Share of the top 20	30.39	31.79	34.91	37.66
Share of all large firms	53.45	54.94	58.59	59.58

Technology class: 9 electronic consumer goods (inc. photography and photocopy)

	69–73	*74–78*	*79–83*	*84–88*
1 Canon (Japan)	0.66	2.33	4.09	6.50
2 Fuji (Japan)	1.96	3.69	3.35	5.40
3 RCA (USA)	2.67	3.34	4.68	3.64
4 Matsushita Electric (Japan)	1.56	2.46	2.52	3.41
5 Sony (Japan)	0.99	1.80	2.43	3.14
6 Toshiba (Japan)	0.43	0.45	1.11	3.13
7 Philips (Neth.)	2.40	2.79	2 47	3.10
8 Xerox (USA)	3.41	6.76	3 23	2.94
9 Eastman Kodak (USA)	6.64	3.31	3.76	2.85
10 Hitachi (Japan)	0.45	1.51	1.85	2.64
11 Konishiroku (Japan)	0.48	0.68	1.48	1.97
12 Ricoh (Japan)	1.05	0.98	1.71	1.63
13 IBM (USA)	2.99	2.39	1.91	1.61
14 NEC (Japan)	0.72	0.80	1.30	1.61
15 Minolta (Japan)	0.73	1.44	1.49	1.44
16 Pioneer Electric (Japan)	0.11	0.38	0.63	1.24
17 Sharp (Japan)	0.00	0.12	0.29	1.21
18 Siemens (FRG)	0.56	0.78	0.96	1.10
19 Polaroid (USA)	2.95	2.30	2.03	0.96
20 Nippon Gakki (Japan)	0.94	1.07	1.24	0.92
Share of the top 20	31.72	39.36	42.52	50.43
Share of all large firms	60.41	62.22	64.23	69.54

Technology class: 10 raw-materials-related technologies

	69–73	*74–78*	*79–83*	*84–88*
1 Mobil Oil (USA)	2.16	2.99	3.37	4.91
2 Exxon (USA)	3.11	2.67	2.53	2.59
3 Standard Oil, Indiana (USA)	2.59	2.41	2.60	1.88
4 Halliburton (USA)	0.53	0.84	1.18	1.77
5 Phillips Petroleum (USA)	2.40	1.70	1.89	1.76
6 Amoco (USA)	0.67	1.07	1.61	1.66
7 Texaco (USA)	2.17	3.11	2.57	1.64
8 Atlantic Richfield (USA)	0.89	0.88	0.87	1.25
9 General Foods (USA)	1.07	1.54	0.93	1.23
10 Shell (NL)	2.26	1.21	0.99	1.22
11 Signal Companies (USA)	2.94	3.02	2.11	1.06
12 Baker International (USA)	0.37	0.47	0.77	0.90
13 Union Oil Company of California (USA)	0.58	0.56	0.42	0.89
14 Deere & Company (USA)	0.98	0.66	0.74	0.78
15 Hughes Tools (USA)	0.15	0.26	0.31	0.69
16 Nabisco (USA)	0.14	0.24	0.41	0.65
17 British-American Tobacco (UK)	0.13	0.31	0.41	0.62
18 E I du Pont de Nemours (USA)	0.82	1.06	0.81	0.61
19 Philip Morris (USA)	0.20	0.16	0.50	0.59
20 British Petroleum (UK)	1.05	0.63	0.35	0.54
Share of the top 20	25.20	25.80	25.37	27.23
Share of all large firms	51.69	50.12	48.96	48.07

Technology class: 11 defence-related technologies	*Years*			
	69–73	*74–78*	*79–83*	*84–88*
1 Boeing (USA)	0.96	2.97	4.28	5.16
2 Messerschmitt-Bolkow-Blohm (FRG)	1.21	1.16	1.76	1.98
3 Oerlikon-Buhrle (Switz.)	0.91	1.05	0.80	1.35
4 Morton Thiokol, (USA)	1.01	1.08	0.50	1.27
5 Flick Group (FRG)	1.69	1.08	0.99	1.24
6 General Dynamics (USA)	0.27	0.63	1.49	1.21
7 Westinghouse Electric (USA)	0.18	0.18	0.38	1.02
8 Imperial Chemical Industries (UK)	1.37	1.21	1.87	0.99
9 Aerospatiale (France)	0.39	0.29	0.53	0.93
10 British Aerospace (UK)	1.05	0.47	0.46	0.85
11 General Electric (USA)	1.44	1.60	1.60	0.85
12 United Technologies (USA)	1.10	0.79	0.88	0.79
13 Olin (USA)	1.28	0.53	0.23	0.76
14 Grumman (USA)	0.02	0.16	0.31	0.71
15 Lockheed (USA)	0.59	0.58	0.61	0.71
16 E I du Pont de Nemours (USA)	1.26	1.21	0.76	0.68
17 Honeywell (USA)	0.37	0.03	0.19	0.68
18 Ford (USA)	0.09	0.13	0.42	0.56
19 Sundstrand (USA)	0.02	0.03	0.00	0.54
20 Halliburton (USA)	0.14	0.13	0.23	0.51
Share of the top 20	15.35	15.30	18.30	22.77
Share of all large firms	32.37	27.47	29.50	32.90

Source: Data supplied to SPRU from the United States Department of Commerce, Patent and Trademark Office

Second, as might be expected, large countries and regions are technologically more self-contained than smaller ones – in the sense that foreign technological activities (either foreign firms at home, or national firms abroad) are relatively less important. The structure of technological activities in Western Europe taken as a whole is very similar to that in the United States, with locally controlled large firms accounting for more than 40 per cent of the total, other national organisations for about 50 per cent, and foreign activities for less than 10 per cent. In Japan, large firms are more important and foreign technological activities even less so.

Strategies

In order to identify the determinants of the differences in countries' technological performance shown in this chapter, we must look beyond these structural characteristics to the technological strategies of large firms. Table 3.7 shows, for the same 11 technological fields as in Table 3.5, trends in shares of total patenting of the worlds' top 20 firms and of all the large firms in our data-base. Four sets of conclusions emerge from these data.

First, the share of our large firms in the world's technological activities varies greatly among technological fields. The share for all fields combined is about 49 per cent, but this increases to more than 70 per cent in electronic capital goods and to more than 60 per cent in other chemicals, materials, motor vehicles and electronic consumer goods. The top 20 firms in motor vehicles and electronic consumer goods have more than 50 per cent of the total. Mechanical engineering (including instruments) is the least concentrated field.

Second, the strengths and weaknesses of each region or country are in general reflected in the relatively large number of nationally based large firms appearing in the top 20. Thus, ten of the top-20 patenting firms in automobiles, and 13 in electronic capital goods are Japanese-controlled, while none are in raw materials and defence. The United States, by contrast, has 17 of the 20 raw materials firms, and the oil companies among them also make a major contribution to American industrial chemical technology; in automobiles and electronic consumer goods, however, the numbers of large American firms are as few as five. Europe's strength in chemical technology is also reflected in the patenting of locally controlled large firms (eight of the top 20 in fine chemicals and seven in other chemicals). However, the same is not true in mechanical engineering, where Europe has only three large firms in the top 20, but where its strength resides in the technological activities of the smaller firms in this much more unconcentrated field.

This pattern is consistent with a result that we have presented elsewhere, and that holds equally well for Japan, the United States and Western Europe: in each region, sectoral strengths of locally controlled large firms are explained statistically by their *number* and not by their *size*. In other words, relative technological strength is correlated with rivalry among the large firms and not with gigantism (Patel and Pavitt, 1990).

Third, we found close correlations between countries' levels, trends and sectoral patterns of technological activity and those of their nationally controlled large firms. Furthermore, the technological activities of these large firms are heavily dependent on their home base, in two senses:

(i) a high proportion of large firms' technological activities are performed in their home country: 90 per cent for all the firms in our data set, although Table 3.6 shows that the share varies widely among countries;

(ii) levels and trends in the home-based technological activities of these large firms are much more closely correlated with those of other home-based technological activities than with those of their foreign-based technological activities. In other words, country-specific features determine large firms' technological performance, rather than the reverse. Globalisation, although much trumpeted, is not yet a decisive factor and national systems are still predominant (Patel and Pavitt, 1991).

Finally, the relatively poor technological performance of the United Kingdom is associated with the following characteristics:

(i) a high level of internationalisation of technological activities compared to other countries of equivalent size(see Table 3.6). Relatively high levels of technological activities outside the United Kingdom by British firms can be found in raw materials-based technologies and in mechanical engineering. Relatively high levels of foreign-controlled technological activities in the United Kingdom can be found in electronics and in some chemical sectors;
(ii) the strong British performance in chemicals and defence-related technologies, reflected in the strong position of large British-owned firms in these sectors: ICI, Beechams, British Aerospace (see Tables 3.5 and 3.7). At the same time,

large British electronics firms have virtually disappeared below the threshold of world-wide visibility. In an electronics category combining the three electronics fields in Tables 3.5 and 3.7, British GEC ranked 58th in the world in American patenting in 1984-8, Plessey 73rd and the two combined still lower than 40th.

V Policy conclusions for Europe

The economic environment: diversity, competition and growth

The policy conclusions emerging from the above analysis are inevitably mixed given Europe's great heterogeneity. In one sense, this is a source of strength, since variety and experimentation are vital for any technological system at the world frontier. In this context, our clear statistical conclusion that rivalry among the big – rather than gigantism and monopoly – are the characteristics of national technological strength, is a clear pointer towards the vigorous enforcement of a strong European competition policy. This is particularly important at a time when the prospect of the Single Market is pushing large firms to hedge their bets (and increase their monopoly power) through mergers, joint ventures, and so-called 'strategic alliances'. Another conclusion is the importance of the macro-economic climate on the rate of growth of technological activities. Both the fact of high growth in the past and the expectation of high growth in the future encourage the allocation of resources to technological activities.

The institutional context: myopic or dynamic?

However, the creation of a favourable economic climate, while necessary, is not sufficient to ensure a satisfactory allocation of resources to technological activities. Earlier research has shown that national institutional features can help or hinder technological activities (Ergas, 1984; Dosi et al., 1990). Given the great variation among European countries in their technological levels, one of the major objectives should be to bring the whole of Europe – including central Europe and the Mediterranean countries – up to international best-practice.

In order to understand these institutional differences, we have found it useful to distinguish between national systems of innovation that we define as myopic and those that we define as dynamic (Patel and Pavitt, 1988). Briefly stated, myopic systems treat investments in technological activities just like any conventional investment: they are undertaken in response to a well-defined market demand and include a strong discount for risk and time. Dynamic systems, on the other hand, recognise that technological activities are not the same as any other investment. In addition to tangible outcomes in the form of products and profits, they also entail important but intangible by-products in the form of cumulative and irreversible processes of technological, organisational and market learning that enable them to undertake subsequent investments. In the European context, the archetypal dynamic national system of innovation is Germany, while the myopic system is the

Table 3.8 Trends in investments in science, technology and skills in France, the Federal Republic of Germany and the United Kingdom

Measure	France	Fed. Republic Germany	United Kingdom
*Govt. exp. on academic and related research** (UK = 100)			
1987	115	144	100
1980	98	147	100
Output of scientific publications (UK = 100)			
1986	59	71	100
1981	60	75	100
*Industry-funded R & D** (UK = 100)			
1988	90	191	100
1981	96	191	100
American patents granted (UK = 100)			
1988	101	268	100
1980	87	239	100
Mechanical and engineering qualifications awarded at craft & technician levels (000s)			
1987	98	134	30
1975	66	103	27
Workforce with vocational or higher qualifications (%)			
1988	47	n.a.	36
1979	38	67 (1978)	31

*Calculated using purchasing power parity exchange rates.

Sources: Irvine *et al.*, 1990 for academic and related research; Martin *et al.*, 1990 for publications; Prais, 1981 and Steedman, 1988, 1990 for skill qualifications; SPRU-OTAF data base for American patenting; OECD for R & D expenditures.

United Kingdom. The essential differences between the two can be found in three sets of institutions:

(i) in the financial system underlying business activity: in Germany, it gives greater weight to longer-term performance, when the benefits of investments in learning begin to accrue; and it provides both the information and the competence to enable the value of firm-specific intangible assets to be evaluated by the providers of finance;
(ii) methods of management, especially in large firms in R & D-intensive sectors: in the United Kingdom, the relatively greater power and prestige given to financial (as opposed to technical) competence is more likely to lead to incentive and control mechanisms based on short-term financial performance and to decentralised divisional structures insensitive to changing technological opportunities;
(iii) the national system of education and training: the German system of widespread yet rigorous general and vocational education provides a better basis for cumulative learning, especially in the engineering-based industries. The British system of higher education, on the other hand, has performed relatively well, which helps explain Britain's strong performance in a field where science and marketing count for much more than production, namely, fine chemicals.

Britain's technological deficiencies are long-standing and well-known. The shortcomings compared to Germany and the United States at the beginning of the century were described at the time by Alfred Marshall (among others) and have since been confirmed by many economic historians. They again became evident in the 1950s and 1960s as British firms were made to compete increasingly in the competitive markets of North America and Europe, rather than in a cosy, protected Commonwealth.

It has been the hope of many, and the claim of a few, that Britain's relative technological decline was stopped in the 1980s. The historically high rates of increase in British productivity in the 1980s were achieved largely by eliminating worst practice in the form of obsolete plant and working practices. Increases in future will require improvements in best practice, which depend on substantial investments in technology and skills as well as in plant and machinery. In this context, the comparisons with France and Germany in Table 3.8 are disquieting. Certainly, it shows that British performance in scientific publications remains impressive, even if – as we mentioned earlier – part of it reflects English-language bias. Otherwise, it shows an enormous gap between the United Kingdom and the Federal Republic of Germany, with the latter committing at least twice the British level of resources to technology and training twice the proportion of its labour force to vocational and higher levels. France has improved its position in American patenting and labour-force qualifications but not in R & D expenditures. There are no convincing signs in the 1980s of any change for the better in the historical trend; Britain continues for the moment to hold its own in basic research, in spite of funding cutbacks; its industrial technology remains in relative decline; and its training stubbornly backward.

It is by no means certain that the functioning of the Single Market will, in and of itself, ensure that the features associated with technologically dynamic systems diffuse throughout the Community. Competition in increasingly international product markets, coupled with the increasing international mobility of labour, capital and know-how, should in principle lead to the diffusion of best-practice systems of management. However, management behaviour is also mediated by both the financial markets on which it depends and the range of skills at its disposal, both of which will be heavily influenced by European governments, acting individually or collectively. Matters of national pride and political ideology notwithstanding, it is important that Europe takes the German systems of industrial finance and training as models and not the British ones.

Electronics

One field where 'Euro-pessimism' in the early 1980s might have been generally justified is electronics: as we have seen in Chapter 2 and in Tables 3.5 and 3.7, European countries have become progressively weaker since the 1960s in electronic capital and consumer goods, and components, and this is reflected in the technological performance of European electronics firms. However, relative performance has improved in telecommunications, and Europe has been effective in exploiting information technology in other sectors, including capital goods, design, banking, retailing and tourism. Furthermore, the relative decline of European electronics firms has been very uneven. Since the late 1960s, one group has increased shares of American patenting: Philips, Siemens, Thomson, Bosch; while others have seen decline: CGE, GEC, Ericsson, Plessey, Olivetti and AEG-Telefunken. In Part II we consider in much greater depth the performance of the ICT industries. And in the next chapter we discuss the background to Community technology policies.

4 The single market and European technology policies*

Margaret Sharp

I Introduction

Community policies towards advanced technologies were for the first time given a legal base in the Single European Act of 1987. Until that time, policies to promote research, development and technology, with the exception of nuclear power (which was covered by the Euratom treaty), and iron and steel (covered by ECSC), were based upon action under Article 235 of the Treaty of Rome. This provides general powers for Community action where no specific powers have been granted, subject to unanimous decision of the Council and consultation with the Assembly.

The Single European Act adds to Part III of the EC Treaty a Title (Title VI, Article 130F) which reads:

> The Community's aim shall be to strengthen the scientific and technological base of European industry and to encourage it to become more competitive at an international level. In order to achieve this it shall encourage undertakings, including small and medium-sized undertakings, research centres and universities in their research and technological development activities; it shall support their efforts to co-operation with one another, aiming, in particular, at enabling undertakings to exploit the Community's internal market to the full, in particular through the opening-up of national public contracts, the definition of common standards and the removal of legal and fiscal barriers to that co-operation.

The Single Act also lays down a two-pronged implementation mechanism: first, unanimous adoption by member states of a multi-annual framework programme, laying out the main scientific and technological objectives and defining priorities; second, implementation of that programme through a series of sub-programmes which can be adopted by the Council, after consultation with Parliament and the Economic and Social Committee, by qualified majority voting.

The purpose of this chapter is to look behind the formal wording of the Single Act at what has been happening in practice. To do so requires going back in time to developments in the 1960s and 1970s, to examine the unsatisfactory nature of some of these developments, the sea-change in policy which took place in the 1980s with the introduction of ESPRIT and other Community programmes and the degree to which the Single European Act made *de jure* what was already *de facto*.

*An earlier version of this chapter was published in the *Political Quarterly* for October 1990. It is reproduced here by kind permission of the editors and Messrs Blackwells Ltd, its publishers.

The ESPRIT programme captured well the spirit of the early 1980s – a spirit, on the one hand, of fear and pessimism as European firms watched American and, particularly, Japanese competitors picking off their traditional markets and, on the other, of a willingness to change and accept change. There was also, at that time, a vacuum in leadership which the Community, in the ESPRIT programme, seized, and in so doing provided one of the catalysts to its revitalisation and recommitment to the Single Market. Part of the thesis of this chapter is that developments in technology themselves underlay and have provided a considerable spur to these other developments.

The second part of the chapter relates to what is described as 'the dialectic of 1992'. Community policies for advanced technology are currently proceeding along a twin track. On the one hand is the Framework Programme, with its series of sub-programmes, many modelled on the collaborative, shared-cost basis that proved so effective with ESPRIT. While these have done much to revitalise European firms in the sectors they cover, they have also implicitly encouraged and promoted the substantial concentration that has taken place in most of the new technology sectors. Many of the sectors are now dominated by a few large firms whose size and global affiliations put them well beyond the control of national competition authorities – arguably also beyond the control of the Community authorities. The other track of Community policy, and one of the central features of the Single Market, is deregulation and liberalisation – a policy at times seemingly at odds with the collaborative ethic of the Framework Programme. The chapter argues that the two tracks need not conflict, and that 'Fortress Europe' is the wrong interpretation of Community technology policies. On the contrary, it is argued, the two complement each other. Only by maintaining an open, liberal trading environment can Europe hope to control its giant firms; yet only by 'breeding' firms of a size to compete with the IBMs or Toshibas of this world can Europe hope to remain competitive in advanced-technology sectors. The dialectic is thus resolved.

II Early moves towards a European Technological Community

The concept of the European Technological Community has never been far below the surface for those seeking to create a united Europe. It formed an explicit part of the programme for Jean Monnet's Action Committee for a United States of Europe and was subsequently translated into an action plan in Christopher Layton's book, *European Advanced Technology: A Programme for Integration* (1969). More recently it has resurfaced in Karl Heinz Narjes's (Davignon's successor as Research and Technology Commissioner) call in 1985 for the creation of a Technological Community to be built upon the Community's research and development (R & D) programmes and the various technology initiatives.

There has, however, always been a degree of schizophrenia in such advocacy, with tension between the free marketeers, whose emphasis throughout has been on competition and the diminution of internal barriers to free competition within the Community, and the mercantilists, who have constantly warned against Europe's increasing dependence on American (and latterly Japanese) technology, and whose vision centred on a more positive and interventionist approach (Servan-Schreiber, 1967; Williams, 1973). The Treaty of Rome exacerbated this schizophrenia, in so far as it did not endow the Commission with explicit powers to promote R & D or, for that matter, industry. It provided for a range of policy powers which could be used to determine the regulatory framework and help shape the market environment: competition policy, freedom of movement for capital and labour, the right of establishment, the regulation and harmonization of state aids. But no general policy framework was established for either industrial or technology policy. This left the Commission to operate only through unanimous decisions of the Council of Ministers and hostage, therefore, to any national interests considered to be of over-riding importance.

The need to operate via consensus among the Council of Ministers helps to explain the stop–go nature of some of the early experiments. DG III, responsible for industrial affairs, was not established until 1967, and then took some time to decide what its proper role should be. Initially it decided on an active stance, encouraging trans-national mergers (CEC, 1970), but became caught between West German liberalism and the more interventionist philosophy of the French. Attempts in the early 1970s to create an Industrial Policy Committee under the joint auspices of the Council of Ministers and the Commission failed, in effect leaving Community industrial policy in limbo until the end of that decade.

Science and technology policy fared little better. The first meeting of the Council of Science Ministers was held in 1967, and in a flurry of activity they commissioned studies of the potential for Community action in six broad areas of technological development – transport, oceanography, metallurgy, environmental issues, data processing and telecommunications – but detailed discussion of proposals was held up by Britain's application for membership. Eventually the initiative led to the establishment in November 1971 of COST (European Co-operation in the field of Scientific and Technical Research), a grouping centred on the Community but comprising, in fact, all 19 OECD Western European members including Switzerland, Sweden and Austria. It has become a useful, if low-key, framework for the preparation and implementation of pan-European projects, although in the latter half of the 1980s, it has been somewhat overshadowed by the EUREKA initiative (Williams, 1973; Sharp and Shearman, 1987).

In the computing sector, where Servan-Schreiber (1967) specifically warned of the growing power of IBM, European initiatives were particularly slow to get off the ground. Several attempts were made during the 1960s to develop cross-national groupings, but most turned out to be inconclusive and the attempt in 1973 to bring Siemens, Olivetti and CII-Bull together under the Unidata umbrella also collapsed amid much wrangling on all sides. While hopes for the success of Unidata were still riding high, the Commission had pushed through, in 1974, a Council Resolution backing a medium-term programme on the application, development and production of data processing systems. The hope had been that this would mark the beginning of a major EC programme in computers and electronics. With the failure of Unidata this initiative fizzled out, leaving nothing but a series of small and isolated Commission initiatives and illustrating well the problems of trying to create a coherent programme when the susceptibilities of national champions were at stake (Sharp and Shearman, 1987).

Even nuclear power, where under the Euratom Treaty there were explicit powers to promote R & D, brought difficulties and disagreements. Four joint research centres were set up – at Karlsruhe, West Germany; Ispra, Italy; Geel, Belgium and Petten, the Netherlands. But these joint research programmes fell victim to both mistaken assumptions about the future role of nuclear power (which many in the 1960s saw as central to the future economic prosperity of the industrialised world) and the nationalistic ambitions of the individual power plant manufacturers, whose traditional linkages were with American not European firms and who had, in any case, licensed American nuclear technology and were anxious to exploit it. Euratom has, however, had an important (if controversial) role in regulating and monitoring the use and transportation of nuclear materials in compliance with safeguards.

Of all the new technologies of the 1960s and 1970s, the one success story of European collaboration – aerospace – owed nothing to the EC and much to the determination of governments and their respective national champions to see collaboration through to success. Admittedly, both for aviation and space, economics rapidly ruled out the individual firm from acting alone unless heavily supported by government subsidy and even, in time, inhibited individual governments from acting alone. The lessons culled from first Concorde and then Airbus, are of the long, slow haul to collaboration, and of the importance of the industry, not governments, defining the scope for collaboration (Hayward, 1986). Similar lessons were learned, too, from the space experience with the European Space Agency (ESA). As a narrow but functionally orientated collaborative framework, it has helped to produce a coherence of objective and purpose in the closely knit community of policy makers, scientists, engineers and industrialists and has ensured that, by the mid-1980s, Western Europe had the range of space capabilities sufficient to give it a degree of autonomy (RIIA, 1987).

What lessons then are to be learned from the 1960s and 1970s? Perhaps above all that experience bred cynicism – doubts about the ability of any bureaucracy to 'pick winners' and fears that expensive mistakes at the national level would be translated into even more expensive mistakes at the supra-national level where accountability is weakened. There were, however, some positive lessons to be gained from the experience. Indeed, part of the success of the EC programmes of the 1980s lay in the degree to which these lessons have been heeded.

The first of these lessons was to recognise the need for flexibility – that not all programmes had to conform to a uniform EC pattern with participation from all EC countries but needed to be tailored to circumstances, with scope for reappraisal and review. The 'variable geometry' that was to become an increasingly important theme of the 1980s was already apparent. Second was the need for functional specificity. Programmes with limited but clear objectives were more likely to meet with success than grandiose programmes which promised the earth but could deliver nothing. Equally, it was crucial that the partners perceived sufficient benefit in the longer term to sustain their commitment through the short run and to provide the incentive to reconcile major differences in outlook and interest. In this respect the Franco-German axis was crucial. Programmes which won support from these two partners were more likely to be successful. Third was the need to align the government and industrial roles into one that was mutually reinforcing. There was little point in governments promoting programmes which were of no interest to the industrial community. In this respect programmes needed to be drawn up in line with the 'grain of the market', not against it.

III ESPRIT – a watershed in Community policies

The issue of technological dependence which had underlain early moves towards the establishment of a Community technology policy (and stimulated a rash of national policies) resurfaced in the late 1970s when, emerging from the traumas of oil crisis and recession and beset by problems of adjustment in older industries such as steel, shipbuilding, even chemicals, Europe suddenly became aware of the very fast advances being made by American and Japanese firms in micro-electronics and associated technologies. European firms seemingly blind to many of these developments were now suddenly faced by the prospect of American and Japanese competitors picking off markets which they had long regarded as theirs. Many businessmen despaired of their ability to regain lost ground – hence the term 'Europessimism' which came to describe the era.

In fact things were by no means as bad as the pessimists made out. As Chapters 2 and 3 demonstrate, in comparison with the United States and Japan, Western Europe's overall competence in advanced technology sectors (including within that definition sectors such as pharmaceuticals, new materials and aerospace), held up well. However, across the board, Europe was, and is, indeed weak in the electronics sectors, showing a poor performance in both capital and consumer goods sectors.

Why should Europe's performance in these sectors have been so poor? Since the 1960s they were sectors picked out for that strategic importance. The French, British and West German governments had all pumped sizeable support into them. The French had run a series of *Plans Calculs* and *Plans Compositants* culminating in their grand strategy for the *filière éléctronique*. The British government, like the French, acted as marriage broker in bringing firms together, on the grounds that size was a key element in competitiveness but in the 1970s had switched emphasis to a series of more generalised programmes aimed at promoting general capabilities and awareness in micro-electronics (Sharp, 1985). In all three countries lucrative public-sector contracts in data processing and telecommunications were steered towards the respective 'national champions'; and in France and Britain the defence sector, as in the United States, was an important public purchaser.

Increasingly, observers began to question whether these policies themselves were not at fault – whether promoting 'national champions' in each individual member state was not diverting European companies from taking advantage of the Community's major asset, its common market of 320 million and instead fragmenting effort between the various national markets. Increasingly, too, the lack of competition was questioned. Large companies might be necessary to achieve the relevant economies of scale and scope, but such benefits were rapidly negated if the companies concerned made insufficient effort to keep up with new technologies, complacent in their protected position in home markets.

It was against this background that Viscomte Davignon, Commissioner for Industry from 1977 to 1985, and for Research and Technology from 1981 to 1985, set in train moves which were eventually to result in the ESPRIT programme. The model was the Japanese Very Large Scale Integration (VLSI) programme of the mid-1970s which had successfully launched the Japanese into making mainstream memory chips. Like this programme it was to be a strategic programme of 'pre-competitive research' – that is research necessary to underpin new developments but not in itself having immediate commercial application. But it was not to be a heavy-handed, top-down programme designed by Community bureaucrats, and it was to be collaborative across national boundaries to get the national champions thinking along Community, not national, lines. (The Commission's definition of pre-competitive research is 'projects which competitors agree to do jointly' on the grounds that competitors will not wish to cooperate on near market R & D. This does, however, beg a number of competition policy issues.)

Davignon played a vital role in developing the ESPRIT programme. In his evidence to the 1985 House of Lords' Select Committee on the European Communities, he outlined the three factors which had motivated his attempts (House of Lords, 1985). In the first place, he had been struck by the 'very distinctive difference in performance' between the industries of the United States, Japan and the EC. Second, he had felt that the time had come for Community competence to be upgraded to reflect more accurately the state-of-the-art; and third, he had become aware of the fact that no real incentive existed for cross-border collaboration. Any EC-level solution therefore needed a new approach to policy development. Until then the Commission had tended to work with research directors or their equivalents and initiatives had come unstuck because they had been unable to carry them higher up the hierarchy. Davignon determined to liaise with only the very highest levels of company management, to define priorities with them and thus secure their commitment to subsequent programmes.

During 1979–80 Davignon therefore invited the heads of Europe's leading electronics and IT companies to a series of Round Table discussions. The 'Big Twelve', as they came to be known, comprised ICL, GEC and Plessey from Britain; AEG, Nixdorf and Siemens from Germany; Thomson, Bull and CGE from France; Olivetti and STET from Italy; and Philips from the Netherlands. Davignon received a more favourable response than the Commission had received from their lower-level counterparts in earlier years. By September 1980 an outline programme began to emerge. The idea was to develop a European strategic programme based on the collaboration among the major European companies, their smaller counterparts and universities and research institutes. This would focus on the pre-competitive end of collaborative research, a stance which from the Commission's viewpoint neatly avoided the issue of competition policy. (Under Articles 85 and 86, collaboration for the purposes of pre-competitive research is granted a block exemption; collaboration at the development stage – competitive research – is not allowed.) By May 1982 these had been worked into a full proposal and the Commission's paper 'Towards a European Strategic programme for Research and Development in Information Technology' was put to the Council and subsequently to the Versailles European Summit in June 1982. The response was favourable and by December 1982 the Commission had the go-ahead for the first pilot phase costing ECU11.5m (£8.5m).

The pilot phase in 1983–4 was a deliberate part of the Davignon strategy. Given the doubts he encountered from the participants in his Round Table, in particular over the capacity of the Commission to mount an effective programme which would not become bogged down in bureaucratic delays, his strategy was one of the 'toe in the water' – see how the pilot phase goes before making a further commitment. A special task-force, mainly recruited from industry, was set up to handle applications and to cut through the Brussels red tape. The call for proposals went out in February 1983, and contracts began to be signed in May that year. By September, 38 projects had been launched which were later to be incorporated into the main part of the programme. Over 80 per cent of the first round of contracts went to the 12 'Round Table' companies (comprising, it has to be said, 70 per cent of the industry).

Encouraged by the success of the pilot phase, the Commission rapidly pushed ahead with its full plans. Those comprised a ten-year programme (1984–93) with an overall budget (50 per cent Community funded; 50 per cent industry funded) of ECU1.5 bn (approximately £1 billion). The first five-year phase was to concentrate on the pre-competitive stage of developing the technology in three areas (microelectronics, advanced information processing and software technology) and two fields of application (office systems and computer-integrated manufacturing). These plans were put to the Council in November 1983, but held up until February 1984 by British and German reservations over budgetary costs.

The first call for proposals under the full programme went out in March 1984 and met with a huge response. The 226 projects eventually selected for the programme's first phase involved 240 firms (57 per cent of these from firms with fewer than 500 employees which count as small or medium-sized business (SMEs)) and 210 research institutions. Three-quarters of the research projects involved collaboration between firms and academic research units. By January 1987, a total of ECU1.36 bn (approximately £900m) had been committed, almost the whole of what had originally been seen as a ten-year programme.

An important feature of ESPRIT has been the openness and commitment it has required of the firms linked into it. Project proposals have to be submitted in reply to open invitation. Each project must involve at least two independent industrial partners from separate member states. Costs are generally co-financed by the EC and industry on a 50/50 basis although academic institutions receive 100 per cent funding. Research results may be patented but have to be shared between all the participants in any given project who are free to apply them commercially, and preferential access is then granted to other ESPRIT participants outside that project. These guarantees are the cornerstones of the whole ESPRIT process.

The success of ESPRIT led to immediate demands for a second and yet more expensive programme. Plans for Phase II were, however, delayed by the protracted negotiations over the Framework Programme and did not get underway until 1988, with a projected five-year life over 1988–92. In fact under the renegotiated Framework Programme for 1990–94, a somewhat lower priority was subsequently given to information technology (see below).

ESPRIT II was roughly twice the size of ESPRIT I with a total spend over its five-year life of ECU3.2 bn, the expenditures being split, as with ESPRIT I, 50/50 between the Commission and industry. (The cost to the Commu-

Table 4.1 The main Community programmes for promoting new technolog

Name	Dates	Budget Cost* (ECUm)	Main objectives
ESPRIT European Strategic Programme for R & D in IT	I 1984–88 II 1988–92	800 1,530	To promote European capabilitie and competitiveness in IT technologies, primarily micro-electronics and systems development.
RACE R & D in Advanced Communications Technologies for Europe	Definition Phase 1985–87 Main Programme 1988–92	21 460	To establish European competen in broadband communications by developing the equipment, standa and technology necessary for an Integrated Broadband Communications (IBC) system.
BRITE/EURAM Basic Research in Industrial Technologies Advanced Materials for Europe	BRITE I 1985–88 EURAM I 1986–88 BRITE/EURAM 1989–92	100 450	Support for industrial R & D wh upgrades technological/materials base of production.
BAP Biotechnology Action Programme	1985–89	75	Support for infrastructure development in biotechnology wi particular emphasis on research training.
BRIDGE Biotechnological Research for Innovation, Development and Growth in Europe	1989–93	100	As for BAP, but emphasis put o larger, more comprehensive proj in areas such as advanced cell culture, molecular modelling etc
ECLAIR European Collaborative Linkage of Agriculture and Industry through Research	1989-94	80	Industrial applications of advanc biotechnologies in agro-industria sector, with emphasis on use of agricultural produce as industria raw material.
FLAIR Food Linked Agro-Industrial Research	1989–94	25	
COMMETT Community Programme in Education and Training for Technology	Phase I 1987–89 Phase II 1990–92	30 30	Cooperation between university industry in training programmes for innovation (a) through estab ment of university/enterprise pa ships and (b) through transnational exchange of stude and staff.

*The figures given as budget cost exclude any industrial contribution for the ESPR other programmes which are shared cost programmes; total expenditures under the programmes are therefore approximately double the amount of the budget cost quo

Sources: EC, Brussels

nity's budget over the five-year period was to be ECU1.6 billion. This was committed by the spring of 1990.) The ESPRIT II programme was streamlined into three main areas of research – micro-electronics, IT processing systems and applications technologies – and there was greater focus on what was called 'demand driven' aspects of the programme, for example greater emphasis on Application Specific Integrated Circuit (ASIC) technology. As with ESPRIT I, considerable emphasis was given to the development of open systems of data-processing such as OSI (Open Systems Interconnect) (see Chapters 11 and 15). ESPRIT II also incorporated a number of demonstration projects known as Technology Integration Projects (TIPs) designed to pull together various strands of other work and show how they interlink. For example, work on desk-top workstations is being linked with theoretical work on parallel architectures to offer greater finger-tip capabilities (DTI, 1990). Although nominally still pre-competitive, there is a good deal in ESPRIT II that in fact comes close to competitive research, as those involved in Phase I seek to develop some of the ideas promoted at that stage. Its popularity among manufacturers and research institutions can be gauged from the excess of applicants for funds available – in the 1988 funding round the budget was ten times over-subscribed. The SMEs in particular are very positive about ESPRIT and the Commission now claim that 70 per cent of the ESPRIT budget goes to SMEs, a considerable turnaround from its early days.

IV The Single European Act, the Framework Programme and other policies for advanced technologies

Since 1985, ESPRIT has served as a model for other European programmes promoting new technologies. There is not space to go into these in detail, but Table 4.1 summarises the most important programmes currently in being. (More are being added all the time.) It is worth noting from Table 4.1 how ESPRIT dominates in terms of spending, being more than twice the value of the next largest programme – RACE (telecoms). Both ESPRIT and RACE are administered by DG XIII which deals exclusively with information technology and telecommunications and incorporates the rump of the old Task Force set up to administer ESPRIT. DG XII, which deals with science and technology, administers most of the other programmes and is generally regarded as more traditional in its approach. (It is perceived to be more bureaucratic, but also more *laisser-faire* compared with an element of *dirigisme* seen within DG XIII.) The largest of the DG XII programmes is BRITE/EURAM, whose aim is to upgrade industrial technology and materials. The biotechnology programmes pale to insignificance beside the expenditures on IT and to date have largely funded academic research. Indeed one of the criticisms of the programmes is that grants have been too small to attract major industrial participation – a criticism which the new BRIDGE programme aims to overcome with its emphasis on industrially orientated projects. More recently, the Commission have begun to embark upon a series of programmes aimed at promoting education and training. COMMETT is the oldest and largest of these, now entering its second phase, but

Table 4.2 1987–91 Framework Programme budget

	Projected expenditures ECUm	% of total
I **Quality of life**		
Health	67	1.2
Radiation protection	60	1.1
Environment	292	5.1
II **Towards an information society**		
IT (ESPRIT)	1,534	27.3
Telecommunications (RACE)	462	8.2
Applications of IT	105	1.9
III **Modernisation of Industry**		
Manufacturing (BRITE)	396	7.0
Advanced materials (EURAM)	205	3.6
Raw materials	65	1.5
Standards	188	3.3
IV **Biological resources**		
Biotechnology	121	2.2
Agro-Industry	88	1.6
Agricultural competitiveness	56	1.0
V **Energy**		
Fission	472	8.4
Fusion	902	16.0
Non-nuclear	190	3.3
VI **S & T for development**	67	1.2
VII **Exploiting the sea-bed**	67	1.2
VIII **S & T Cooperation**		
Stimulation	176	3.1
Large Facilities	25	0.4
Forecasting	22	0.4
Dissemination	56	1.0
Total	5,617	100.0

Source: DTI, *Guide to European Community Industrial R & D Programmes*, July 1990 edition

programmes such as SCIENCE (to promote collaborative research and exchanges with the scientific community) and SPES (to promote similar objectives in economics) are as yet very new and take the Commission into new territory (the promotion of basic research and education). There are also a series of projects aimed explicitly at developing applications of technology – hence AIM (medical technology), DRIVE (automobile and IT), DELTA (educational technology). None at present involve expenditures on a par with those identified in Table 4.1.

Although the programmes all differ in aims and levels of expenditure, all share four common characteristics:

(i) they are collaborative between organisations and between countries – to qualify all applications must involve firms and/or institutions from two or more Community countries;
(ii) where there are industrial partners (all except the educational programmes), funding is on a shared cost basis with industry meeting 50 per cent of the research costs;
(iii) while the Commission identifies priorities and broad areas of research, actual research projects are chosen by the applicants and hence the programmes are to some degree demand driven;
(iv) each project, once agreed, is subject to a tight timetable and monitored through a system of programme management. Those dealt with by DG XII (Science and Technology) are subject to peer review assessment prior to project approval.

The basis for these policies is now provided by the 1986 Single European Act which, given that ESPRIT I was well underway in 1985 when the Act was first mooted, effectively made legal what was already happening in practice. The Single Act also introduced the notion of the Framework Programme – an overall programme looking five years ahead, clarifying objectives and identifying, with funding provision, general expenditure priorities. The Framework Programme adopted in 1987 related in fact to the years 1987–91 (inclusive) and had been under discussion for nearly two years, the original hope being that it would be agreed by the Council of Ministers in late 1986. British objections to the level of funding (influenced partly by the wish to use this as a counter in the wider budget discussion on the funding of agriculture) held up agreement on the budget until September 1987 (Sharp, 1989c). The final sum agreed amounted to ECU5.6 bn, although Narjes' original bid in 1985 was for funding at the level of ECU10 bn (*Financial Times*, 7 March 1985), with the British holding out for spending at ECU3.6 bn, little above the spending levels of the years 1983–86 (*Financial Times*, 9 April 1987). Even at the higher level finally agreed, spending amounted to only 2 per cent of the total EC budget, whereas agriculture consumed over 70 per cent. It is not surprising that the recent review of the Framework Programme commissioned, as required, in its third year of operation, was at pains to stress that the Framework Programme itself needs to be considered in the light of overall community priorities (Aigrain, 1989).

The first Framework Programme (for 1987–91) is summarised in Table 4.2 which identifies both broad objectives (for example, quality of life etc.) and more specific programmes. It will be noted that the IT budget (line 2), at over ECU2,000m, was very considerably larger than any other, with ESPRIT and RACE the two main items in this programme. But note also that the second largest expenditure went not to the BRITE programme or biotechnology, but to energy, where fusion and fission power together still consumed resources equivalent (almost) to the total ESPRIT programme. The fusion expenditures under the JET (Joint European Torus) and the NET (New European Torus) programmes funded research centres at Culham, near Oxford in the United Kingdom, and at other smaller research centres in France, Italy and the Federal Republic of Germany. The fission

Table 4.3 1990–94 Framework Programme budget

	Projected expenditures ECUm	% of total
I **Enabling technologies**		
1 *Information and communications technologies*		
Information technologies	1,352	23.7
Communication technologies	489	8.6
Development of telematics	380	6.6
2 *Industrial and material technologies*		
Industrial and materials	748	13.1
Measurement and testing	140	2.5
II **Management of natural resources**		
3 *Environment*		
Environment	414	7.3
Marine Sciences	104	1.8
4 *Life sciences and technologies*		
Biotechnology	164	2.9
Agriculture and agro-industry	33	5.8
Biomedical and health	133	2.3
Life sciences for Third World	111	2.0
5 *Energy*		
Non-nuclear	157	2.9
Fission safety	199	3.5
Controlled fusion	458	8.0
III **Management of intellectual resources**		
6 *Human capital and mobility*	518	9.0
Total	5,700	100.0

Source: DTI, *Guide to European Community Industrial R & D Programmes*, July 1990 edition

expenditures reflected mainly continuing expenditures on the Joint Research Centres (JRCs) set up in the 1960s. Attempts to diversify the JRCs into, for example, new materials and ceramics, have not been very successful and gave continuing cause for concern (Aigrain, 1989). The BRITE/EURAM programme came third in terms of expenditure, after energy, then the quality of life (mainly environmental R & D) followed by biotechnology and the agro-industrial technologies and the programmes aimed at encouraging cooperation in science and technology, including the promotion of basic science.

One of the main conclusions to emerge from the 1989 review of the Framework Programme (Aigrain, 1989) was the need to promote what was described as the 'scientific humus' of the Community which included not merely exchange programmes but funding European centres of excellence. The review board generally endorsed the Framework Programme and reiterated the commitment of the Community to fund only pre-competitive R & D, although they warned against the inevitable shift of the programmes, as they matured, towards more competitive elements of R & D. This, they

Table 4.4 EUREKA funding commitments, September 1990

Technology	ECUm.	%
Medical and biotechnology	499	6.5
Communications	1,448	19.0
Energy	548	7.2
Environmental	876	11.5
Information technology	1,755	23.0
Lasers	305	4.0
New materials	238	3.1
Robotics and automation	1,218	15.9
Transport	750	9.8
Total	7,637	100.0

Source: EUREKA Secretariat, Brussels, 1990

argued, should either be funded by industry or, where continuing assistance was required, by EUREKA. (See below.)

In the light of the Review Panel's assessment, a new Framework programme, Table 4.3, was negotiated for the years 1990–94 with funding at much the same levels (ECU5.7bn or £4.2m over the full five years) but with a major revision in priorities, in particular a reduction in the sums devoted to nuclear energy (fission and fusion) and a major increase in the sums being devoted to environmental research and academic science and technology. In addition, instead of the 'Action Line' budget of the 1987–91 programme, the budget is now divided into three main categories – Enabling Technologies (IT; telecoms; BRITE); Management of Natural Resources (environment; biotech; agriculture and food; energy); and the Management of Intellectual Resources (academic programmes). This new budget is now operational, with concomitant changes in, for example, resources for the energy sector, where the previous programmes are now being run down. The Framework Programme is seen to be a rolling programme with forward budget commitments which can be adjusted over time to take account of changing priorities. It will be interesting to see how smoothly the adjustments in the energy sector are effected. As we noted earlier, it is the only sector where the Community had actually established its own research institutes.

The relationship between the EC programmes and EUREKA is complex. Originally EUREKA was proposed by President Mitterrand in April 1985 as a counter to the American Star Wars programme when it looked as if the latter might poach Europe's ablest scientists. As it has worked out the EUREKA programme is far from the flamboyant programme envisaged by Mitterrand and instead constitutes an umbrella mechanism for encouraging firm-to-firm collaboration. It applies to all Western European countries, not just the EC, and, unlike the Framework Programme, has no central funding. Projects 'badged' under EUREKA (to verify that they constitute a *bona fide* collaboration) instead qualify for national R & D support funding, the terms of which vary considerably from country to country.

As Table 4.4 indicates the total financing cost of projects agreed under EUREKA by September 1990 amounted to ECU7,637m (£5.34 bn) – more than the total five-year spending commitment under the new Framework Programme. However, this figure should be taken cautiously. First, it is the *gross* cost, including both government and industry expenditures on projects, whereas the Framework budget only shows the Commission costs, not industry's matching expenditures which double the value of the industrial parts of the programme. Second, the Eureka figures reflect commitments to expenditure and there is no guarantee (and no check kept) on whether the project actually goes ahead. It should be recognised that the EUREKA secretariat is primarily a match-making body and has no funds and no monitoring or evaluative role in relation to programmes. In addition, in sharp contrast to the Framework Programmes, there are no rules for sharing IPR, it being up to partners to negotiate their own agreements (which are confidential).

EUREKA has, however, provided the home for a number of major European projects which combine public- and private-sector funding. The most notable (and expensive) of these are PROMETHEUS, a ECU600m programme aimed at developing technology for the 'intelligent' motor car (see Chapter 16); the HDTV (High Definition Television) project, a ECU400m project among Europe's main consumer electronics firms aimed at developing a common standard for the new generation of television transmitters and receivers (see Chapter 11); and JESSI (Joint European Structures on Silicon Initiative), the ECU3bn project aimed at developing the next generation of semi-conductors (see below and Chapter 5). Although initially EUREKA was seen as a competitor to the EC programmes – indeed much of the initial British support for EUREKA stemmed from a desire to counter Community ambitions on new technologies. Over time, however, the two have become complementary with EUREKA concentrating on the competitive end of R & D and the EC programmes, pre-competitive R & D. In JESSI, the EC have actually joined forces with EUREKA to fund the programme jointly with the large electronics firms and national governments.

There remains, however, an inevitable element of overlap between the various cooperative R & D ventures promoted on a European basis. Besides EUREKA, the COST programme still exists (and is funded separately from the Framework budget) and embraces non-EC states. There are also the various specialist initiatives such as CERN (particle physics), EMBL (molecular biology) and ESA (space), all of which are funded by different groups of European countries and have nothing to do with the EC – variable geometry at its best. However, as the EC increases its presence in these areas, relationships have become more complex. Indeed, yet another of the Review Board's recommendations was that the Community should avoid duplicating such programmes, but should aim to establish 'synergistic' relationships.

V Collaboration, confidence, concentration and the Single Market

ESPRIT has proved a turning point in policies towards advanced technologies in more than one sense. As has already been noted it marked a new departure in style for EC policies with the collaborative, decentralised, project-based programme replacing the top-down style of earlier years. It also, as we have seen, provided a model for a whole new range of Commission-based policies latterly brought under the aegis of the Framework Programme. It even, arguably, stimulated the setting up of a competitor programme – EUREKA – which was fairly rapidly tamed and has become a complement rather than a competitor.

Psychologically, however, ESPRIT has proved far more potent in impact than either its example, or, for that matter, the projects it sponsors would imply. (Indeed, there are those who would argue that the projects sponsored were of relatively low-technological value.) In the first place, it proved to be a turning-point in the confidence of the European electronics industry. Its collaborative framework perforce brought the main firms in this sector – the old national champions – together and provided them with a channel for exchanging views and information. (It was surprising how few of them previously had had any regular means of doing this.) In so doing it also provided a mechanism for arriving at a common view of what the future held in store. That common view, around which there proved rapid convergence, was that in the long run there was little protection to be gained from tariffs or other protective devices given that foreign competitors could rapidly circumvent them via inward investment: moreover, with the swing to neo-liberalism, even long-established public purchasing monopolies, such as telecommunications and defence, were vulnerable.

The logic led inexorably to the view that to compete successfully, even *within* Europe, these erstwhile national champions needed to set their sights on global markets and global competitiveness. As Davignon had hoped, ESPRIT proved an important stepping stone in the route from national champion to global competitor. In turn this created an important and vocal constituency in the Community pressing for the completion of the internal market for once these firms had lost their national champion status, it was imperative that they maximised the advantage to be gained from the single market. Behind Delors and Cockfield in 1985, therefore, were the leaders of Europe's major firms, impatient for the end to such things as customs' delays at borders, conflicting national standards in data processing or arcane rules on property ownership.

ESPRIT has also proved to be a catalyst to a far wider reorganisation of European industry, affecting both high-tech and low-tech industries. Collaboration was infectious. Admittedly something of the trend was visible before ESPRIT, but the period since 1983–4 has seen a substantial rise in joint ventures and other forms of strategic partnership both within the Community and with firms outside the Community (Sharp, 1989b). By its nature, collaboration is often temporary. Firms collaborate over R & D for the development of a specific product; once that product is in the market place the collaboration often ends. Successful collaborations frequently result in merger (as indeed sometimes do unsuccessful collaborations). It is not therefore surprising to find European industry in the latter half of the 1980s in the midst of a merger boom affecting all sectors of industry – from food-processing (Nestlé–Rowntree) to heavy electrical equipment (ASEA–Brown Boveri). What the Colonna Report of 1970 tried ineffectually to promote – the cross-country European merger – is now happening seemingly every day.

The electronics sector in particular has experienced a major concentration, reflecting in most sectors the increasing importance of economies of scale and scope. The chapters in Part II detail these developments. Suffice it to say here that in all the main sectors of electronics – semiconductors, computers, telecommunications, consumer electronics – where there were in 1980 six or more companies, there are now usually only three major operators – a German and a French-based firm and one other (although this pattern is not invariable). The most dramatic moves came in telecommunications where the French firm Alcatel – the telecom subsidiary of the French electrical and electronics giant, CGE – in 1987 took the ITT's interests in Europe, in the process converting what had been in 1975 a comparatively small company into the second largest telecommunications equipment company in the world. The French firm Thomson, likewise, by acquiring the American RCA's manufacturing interests, became one of the world's largest manufacturers of TV sets. The moves have also consolidated Siemens dominant position in electronics and telecommunications in Germany. The national champions of yesteryear have thus been transformed into major European-based multinationals. How far they have really shed their national champion mentality is a topic to which we shall return in Part II of this book.

Collaboration may have led the way, but there is little denying that major impetus to concentration has also come from competition, or more specifically from the threat of competition likely to emerge in the wake of 1992. The nexus is an involved one in which cause and effect are intertwined and it is not always possible to disentangle. Briefly, the logic and the thesis developed in this chapter are as follows. World developments in technology, and particularly in information and communications technologies, led to the emergence of the United States and Japan as the strongest players in the global context, threatening the fiefdoms established by the European national champions; this led to Europessimism, which in turn caused Davignon to promote ESPRIT; ESPRIT (combined, it has to be admitted with recovery from a major recession) worked like a dream, restored the confidence of the European firms in their own abilities, got them working together and promoting, on the one hand, the completion of the internal market and, on the other, further collaborations and mergers; this latter trend was then substantially reinforced by moves towards the Single Market, with the result that, paradoxically, what was intended to be a major stimulus to competition has ended by promoting a substantial increase in concentration.

VI The dialectic of 1992

Where does this get us? To outside observers there appears growing conflict between the main objective of the Single European Act – more competition in the internal market – and what has always been seen as a subsidiary objective, namely the strengthening of the supply-side of the European economies. The Single European Act and 1992 is supposed to be all about competition. Consider the following extracts from the Cecchini Report, taken from the section headed 'The Competitive Vista of 1992':

> 'A dramatically new environment awaits consumers and producers alike in the integrated Community market post-1992. The removal of a whole range of non-tariff barriers – frontier red tape, closed public procurement, a plethora of differing product standards – leads to an immediate downward impact on costs. But this is merely the primary effect ... Much more substantial gains will be generated by ... a new and pervasive competitive climate. One in which the players of the European

economy – manufacturing and service companies, and consumers of their output – can exploit new opportunities and better use of available resources ... For firms the era of the national soft option will be over.' [Cecchini, 1988:73–4.]

Admitted, Cecchini is looking to dynamic as well as static gains from increased competition, above all to the removal of the inefficiencies that stem too readily from the quiet life of protected monopoly profits. But is this really what is happening? Or is the process being dominated by the other 'track' of the Single European Act – that which seeks to promote competitiveness by the more traditionally mercantilist route of promoting R & D in strategic sectors? Indeed, is the mercantilism dominating the process and creating instead a Fortress Europe?

The existence of the 'twin tracks' of 1992 is apparent in many sectors, with simultaneous pressures towards liberalisation and deregulation on the one hand and towards collaboration and concentration on the other. Among advanced technologies, it is best illustrated by developments in telecommunications. (See Chapter 8 for fuller details.) The emergence of the RACE programme with its commitment to the integrated broadband network has already been mentioned, a development very much in line with the traditional, ordered public-utility approach to network provision which seeks to exploit the maximum *social* benefit (externalities) to be gained from an extensive network. Simultaneously, and anticipating the liberalisation moves of 1992, there has also been major concentration and restructuring, with the emergence of three leading equipment firms within Western Europe – Alcatel (CGE), Siemens and the Swedish firm Ericsson – all of whom are collaborating extensively within the RACE and ESPRIT programmes.

Alongside this strengthening of the supply side, the EC Commission has issued a series of directives, using for the first time powers under Article 90 of the Treaty of Rome (which empowers the Commission in certain circumstances to take action to prevent undue preferential treatment of public enterprises or similar establishments) and seeks to force the pace of liberalisation. In the summer of 1988, it issued its first directive setting a timetable for the liberalisation of (telecommunications) terminal equipment. In October 1988 came another directive aimed at opening the public procurement of network equipment to competitive tender. In December 1988 came a directive calling for the liberalisation of network use for all except voice services by April 1990. Further measures, dealing with access to the publically owned voice networks, are addressed by the Commission in its 'Open Network Provision' directive, a draft of which was issued in April 1989. This lays down rules for tariffs, standards and the use of private networks – issues in the heartland of any telecom authority. The strong stance being taken by the Commission is an indication of the importance they attach to these issues – the 'competitive' track of the Single Act – and the degree to which they wish to force the pace on liberalisation and deregulation. (Equally, it has infuriated some member states, notably the French, who are taking the Commission to the European Court over their use of Article 90.)

The 'twin tracks' of the Single Act in many respects echo the tension that has been implicit within the Community since its foundation, between the minimalists who saw competition policy as the only necessary tool of industrial policy and the maximalists who from the start argued for more interventionist supply-side powers. With the Single European Act the maximalists' stance has been legitimised, but the tension has not been reduced. On the contrary, in many senses it has surfaced with greater intensity; witness the developments in telecommunications.

Yet it can be argued that the tension is creative rather than destructive – that the twin tracks of the Single Act represent a dialectic whose resolution explains all. Fortress Europe is the wrong interpretation of 1992. The logic is that to be competitive, Europe had to transform its national champions into firms which could compete against all-comers in world markets. This inevitably meant firms of a size and influence beyond the control of the nation state, even beyond the control of the European authorities. To protect such firms – to turn them from national into European champions – risked defeating the whole purpose of the exercise which was to raise and maintain their competitiveness. 'Control' therefore has to come from the market, from these firms having to compete with their Japanese and American competitors in home markets as well as in foreign markets. Liberalisation and deregulation therefore become an essential complement to collaboration and concentration – the twin tracks to competitiveness. This is summed up by another quotation from the Cecchini Report:

> 'In short, strengthening the European competitivity leads, so to speak, to the reconquest of the European market. Failure to meet the demands of competitivity does not mean that the challenges of the (Single) European market will not be mastered. They will. But not by Europeans.' [Cecchini, 1988:75].

There is, of course, another point of view. Cynics can view developments as a relatively smooth transition from national to European champions and view the behaviour of the European-based multinationals in relation to, for example, voluntary export restraints, as highly opportunistic (see Chapter 10). The issue of control is an important one – even the largest superstates are no longer able to cope with some multinationals – and we shall return to it in the final chapter of the book.

In the meantime we prefer to stick to our more positive view of the European technology programmes. In themselves, the various projects may be relatively unimportant: what has been important has been the degree to which they have acted as a catalyst in getting industry in the various member states to 'think European', helping to form linkages, networks and groupings and to accelerate moves towards the Single Market. It may be that there is no further requirement for such pressures – that they have achieved what was needed and that now is the time to dismantle these programmes. Part II of this book looks in greater detail at the electronics sectors and at the twin tracks of de-regulation and liberalisation on the one hand and support and protection on the other. In Part V we return to the policy issues and in particular to the question of whether Europe needs a technology policy.

Part II
Information and communication technology

Preface to Part II

The chapters in Part I have reviewed the competitive position of the EC in the context of world-wide developments in technology, economic structure, international trade and the global environment. From this review it is clear that although the EC is one of the richest regions of the world and has many advantages, it is faced with a serious erosion of competitive strength in some of those industries which are most crucial for future competitive performance in the 1990s.

The information and communication technologies (ICT) are important not only for global competition but also because they offer some possibility for a switch away from an energy- and materials-intensive pattern of production to one which is less degrading and threatening to the environment.

As Chapter 4 has shown, the recognition of the European lag in ICT and of the need for cooperative action to retrieve the situation was a major spur to the negotiations leading to the SEM in 1992. It is still possible to argue about the technical advances which can be attributed directly to the ESPRIT or EUREKA programmes. But there is no doubt that these and other similar programmes played a very important part in bringing EC firms, universities and governments together to achieve certain common objectives. They also stimulated steps towards the re-structuring of the ICT industries.

The chapters in Part II examine in detail the progress of this re-structuring. In particular, they confirm that in each of the main areas (components, computers, software, consumer goods, telecommunications) there was indeed a serious problem of relatively weak 'national champions' which were too small to compete in the big league and extremely vulnerable to Japanese and other competition. The EC ICT industries were losing ground almost everywhere in world markets but especially outside Europe. Only voluntary export restraints and, in some cases, non-tariff barriers prevented a still more serious loss of market share in the EC itself.

However, Part II also shows that significant progress has been made in the re-structuring of the EC ICT industries during the 1980s. Especially in components Europe does now have firms which have the strength in R & D and the financial resources to compete in the global first division. There are also some encouraging developments among small and medium-sized firms in components. The first chapter describes these developments. In this and the other chapters of Part II it has been necessary to make some use of secondary and unofficial statistical sources. Official statistics still lag too far behind the pace of change in these rapidly growing industries.

5 The European semi-conductor industry; resurgence and rationalisation

Mike Hobday

Introduction

Semiconductor (SC) technology is at the heart of the telecommunications, computers, consumer electronics and other information technology (IT) industries. Sustained advance in SC technology has led to continual reductions in the cost of IT equipment. Some firms, including NEC and Fujitsu of Japan, believe that market performance in IT generally depends on a strong in-house capacity in SCs.

During the 1980s Western Europe produced around 12 per cent of world SC sales and this share was slowly declining – the bulk being produced by American and Japanese firms. The United States and Japan each accounted for over 40 per cent of world sales. However, during the mid-1980s, the European SC industry appeared to gain a new lease of life. Governments began to coordinate their strategies at a pan-European level. Major IT producers formed strategic alliances to rationalise R & D investments and overcome the fragmentation of the European market. Europe committed itself to a massive investment in SC technology in an effort to catch up with the United States and Japan.

This new commitment to SCs was based upon the belief that the ability to make and design highly integrated semiconductors such as DRAMs, ASICs and microprocessors (see glossary) was the key to globally competitive European IT firms. Conversely, a continuing weak and fragmented European SC industry could well result in competitive disadvantage across the entire IT complex.

The purpose of this chapter is to review the upheavals in the European SC industry and to assess how far innovative and market performance has improved as a result of ESPRIT and other EC programmes. Section I examines the poor historical performance of European chip producers and illustrates the pressing need for a major European initiative in the mid-1980s. Section II describes the main features of Europe's resurgence in SCs: pan-European government IT initiatives; strategic alliances among the largest SC producers; and the emergence of innovative start-up companies. Section III turns to the challenges confronting European producers. These include: the survival of the start-up companies; strong competition from Japan, the United States and Korea; conflicts between pan-European objectives and firm strategies; and possible ill-effects of subsidy. Indications of improved European performance are also touched upon.

The chapter argues that, partly as a result of EC initiatives, large indigenous European firms for the first time have a chance of keeping up with the international frontier in SC technology. This in turn has improved their potential in world SC and IT systems markets. A second argument put forward is that Europe's revitalisation is partly owing to the professionalisation of European SC management – a side-effect of inward investment by American firms during the 1960s and 1970s.

One bleak note concerns the United Kingdom. British-owned industry has been overtaken by events in Europe. Indeed, the United Kingdom is fast becoming a 'second-tier' European economy in SCs. In contrast, major continental European firms such as Siemens (The Federal Republic of Germany) and SGS-Thomson (Italy/France) have made massive investments in leading-edge SC technology, re-organised to meet customer needs and stepped up marketing operations outside Europe, particularly in the United States.

I The European SC industry: historical performance

Europe's poor performance

Europe's competitive performance in SCs has been poor by any standards. As Table 5.1 shows, between 1980 and 1988 the share of Europe in world SC sales fell from 16 per cent to 10 per cent. European consumption of SCs has been satisfied largely by imports and inward investment from the United States and more recently Japan. Despite recent European support programmes Europe's overall share continued to fall in the 1985 to 1988 period but has become rather more stable.

Over the past decade Philips, the Dutch consumer goods manufacturer, was the only European SC producer consistently in the top-ten list of suppliers. With an output of around $1.7bn in 1989, Philips was ranked tenth in the world, but has recently been experiencing some major problems. As Table 5.2 shows, in 1989 11 out of the 20 largest producers were Japanese, five were American and three were European.

Table 5.1 European manufacturers share of the world semiconductor market (%)

Year	%
1979	16
1980	13
1981	12
1982	12
1983	11
1984	11
1985	12
1986	11
1987	11
1988	10

Source: Dataquest, cited in *Wall Street Journal*, 19 June 1989

Table 5.2 Leading semiconductor manufacturers world-wide, 1988–89

Rank 1989	1988	Company	1989 SC revenues ($m)	Change from 1988 (%)	Market share 1989 (%)
1	1	NEC	4,964	+9	8.9
2	2	Toshiba	4,889	+11	8.8
3	3	Hitachi	3,930	+12	7.0
4	4	Motorola	3,322	+9	5.9
5	6	Fujitsu	2,941	+13	5.3
6	5	Texas Instruments	2,787	+2	5.0
7	8	Mitsubishi	2,629	+14	4.7
8	7	Intel	2,440	+4	4.4
9	9	Matsushita	1,871	–1	3.4
10	10	Philips	1,690	–3	3.0
11	11	National	1,618	–2	2.9
12	12	SGS-Thomson	1,301	+20	2.3
13	18	Samsung	1,284	+42	2.3
14	15	Sharp	1,230	+19	2.2
15	20	Siemens	1,194	+32	2.1
16	14	Sanyo	1,132	+5	2.0
17	17	Oki	1,125	+19	2.0
18	13	AMD	1,082	0	1.9
19	16	Sony	1,077	+13	1.9
20	19	AT&T	873	+2	1.6

Source: Dataquest, cited in *Electronic Business*, 16 April 1990

Within Europe, indigenous producers have been far more significant than at the world level. Major European suppliers other than Philips include SGS-Thomson and Siemens, both of whom rank in the top-20 list of world suppliers. No British firms rank in the top-20 listing. Throughout the past three decades, American manufacturers supplied the bulk of Europe's SC requirements. Of the total European chip market of $8.5bn in 1988, American firms supplied 44 per cent, European firms 38 per cent and Japanese firms 16 per cent (Dataquest, cited in *Electronics*, July 1990). The reasons for Europe's poor performance in chips relate both to firm strategies and government policies.

Inward-looking, nationalistic strategies

Until the mid-1980s, European companies' strategies were, relative to those of American and Japanese firms, cautious, inward-looking and nationalistic. Most European manufacturers remained outside the commodity chip market, concentrating instead on low-volume, domestic niche markets. A fairly large proportion of Europe's SC production was in mature technologies destined for in-house consumption, rather than the traded (or so-called merchant) market. In 1980, for example, the share of external component sales in total revenues was less than 5 per cent for Philips, Siemens, Thomson, AEG-Telefunken and Plessey (Mackenzie, 1985).

In contrast, during the 1960s and 1970s, American multinationals followed aggressive strategies within Europe, maintaining a persistent technological lead. Tariff barriers erected by European countries encouraged American producers to increase investment in France, Germany, the Netherlands, Italy and the United Kingdom.

The American lead in commodity chip technology (especially memory devices such as DRAMs) spilled over to other segments of the industry. With each successive generation of SCs technological and financial barriers to entry increased, making it more and more difficult for European producers to recapture lost markets.

Compounding these weaknesses was the fragmentation of the European SC market. European firms tended to supply their home markets which are considerably smaller than those of the United States and Japan. In 1986 for instance, the American market for integrated circuits was $8.4bn and Japan's was $9.9bn. In contrast, the market size of the former West Germany was $1.1bn, the United Kingdom $1.0bn, France $0.7bn and Italy $0.4bn (Dataquest, Mackintosh, Butler Cox: cited in the *Financial Times*, 31 March 1987). The American and Japanese home markets were useful for building up economies of scale in R & D, production and marketing. In comparison, European-based firms were at a distinct disadvantage.

Government policies of procurement and subsidy encouraged inward-looking strategies, especially in SCs destined for the protected markets of telecommunications and defence. Governments tended to support national champions via protection, subsidy and procurement policies. Parochial, nationalistic policies were a constant barrier to the rationalisation of the European market.

Europe also lacked the dynamic start-up firms which pioneered the development of Silicon Valley and populated the American SC industry with capable and creative managers and engineers. Dorfman (1987) explains the important role played by small, dynamic firms in the development of the American SC industry. Successful start-ups pressurised larger firms to become more efficient and innovative. Some, like Intel, grew to become large multinationals. This virtuous cycle of innovation was almost non-existent in Europe during the 1960s and 1970s.

In summary, rising technological barriers to entry, the inertia of European firms and the segmentation of European markets, bedevilled the industry's performance. Nationalistic policies encouraged corporate parochialism and generated protected, subsidised markets. By the mid-1980s, Europe's SC industry was technologically backward, fragmented and uncompetitive, especially in mainstream markets.

II The revitalisation of Europe's SC industry

A new mood of optimism

In spite of Europe's track record, in the mid-1980s several observers believed that with the correct strategy, Europe could stage a comeback in SCs. This optimism was partly due to the rapid growth of the European SC market. Dataquest (1987), for example, predicted that the European SC market would more than quadruple, from a level of $5.6bn in 1986 to around $25bn in 1996. Optimism was also the result of forceful actions taken by European firms and governments.

Japanese firms had demonstrated that the American lead was far from impregnable. With some government coordination through MITI's VLSI Project, the large Japanese IT corporations integrated backwards into commodity SC production (Sigurdson, 1986). To the alarm of American producers, Japan quickly gained the lion's share of the international market for DRAMs. Following Japan, Korea and Taiwan began producing leading-edge commodity chips. These events showed that with sufficient financial and industrial commitment, and with the correct strategy, lagging economies could catch up. Another factor was the financial strength of some European chip makers. Unlike their American counterparts these European companies were insulated from the worst effects of the severe 1985–7 SC recession, precisely because of their low share of the commodity chip market. They were not forced into investment cutbacks or redundancy programmes and remained financially strong throughout the recession.

Pan-European IT initiatives

During the mid-1980s, as described in Chapter 4, national governments and the European Commission took a leading role in formulating R & D recovery programmes in SCs. National IT programmes were set up to support R & D and, to some extent, to complement the European-wide programmes. France, the United Kingdom, Italy, the Federal Republic of Germany, Sweden and the Netherlands began to channel resources directly to their leading electronics manufacturers. The British Alvey Programme, for example, included government funding of £73.8m for R & D, under its VLSI Programme.

New pan-European initiatives were set up to strengthen capabilities across the spectrum of IT and other high-technology industries. These included ESPRIT, RACE, BRITE and EUREKA. The ESPRIT 1 Programme included five IT sub-programme areas, the largest of which was the Advanced Microelectronics Programme (AMP), which accounted for about 25 per cent of allocated funds (Hare *et al.*, 1988). The AMP's objective was to ensure that Europe maintained a competitive position in the supply of microelectronic components for the domestic IT industry, both in sufficient quantity and at competitive prices (CEC, 1987b). The number of approved projects in ESPRIT 1 was 226, of which 49 (22 per cent) were microelectronics (mostly SC) projects.

The second phase of ESPRIT (ESPRIT II) was agreed in 1987 and committed a total budget of ECU3.2bn, of which ECU1.6bn was to be provided by the EC, the balance by EC firms. Within ESPRIT II, a newly constituted microelectronics and peripherals sub-programme took over R & D in SCs. One of its main objectives was to develop advanced Application Specific Integrated Circuits (ASICs). Another was to ensure the commercial exploitation of R & D results achieved in ESPRIT 1.

As Sharp argues in Chapter 4, ESPRIT played an important part in influencing the SC strategies of Europe's IT corporations. ESPRIT provided an effective mechanism for enabling cooperation among European companies and for creating convergent expectations about the future. Most importantly, ESPRIT provided a platform for European companies to decide how they could best meet competition from the United States and Japan. The programme pressed for an integration of the fragmented European market, common standards in IT and the abolition of remaining barriers to trade. In the field of SCs, ESPRIT was inextricably bound up with strategic alliances among the largest European IT firms.

Technology catch-up strategies

During the 1980s, major European electronics firms embarked on strategies to catch up with their Japanese and American competitors. Catching-up involved not only SC technology but also a thorough business restructuring across all electronics sectors. Partly influenced by ESPRIT, the largest European chip makers engaged in takeovers and strategic partnerships in order to rationalise R & D investments and to position themselves for the opening of the European market in 1992.

The strategies adopted by European firms in SCs resembled that of their Japanese, rather than their American competitors. In Japan, the large IT firms are horizontally integrated across market segments. They are also vertically integrated, supplying the large proportion of their SC needs in-house. In contrast, most American chip firms see virtue in remaining 'merchant' chip suppliers – independent, relatively small firms which concentrate largely on SC components.

The catch-up model adopted by the larger European firms was to integrate backwards into chip manufacturing and design technology in order to supply their in-house needs, as well as the merchant market. This strategy runs risks. These risks were emphasised once again when Philips was obliged to withdraw from some of its major partnerships in 1990. In the past it had led to inefficient and poorly run chip operations in Europe. Nevertheless, even though the 1980s' version was a renewal of an old formula, it had several new features: heavy although risky R & D investments to match the world competition; outward-looking strategies; and pan-European strategic partnerships.

Strategic partnerships in semiconductors

Table 5.3 outlines some of the main takeovers and partnerships among European firms which took place during the latter half of the 1980s. One of the most notable partnerships was the Siemens/Philips Megaproject. Siemens of the Federal Republic of Germany and Philips of the Netherlands together committed roughly $1bn to leading-edge (sub-micron) SC technology. The Dutch and (former West) German governments financed approximately one-third of the total cost, as was the norm under EUREKA arrangements.

The Megaproject was a four-year joint R & D venture in memory chip technology (one megabit and four megabit DRAMs). The aim was to produce sub-micron memory circuits and ASICs for in-house use and the world market. As well as addressing the important DRAM market, Philips was to

Table 5.3 Major European partnerships, takeovers and mergers in SCs

Date	Venture	Firms	Cost*
1983–89	Megaproject	Siemens/Philips	\$930m**
1986	Mostek acquisition	Thomson	\$120m
1986	Megaproject 2	SGS/Thomson	\$655m
1987	Merger	SGS/Thomson	
1988–95	JESSI	Siemens/Philips Thomson/SGS/Plessey	\$3,700m
1988	Takeover	Ferranti SC by Plessey	
1990	Takeover	Plessey by GEC and Siemens	
1989	Takeover	Inmos by SGS/Thomson	

* Cost of programmes only
**Exchange rate \$1.7 = £1.0

Compiled from various sources

gain access to advanced ASICs for consumer electronics; Siemens would concentrate on ASICs for telecommunications, automobiles and factory automation systems. The venture promised to help overcome the segmentation of the European market and to provide the economies of scale necessary for the large-scale production of commodity memory chips.

In 1986, Thomson, the state-owned French electronics group, invested roughly \$120m to acquire SC (SRAM) production capacity from Mostek, the American chip company. Later that year, SGS and Thomson jointly applied to EUREKA for \$255m to support a new project in standard memory chips (four megabit and 16 megabit EPROMs) and ASICs. In January 1987, the so-called 'Megaproject 2' was approved by the EUREKA Commission. Each company was reported to have committed \$200m to the venture. In April 1987 the two firms launched a jointly-owned pan-European chip company (SGS-Thomson). The headquarters of the new company were to be set up in the Netherlands, adding a further cross-European dimension to the enterprise.

A further pan-European initiative was JESSI (the Joint European Submicron Silicon Initiative), which was to run from 1988 to 1995. Initially, JESSI was a Siemens/Philips initiative designed to follow on from Megaproject 1 (due to end in 1989). Its aim was to develop sub-micron SC technology suited to the production of 16-megabit DRAMs and advanced ASICs. JESSI's total estimated expenditure was reported to be in the region of £2.2bn (roughly \$3.7bn). Thomson/SGS also joined the consortium. From the United Kingdom, Plessey joined, but GEC did not (*Electronics Times*, 12 January 1989). Because of British defence policy, the British government insisted on GEC acquisition of Plessey's semiconductor operations when Siemens and GEC divided up Plessey between them later in 1989. GEC has inherited a renewed commitment to semiconductor activities despite a track record of indecision and withdrawal.

Although the withdrawal of Philips in 1990 was a serious blow, the Megaprojects and JESSI undoubtedly boosted Europe's SC prospects both in terms of the EC market and beyond. Strategic partnerships helped overcome the segmentation of the European market. At the same time, acquisition of leading-edge process technology gave new impetus to the firms' world market efforts in IT systems. Siemens and SGS Thomson both showed major gains in SC sales (Table 5.2). Indeed Siemens themselves report the increase even higher than the 32 per cent estimated by Dataquest. Siemens' *Presse Meldungen* of 29 May 1990 put the increase at 52 per cent.

New European ASIC start-ups

As noted earlier, one of the main problems facing Europe's SC industry in the 1960s and 1970s was the absence of 'Silicon Valley-style' start-up companies. In the United States, these companies pioneered many leading innovations in SCs. They were also a constant source of competitive pressure, forcing the larger companies to improve their innovation and marketing performance. During the 1980s, for the first time, Europe began producing 'home grown' start-up companies, mostly responding to opportunities in the ASIC market. This less-visible phenomenon could well prove to be important for Europe's long-term prospects in SCs for two sets of reasons: first, the general importance of small dynamic firms; second, the special role of small firms as carriers of SC innovations.

The importance of small SC firms goes beyond their market share. (The 200 or so small firms in the industry world-wide account for only 20 per cent of the world SC market, whereas the leading 20 firms supply 80 per cent). First, small firms are very active in the design-intensive ASIC market. ASICs are a very rapidly growing segment of the world SC market. Narrowly defined (that is, gate arrays, cell-based and programmable logic circuits), ASICs accounted for more than 20 per cent of the world integrated-circuit market in 1988 (roughly $7.4bn, Dataquest, cited in *Electronic Business*, 18 September 1989). Through the early 1990s, the ASIC market is forecast to grow at rates of 20 per cent to 25 per cent per annum. In Western Europe the ASIC market is forecast to nearly double from $731m in 1989 to $1,357m in 1992. (Dataquest, cited in *Electronic Business*, 4 September 1989).

Second, with ASIC technology, chip design becomes gradually de-coupled from SC process technology. Design technology becomes more important, increasingly sophisticated, specialised and commodified. As a result of innovations and falling costs in automated chip design systems, many of the design functions which were previously integrated, organisationally and technologically, within large SC manufacturing companies become de-coupled. With a comparatively small investment in design systems, a user, a distributor or an independent design firm can now create designs relatively quickly and cheaply.

Manufacturers and users of SCs are testing out new organisational forms to control the new technology and to position themselves in the market. New firms have entered an independent design sector for ASICs, acting as carriers of new design innovations. These firms can react quickly to market niches and supply design technology to users, especially small and inexperienced systems firms. More generally, technological specialisation (or de-coupling) is providing a range of new opportunities for specialist, flexible firms involved in areas such as: ASIC design; cell library development design-process integration; speciality memory and logic production; high-performance products; fully customised devices and special packaging.

As Saxenian (1990) shows for the United States, new start-up firms led a recovery in the fortunes of Silicon Valley in the 1980s. Eighty-five new firms were generated, creating 25,000 jobs and more than $2bn in annual sales. The new firms tended to form close, service-orientated relations with users. They often utilised new technologies such as small-scale manufacturing lines (so-called Minifabs, a fraction of the cost of the standard production lines, Megafabs) and electron-beam systems which 'write' chip designs directly onto wafers.

In Europe, during the mid-1980s, small and medium-sized start-up firms began to enter the industry. These firms supplied small volumes of finished ASICs, other specialist products, training in SC design and various engineering and consultancy services. Examples are firms such as ES2 (European Silicon Structures), Mietec of Belgium and Advanced Silicon Corporation which started life in Europe in order to meet the needs of European users of specialist chips. ES2 benefited from a large injection of funds from the EUREKA project. If they continue to be successful, these firms (and new American start-ups in Europe such as LSI Logic) will provide healthy competition for the larger firms as well as advanced SCs for European users and design training for systems firms.

The take-up of ASICs during the 1980s also enabled SC distributors to enter the design sector. In the United Kingdom, for example, around 12 distributors entered between 1985 and 1987 (*Electronics Times*, 28 May 1987). Plessey Semiconductors of the United Kingdom could be added to the above examples. Although a well-established firm, it adopted extremely ambitious objectives towards the ASIC market, acting like a start-up company. Plessey invested around £50m in a new SC plant and supplied a wide range of high-profit, relatively small niche markets, mostly in ASICs. However, in 1989 Plessey Semiconductors was taken over and integrated into GEC. Given GEC's wavering commitment to chips in the past, Plessey's future in SCs is uncertain, but the new head of the company has indicated that they aim to increase sales from £200m to £500m by 1993 (*Financial Times*, 16 October 1990).

Whether or not the new European start-up companies succeed in adding significantly to the technological base of the European SC industry depends to some extent on the abilities of the managers of these firms. Turning to the management of Europe's SC industry, there is good cause for optimism.

Professionalisation of European management

Tracing the origins of SC managers in Europe shows that many of the key players were drawn from American multinational firms which invested in Europe in the 1960s and 1970s. For example, the president of Thomson's components group, Jacques Noel, was a former Texas Instruments (TI) executive. Other top TI managers such as Carlo Zanni joined the Thomson management. SGS, the largest Italian SC manufacturer, recruited American-trained executives with considerable experience in the SC industry. The president of SGS, Pistorio, responsible for restructuring the company, was formerly a Motorola executive. In the 1980s, under Pistorio's stewardship, SGS made a profit for the first time in a decade, set up production facilities in Singapore and began producing high-volume, specialist chips for automotive applications (*Business Week*, 30 June 1986).

Doug Dunn and other managers responsible for Plessey's turn-around were also recruited from Motorola. Dunn's aggressive style of market targeting and public announcement of strategy, followed the typical moves of American SC start-up companies. Plessey's actions were viewed by the British electronics press as highly unusual for a British company, but he apparently now has the full backing of GEC.

ES2's management team originated from TI's British (Bedford) operation. After leaving TI, Rob Wilmot successfully restructured ICL, the British computer company and then became ES2's founding president. Within ES2, key management positions were taken up by other ex-TI employees, including Rod Attwooll, the company's British managing director.

The move of top SC business managers from American multinationals to European firms reflected the sea change occurring in the SC industry in the mid-1980s. Three decades or so of American inward investment had produced layers of professional European SC managers to draw upon. The training of foreign nationals within American subsidiaries in areas such as SC marketing, production, management, engineering and international planning represented technology transfer in the real sense of the words. This not only applied to the industry leaders mentioned above but also to large numbers of less well-known managers, engineers and technicians. There can be little doubt that American direct foreign investment left Europe with a substantially improved indigenous base of management and marketing ability.

III Assessing Europe's Prospects

The survival of European start-up firms

Small start-up firms of the type described in Section II above could play a vital part in Europe's resurgence in SCs. One or more could grow to join the 'big league' of SC suppliers. However, it is possible that the small companies will be taken over by large SC suppliers or simply fail in their aims (as argued by Leibowitz, 1987). During the 1984–6 recession, it was relatively easy for start-up companies to obtain leading-edge ICs from large producers faced with overcapacity. In periods of boom, limits on SC capacity would make it extremely difficult for start-up companies to access world-class fabrication technology. Also, if start-up firms are to join the big league, they will need to make massive investments in volume fabrication facilities.

While it is impossible to predict the future, there are reasons to believe that Europe's small ASIC firms will survive and that new companies may start up in the medium term. The de-coupling of design from manufacturing is likely to continue to generate new demands for engineering services, design supplies and customer training in ASICs and other specialist SCs. The large SC corporations are ill-equipped to meet the demands of small users (see Shortland, 1987). In contrast, small flexible design firms and specialist chip makers could help meet the demands of these users.

According to this view, the survival of the independents is not the main question. The main question is whether any of the small firms can realise their ambitions of becoming large SC corporations. This would require massive investments in fabrication technology and the ability to compete head-on with the major Japanese and American suppliers. This transition has been made in the past by American firms such as Intel and LSI Logic (now the largest supplier of ASICs), and it is not inconceivable that the most dynamic European firms may enter the big league of SC suppliers in the longer term. However, this will in large part be determined by the ability of American and Japanese firms to dominate SC sales and meet user needs within Europe.

American and Japanese competition

The most significant problem confronting European firms is competition from American, Japanese and, more recently, Korean firms. During the mid-1980s, Europe became a major competitive battleground for the United States and Japan. The American firm LSI Logic, for example, claimed to have captured 20 per cent of the total European ASIC market in 1988 (*Electronic Business*, 4 September 1989). Other American firms including Texas Instruments, Intel and Sierra invested ahead of most European manufacturers. Intel set up an advanced ASIC design centre in Swindon in the United Kingdom as early as 1987.

Japanese and Korean companies also stepped up their efforts to gain European market shares in line with their greater shares of American and Pacific Basin markets, but to date, European firms lead Japanese and Korean firms in the scale of their manufacturing facilities within Europe. However, Japanese corporations such as NEC and Toshiba have set up facilities within Europe and have begun to produce and design SCs tailored to European users' needs. Fujitsu, one of the largest producers of ASICs, has a well-established link into Europe through its joint venture with and later takeover of ICL.

Competition for the European commodity chip market will be extremely fierce as the European market is central to the strategies of leading American and Japanese firms. In order to stand a chance of success, European producers must be equal to their competitors in R & D, manufacturing, management and marketing ability. Only outward-looking, global firms prepared to meet the huge investments required in SC technology will be equal to the challenge.

Pan-European programmes versus firm strategies

The main policy aims of European programmes such as EUREKA and ESPRIT are:

- to strengthen indigenous technological capabilities in IT;
- to improve the IT balance of trade;
- to provide necessary defence-related technologies;
- to improve the competitiveness of European firms.

However, European firms are also faced with the necessity sometimes to make strategic partnerships with Japanese and American technology suppliers. Although European programmes obviously favour pan-European partnerships to promote perceived 'European' interests, it may well be that partnerships with Japanese and American corporations are necessary, for instance, to gain access to advanced manufacturing technology. For example, a major agreement in this area is the Siemens/Toshiba agreement on one megabit DRAM technology.

A possible tension therefore exists between EC-driven partnerships on the one hand and firm-level 'best practice' partnerships on the other. Through political persuasion and financial support the EC programmes might promote inefficient or sub-optimal partnerships.

Also, the outcome of existing Japanese–European strategic alliances might be at odds with the objectives of pan-European programmes. As Hamel *et al.* (1986) argue, strategic partnerships can be used as a mechanism for the eventual domination of one partner by another. It is prudent to assume that Japanese corporations treat partnerships as a method of capturing overseas' markets. Partnerships among unequals could eventually lead to the control of the weaker (European) partner. For example, in 1990, the long-standing partnership between ICL and Fujitsu led to a take-over by Fujitsu (*Financial Times*, 19 July 1990). Similar concerns apply to NEC's 15 per cent stake in the French Bull-Honeywell computer corporation.

Programmes such as ESPRIT are not necessarily at odds with strategic alliances. They could help to strengthen the position of European firms in relation to foreign partners and bring about mutually beneficial long-term alliances. However, it is important to recognise that such tensions exist and have not yet been resolved.

Adverse effects of subsidy

A final potential problem facing European firms is dependence upon government and EC subsidies. Through IT programmes, Europe's SC makers received substantial support for pre-competitive R & D as well as close-to-market activities such as fabrication and CAD development. The scale of this European support for firms via direct loans, R & D grants, procurement policies and collaborative programmes is very large.

Given the weak starting position of European SC makers in the early 1980s, subsidies were probably a necessary condition for building up competitive capabilities. Certainly, American manufacturers view government support as legitimate and necessary for their own survival. This belief, for instance, led in 1986 to the American Sematech Programme, a $1.5bn government-funded collaborative SC venture. Japanese firms have also received support although on a much smaller scale.

However, subsidy cannot be a substitute for genuine improvements in management, productivity and competitive behaviour. If subsidies serve to camouflage poor R & D practices, production inefficiencies and antiquated organisational structures, then the large-scale finance being chanelled to firms would only postpone the decline of Europe's SC industry. An unhealthy continuous dependence on state support for finance, combined with the bureaucratic influence of government upon firms, could be debilitating in terms of competitive performance, dynamism and the ability to respond to market requirements.

The impact of subsidy and, for that matter, the success of Europe's resurgence, depend first and foremost on the capabilities of Europe's IT business managers. No amount of subsidy can compensate for second-rate management. In this respect there is cause for optimism. As argued above, European management underwent a regeneration in the 1980s. In the hands of these 'professionalised' managers, the subsidies channelled to Europe's major firms could well have contributed to a long-term revitalisation of the industry. The acid test will be if firms' forward investments in SC technology continue into the 1990s – without the support of government subsidy.

Signs of improving performance

To conclude this section it is helpful to discuss briefly European firms' performance in the latter half of the 1980s. Despite the difficulties highlighted above, the evidence suggests that large European IT firms have committed themselves to outward-looking strategies, very heavy investments in leading-edge technology and global marketing in electronics systems and SCs. Whether the commitment is translated into sustained, long-run, profitable achievements remains to be seen. However, there are signs that the new, positive approach by European firms is reaping benefits.

As far as restructuring is concerned, Siemens underwent one of the most thorough reorganisations of its history. The company reorganised from seven to 16 divisions and decentralised its management and R & D in order to be closer to final goods markets (*Electronic Business*, 20 March 1989). In fiscal year 1988–9 Siemens spent $3.36bn on R & D, massive by any international standards and ahead of all Japanese and American electronics firms except NEC (Japan) and IBM (United States).

Much of Siemens' R & D effort was aimed at penetrating the American market. A substantial proportion was allocated to SCs. In January 1990 Siemens announced a very large partnership with IBM to develop 64-megabit DRAMs by 1994 (costed at over $1bn). This would place Siemens ahead of most SC firms world-wide and could enable them to become a world leader in DRAM technology. The project confirmed Siemens new-found confidence in SCs (*Electronic Business*, 28 May 1990).

Following on from EUREKA, Thomson and Philips also began a huge joint-investment programme (in the order of $3.5bn) in high-definition television, part of which involved developing integrated circuits (*Financial Times*, 16 May 1990) (see Chapter 9). Thomson-CSF became a European giant in defence systems and one of the largest defence companies in the world in 1989 (*Electronic Business*, 27 November 1989). SGS-Thomson, the Franco–Italian chip operation, purchased the innovative British company, Inmos, in order to enter the advanced microprocessor business (by acquiring the so-called Transputer). SGS-Thomson increased its share of the American SC market in 1988 (the only European firm to do so).

On the negative side, Philips recorded a loss in fiscal 1989–90 and was obliged to withdraw from computer activities and various other commitments including some mega-chip programmes. On the positive side, there is evidence of the resurgence of the large chip companies of continental Europe in terms of: technology investment commitment; global marketing and production; strategic partnerships and restructuring to meet user needs. In the latter half of the 1980s the largest European IT firms successfully reversed their 1970s' image of inertia and technological backwardness.

It is not possible to attribute this turn-around to EC programmes alone. Other major influences on corporate strategy included preparation for the Single Market of 1992, reactions to Japanese competition and the emergence of strategic partnerships on the part of IT firms world-wide. However, without doubt the EC programmes were a major factor in the resurgence of Europe's SC industry in the 1980s.

IV Conclusion

Following a long period of inertia, Europe has just begun to close the technology gap between itself and the world leaders in SCs. In the mid-1980s, governments began to coordinate their strategies at a pan-European level through collaborative programmes such as ESPRIT and EUREKA. Some major European players embarked upon offensive technology and marketing strategies both within Europe and beyond. Partly as a result of the EC policy, Europe underwent a major industrial restructuring, backed up with a massive investment commitment to world-class SC technology.

In mainstream SCs such as memories, the strategic partnering between Europe's four largest producers enabled firms to rationalise their R & D activities and surmount the large investment thresholds in fabrication technology. For the first time the technological level of European firms in SCs is comparable to (if not on a par with) that of leading American and Japanese companies. Cooperation between companies helped overcome the segmentation of the European market both in SCs and in IT systems.

The large European firms now have the opportunity to apply advanced SC manufacturing technology to complex, design-intensive chips such as ASICs. In-house capabilities in leading-edge SC technology should lead to improved performance in computers, telecommunications, consumer electronics (notably high-definition television), as well as office and factory automation systems.

The 1980s also witnessed the emergence of small SC firms within Europe for the first time. Although not comparable in scale or scope with American start-ups, the emergence of 'home grown' companies was a significant event of the 1980s. Continuing technological diversification and specialisation in areas such as ASICs suggest that there will be substantial opportunities for creative small and medium-sized firms in the future.

Effective world-class management is the key to Europe's long-term future in SCs. Here there is also room for a little qualified optimism. During the 1960s and 1970s many Europeans were trained within American multinational subsidiaries operating in the United Kingdom, France, Italy, Germany and elsewhere. Several of Europe's leading SC executives were recruited from American firms such as Texas Instruments and Motorola. Larger numbers of less well-known engineers, managers and technicians added significantly to Europe's human resources in SCs.

Nevertheless, it is important to be aware of the problems facing Europe's SC industry. Europe has been targeted as the next competitive battleground by American and Japanese corporations. Japanese firms have stepped up inward investment in preparation for the unification of the European Community market. Competition will undoubtedly be fierce. The difficulties experienced by Philips illustrate its intensity. The European firms are probably still losing in market share within Europe.

Only by continually upgrading their technology and by pursuing global strategies can European firms hope to compete with the world leaders. Failure to compete in the technologically advanced markets of the United States and East Asia will almost certainly lead to a gradual deterioration in competitive performance.

Another cause for concern is the British economy. Although the second largest consumer of SCs in Europe, the United Kingdom has no domestically owned producers comparable with those on the continent. The British government's industrial policy is essentially one of *laissez faire*. As far as domestic firms are concerned, the United Kingdom is fast becoming a second-tier European economy.

Finally, it should be stressed that while it is legitimate to analyse European prospects, today's SC industry is intrinsically global in nature. Europe can and should benefit from direct foreign investment and collaboration with firms from the United States, Japan, Korea, Taiwan and elsewhere. If the unified European market is used as a mechanism to protect less-efficient indigenous firms and to encourage 'Euro-parochialism', then the industry will resume its decline in the medium term. Outward-looking strategies are essential to Europe's long-run prospects in semiconductors. The presence of Japanese, American and other foreign firms in Europe will help to stimulate beneficial technological rivalry provided the European firms can sustain the early indications of their resurgence and translate it into greater market strength with the new products of the 1990s.

6 Electronic computers

Franco Malerba, Salvatore Torrisi and Nick von Tunzelmann

Introduction

Since the earliest days of the electronic computer, European countries have struggled to create dynamic and lasting computer industries. By and large, they have had only limited success. Although European universities were among the leading groups in the 1940s and computers were used in both Britain and Germany during World War II, few of the innovations that later led to today's extraordinary range and power of computer technology have come from Europe; no European company has offered a serious challenge to the leading American computer firms and the European share of world computer markets remains small. The continuing existence of an indigenous computer industry in Europe, and its modest increase in world market share in the 1980s, may be considered to provide some grounds for optimism. However, events in 1990, such as the withdrawal of Philips from computer manufacture, Fujitsu's takeover of ICL and Digital's takeover of Mannesmann Kienzle, have again demonstrated the fragility of Europe's position, particularly in view of the formidable challenge that Japanese firms are now mounting to American supremacy in world markets, including the European market.

So while hopes remain vested in firms like Siemens (Germany), Olivetti (Italy) and Bull (France), Europe's involvement in computer development and manufacture may increasingly depend on the activities of the more dominant American and Japanese multinationals within the enlarged European market. As in other areas of electronics, policy at European, national and regional levels will be orientated towards capturing substantial shares of foreign R & D and production, as well as trying to strengthen Europe's own presence in the industry. Achieving a balance between these two objectives will, however, be no easy task, as the outcry over Fujitsu's takeover of ICL has recently shown.

During the 1980s, the computer industry experienced many changes, extending to the manner in which its field of activity was defined. A definition of the computer industry which considers hardware to the exclusion of its information context is now obsolescent. As already pointed out in Chapter 2 the distinction between manufacturing and services is itself becoming blurred, and this is especially true in this context. Increasingly even the firms manufacturing computer equipment are developing and supplying 'systems'. It is in the ability to tie together systems into packages usually known in the United States as information systems (IS) or in Europe as informatics that the heart of the competitive and innovational struggle now lies. The chapter should thus be read in conjunction with Chapter 7 on software by Brady and Quintas and the wider discussion of informatics in Malerba and Torrisi (1989).

The first section of this chapter summarizes the recent technological and market trends in the computer industry at a global level. The following deals in greater detail with its product structure. The next provides an appraisal of the technological and commercial performance of the European industry, while the one following briefly examines the fortunes of the major European firms since the mid-1980s. Thre final section presents brief conclusions on the future of the industry.

For reasons of space, we are unable to survey public policies for the computer industry, including EC programmes such as ESPRIT (but see Chapter 4 and also Part IV; and Jowett and Rothwell, 1986; O'Brien, 1989; Ferné, 1989; Sharp, 1990), though we do briefly consider the policy implications of our analysis. As in Chapter 5 the nature of the topic makes the use of some technical jargon and acronyms inevitable and the reader is referred to the glossary in case of problems.

I General characteristics of the computer industry

Global size

The share of the electronics industry as a whole in total manufacturing output of OECD countries rose from 3 per cent in 1975 to about 6 per cent by 1988. It is highest in the United States and Japan. Computers, components and telecommunication equipment each accounted for about a quarter of total electronics output in the mid-1980s. But the rate of growth of the computer segment over the years 1980–87 was faster than for other branches of electronics (12.3 per cent p.a. compared to 9.3 per cent in the related activity of office equipment and 4.5 per cent for telecoms equipment). Investments in computers in the six largest countries were over 40 per cent of total IT investments.

The proportion of innovative activity accounted for by electronics within manufacturing, and of computers within electronics, has also been increasing. The electronics industry is characterised by a very high intensity of R & D. Typically, R & D represents 10–15 per cent of value added, against 3–5 per cent for manufacturing generally. Outlays on electronics R & D were estimated at about 30 per cent of total industrial R & D expenditure in the principal OECD countries (Chapter 4). While it is not possible to separate computer from general electronics R & D, patent statistics attest to the dynamism of the computer sector. Between 1969–71 and 1984–86, patenting in computing as a proportion of total IT patenting grew from 19.9 per cent to 28.2 per cent, to become the largest segment (Malerba and Torrisi, 1989, from the OTAF–SPRU database).

Recent historical evolution

During the tempestuous 40-year history of the electronic computer industry, two periods of structural uncertainty and fragmentation have been followed by two periods of oligopolistic concentration. Thus, after a first phase of considerable turmoil, involving many start-ups, take-overs and departures in the 1950s, an international and largely American-based oligopoly in general-purpose mainframes dominated the structure of production from the late 1950s to the mid-1970s. By 1975, the leading American mainframe producers – IBM at their head, followed by the 'Bunch' (the initials of Burroughs, Univac/Sperry, NCR, Control Data and Honeywell) – supplied around 80 per cent of the world computer market. From about that date new market segments and firms arose.

From the mid-1970s to the early 1980s there was a phase of renewed turmoil, leading to destabilisation of the oligopoly and to sectoral and regional 'deconcentration'. The origins were both technological (particularly the introduction of the microprocessor by Intel in 1971) and organisational (for example, the 'unbundling' of software/services from hardware, led from within by IBM in 1969). The movement towards smaller systems, particularly the minicomputer, introduced by Digital (DEC) in 1965, and towards new market niches (for example, office automation), gave rise to fast-growing entrants like Hewlett-Packard and Wang Laboratories, and to crises among many of the established mainframe producers (the 'bunch'). Another wave of new firms in the late 1970s producing microcomputers (Apple, Commodore etc.) further destabilised the industry. In most segments, Japanese firms began to show rapid growth.

In the 1980s there was again a phase of restructuring and oligopolistic reconcentration. As in some other branches of electronics, the economic rationale for concentration included the effect of high R & D 'overheads' and marketing costs coupled with short product life-cycles, in a context of continuing technical change and fierce international competition (now including that from Japan). It was further advanced by organisational factors: as the 1980s progressed, the significance of niche strategies waned and the importance of offering integrated and compatible systems of high value-added and of specific customer-orientated services expanded. Some see the effect of this shift to systems integration and services as further destabilisation, especially in the European industry (for example, O'Brien, 1989), others as the working out of restructuring. By the end of the 1980s the structure of the mini- and micro-computer branches of the industry was becoming more stable, facilitated by their incorporation into the prevailing pattern of systems integration, for instance, through the networking of workstations and PCs. These organisational changes were grafted on to existing technological trajectories, to which we now turn.

Technological trajectories

As shown in Chapter 5, technological advance in components in the 1980s continued the well-established trend to ever-increasing speed and memory capacity. This has been the technical basis for the rapid succession of new generations of equipment. There is considerable debate over which benchmark is most appropriate for assessing the technical performance of computers and their components, but the most straightforward and commonly-cited is 'millions of instructions processed per second' (Mips), or for large machines the related 'Megaflops' (millions of floating-point operations per second). Figure 6.1 shows trends over time of Mips in the dimensions of speed

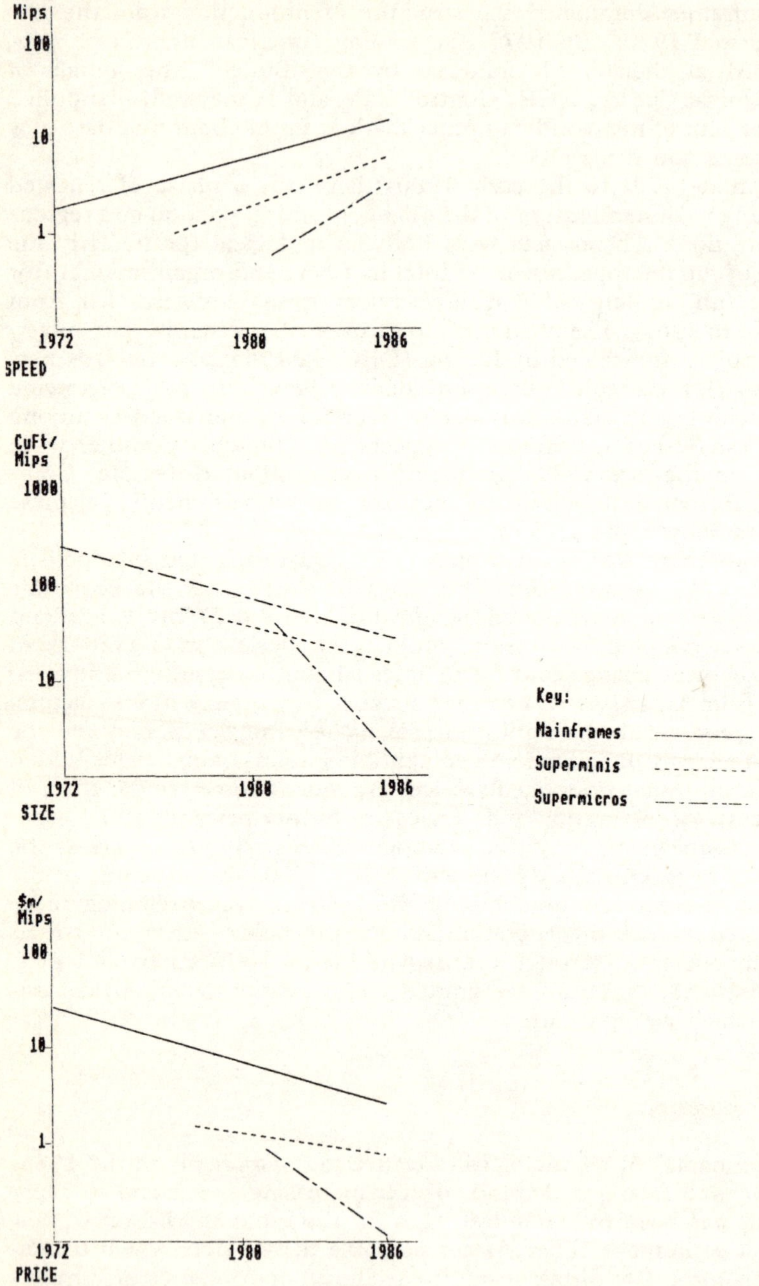

Figure 6.1 Trends in speed, size and price of computer products, 1972–86

Cost comparison of computing power
1981–1993

	1981		1993		
	Cost ($/MIPS)	Cost ratio vs. PC	Cost ($/MIPS)	Cost ratio vs. PC	Growth p.a. in MIPS/$
Mainframes	$419,000	27.9	$54,000	385.7	18.6%
Minis	$168,000	11.2	$6,000	42.9	32.0%
PCs	$15,000	1.0	$140	1.0	47.6%

Source: Gartner Group

Figure 6.2 Cost of computing power, 1981–93

(top panel), size (middle panel) and cost (bottom panel) from 1970 to 1986 and the projected costs to 1993 (Figure 6.2). As the figures imply, the faster rate of growth in speed of smaller systems, coupled with their faster rate of decline in price, has undermined the markets for larger systems – first minis undercutting mainframes, more recently micros (workstations) undercutting minis.

Fundamental changes in computer architecture in terms of either hardware (for example, parallel processing, optical processing) or software (e.g. neural networks, multimedia) have long been forecast (for example, Arnold and Guy, 1986) but mostly remained niche products or experimental operations during the 1980s. This is beginning to change and will of course change further in the 1990s. Parallel architectures, which use more than one central processing unit (CPU), are being extensively applied in supercomputers, At the present time, the obstacles are as much in software as hardware such as how to sub-divide operations for parallel treatment. Applications drawn from, or analogous to, parallel computing that have found wider usage include vector processing (running various sub-divisions of a task simultaneously), fault-tolerance (duplicating facilities to minimize risk of a breakdown) and machine clustering.

Associated with the rise of medium and smaller systems was a shift away from centralised data-processing on large mainframes to distributed processing by end-users. The shift of emphasis from 'stand-alone' machines to computer systems and networks is studied at greater length below because of its central importance to present-day innovation and competition. From the demand side this shift to distributed architectures and systems was accompanied by 'disintermediation' as the man–machine interface evolved, allowing final users to solve their problems independently of their awareness of the network. Examples include Automatic Teller Machines (ATMs) in banking or the use of expert systems in medical diagnostics (see Chapter 9).

Despite the arrival of networking, the 'convergence' of computers with telecommunications has been much slower than originally expected. Advances on the side of telecommunications towards the Integrated Services Digital Network (ISDN) enacted through PTTs have encountered technical and marketing problems (See Chapter 8). The equipment suppliers, either of telecommunications equipment or computers, have for their part scarcely 'converged' in their production patterns, except in some well-advertised Japanese firms (Soete and von Tunzelmann, 1988). These problems of convergence have stemmed principally from the difference between markets catered for by computing and telecoms and from the high cost of associated R & D and especially the differing software needs. However, from the consumer's perspective, some 'convergence' has indeed occurred such as the trends towards computer networking, workstation provision and industry standardisation. Conceivably 'convergence' may ultimately take a different direction from original expectations (for example, through 'intelligent networks'); with a demand-led approach benefiting the computer as much as the telecoms industry.

All of the above trends have implied a considerable and often dominant role for software. Apart from resolving technical problems like systems integration, software is of course of increasing importance for user operations like managing data bases and utilizing expensive support services. The growing predominance of software relative to hardware is analysed in more detail by Brady and Quintas in the following chapter. It is important to stress that hardware and software are partly complements and partly substitutes. Software is important for linking hardware together, but it is also increasingly being embedded in the hardware itself, not least in the latest generation of semiconductor devices (VLSI). This is one reason why computer systems manufacturers increasingly regard access to semiconductor chip technology as strategic to their long-term interests, even where they are not themselves mainline chip producers.

Integration and standards

Integration (of various computer operations) began on individual machines, with the needs for 'portability' (the ability to transfer programmes and data between systems), consistency (homogeneity of interface styles), multiple tasking (running different operations on the same machine at the same time) and multiple use (Fertig, 1985). Incorporation of a variety of machines of differing sizes and capabilities into whole systems involved advances in networking (local area networks, etc.) and more complex changes towards modularity (to increase flexibility in the area of machine upgrading) and compatibility (interoperability of equipment) as for example in the case of firms wishing to link their design, production and marketing activities. As in other compatibility cases, the issue boiled down to whether the equipment itself would become more standardised (for the suppliers, this meant whose

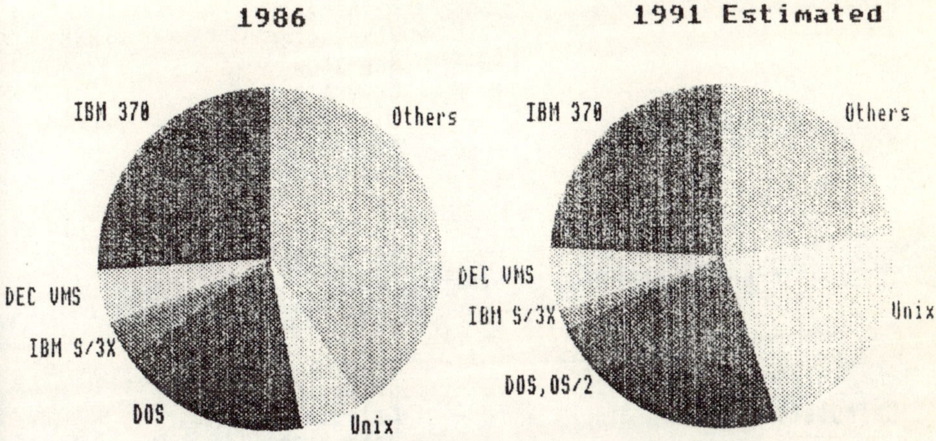

Figure 6.3 Market shares of operating systems, 1986 and 1991
Note: The figures are world-wide sales of American vendors only
Source: IDC, in *Electronic Business*, 15 April 1988

equipment would become the standard) or whether instead interfaces could be designed to permit communication between ostensibly incompatible equipment. Both solutions had high costs, both for individual producer firms and customers (See Chapter 11 on standards).

In terms of hardware, there are significant differences in the processors used, the graphics and the peripherals. But the most difficult questions relate to the operating system. Some hardware and software allow particular machines to 'emulate' other operating systems, but such solutions increase costs and typically reduce available memory and operate very slowly. The Unix operating system invented by AT&T in the late 1960s has increasingly been resorted to as a compromise between two unattractive extremes. See chapter 11 for the reasons why Unix has become so widespread. Unix itself has suffered from a range of technical limitations. It had a long reputation for user-unfriendliness, which is currently being solved by the adoption of 'windowing' along similar lines to modern PCs. It requires a large amount of memory, which is wasted on users having no need for Unix, and partly for this reason operates slowly, although its relative speed increases with the size of the operation and the network. For the latter reasons, it is unlikely to be adopted on PCs for some time, although Hewlett-Packard has recently announced one possible solution by putting Unix on compact disc (CD-ROM).

The problems of Unix are also, however, organisational: it has appeared in a sequence of releases, gradually extending across the product range, so that its adoption has been relatively slow (Figure 6.3). Computer companies were divided into two camps and the detailed story of the conflict between

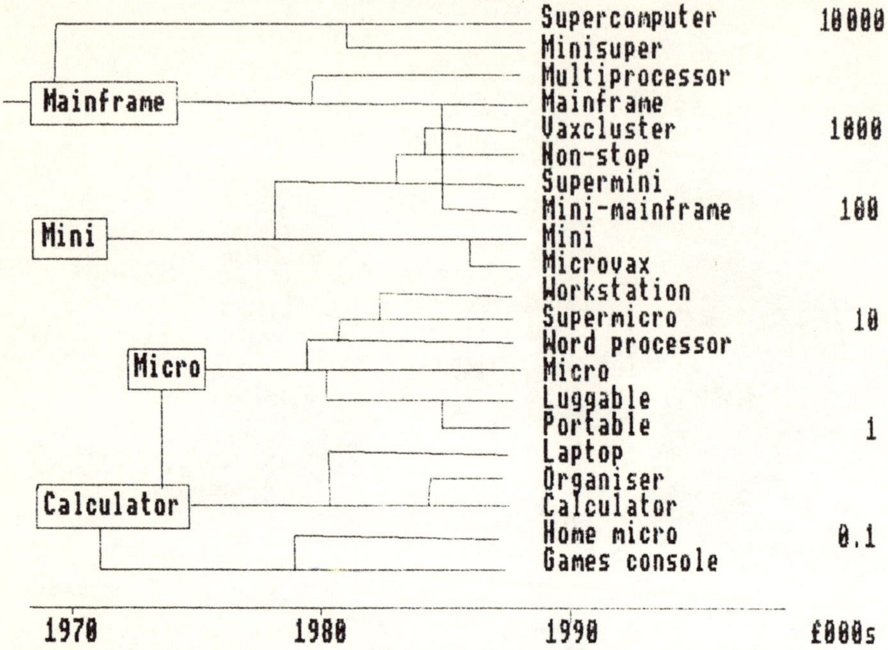

Figure 6.4 Evolution of computer products
Source: Schofield, 1988

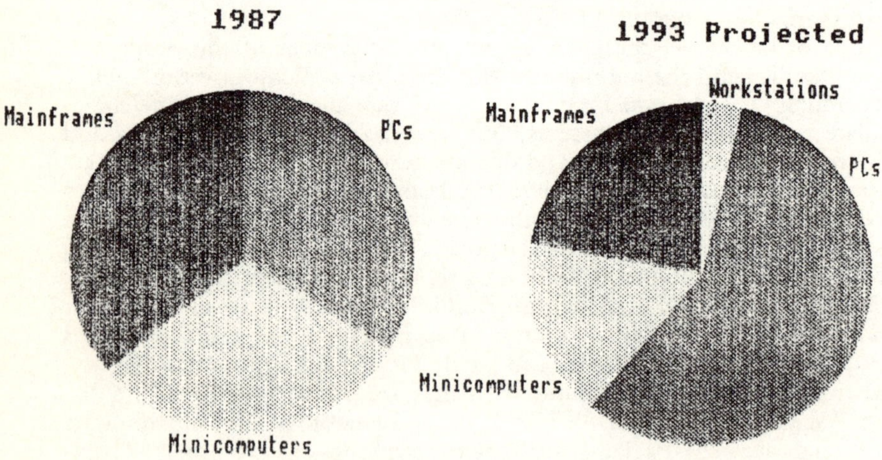

Figure 6.5 The computer market in Western Europe, 1987 and 1993
Source: Dataquest

supporters of different regimes for Unix is told in Chapter 11. The biggest losers from the spread of Unix are the smaller firms with proprietary operating systems (the 'others' in Figure 6.3).

The *network* can also be proprietary, semi-open, or fully open – the last-named in the form of Open Systems Interconnection (OSI) being somewhat nearer to general compatibility than the operating systems. The Japanese government's decision in 1988 to make the OSI the standard for government information systems has provided a boost to its becoming the leading international standard for interfacing systems of different makes. The British government also introduced this procurement compliance policy in 1986 and the EC gave its blessing to it in 1988. The European group for promoting OSI, known as SPAG (Standards Promotion Action Group), includes most of the large European computer and telecoms equipment companies, as well as the American multinationals, and its American equivalent (COS) is similarly encompassing. Because of their complexity and their policy importance, there is also a special chapter on standards (Chapter 11).

II Product structure

The analysis here takes the producer point of view and is necessarily sketchy (see Malerba and Torrisi, l989 for a more extended appraisal). The differentiation by computer type is becoming highly complicated but an attempt is made in Figure 6.4. The relationship between categories and prices on the right-hand side is constantly altering. The schema is explicitly evolutionary (Schofield, 1988). For example, the IBM 9370 'mini-frame', which evolved from the IBM Series 370 mainframes, appears well below DEC's clustering of VAXs in terms of performance and price because the latter evolved from DEC's original Minis, like the PDP-8. The basic pattern of change has been a relative decline of mainframes and minis and a rising share of PCs and workstations (Figure 6.5).

Large systems

This sector is composed of general-purpose mainframes and specialised computers for scientific work (supercomputers and mini-supercomputers). The total large-system market was estimated at just over $30bn in 1988, of which only 5.6 per cent represented the specialised computers. However, general-purpose mainframe markets have been growing at little over 6 per cent annually in the 1980s, as compared with 40 per cent growth of the specialised systems.

General-purpose mainframes were long the foundation of the computer industry. The equally long American leadership has, however, declined in the 1980s, mostly to the benefit of the Japanese (see Table 6.1 for the shares of leading firms identified by *Datamation* – it should be stressed that these figures can be much influenced by changes in the composition of the top 15 in each category). Most firms are global competitors offering across-the-board solutions, with a few more specialised producers like Amdahl in the United States and Comparex in Europe.

Table 6.1 Market shares of leading 15 firms in each segment (%)

	1984	1985	1986	1987	1988
Mainframe:					
EC	5.5	5.3	8.2	8.6	8.3
USA	78.7	79.0	63.6	58.5	54.1
Japan	15.9	15.7	28.1	32.9	37.6
Minicomputer:					
EC	7.6	7.5	7.9	9.6	8.2
USA	82.0	81.5	76.1	74.0	74.9
Japan	10.5	11.0	16.0	16.4	16.9
Microcomputer:					
EC	5.1	8.1	11.4	11.8	11.7
USA	88.3	83.9	76.6	74.6	74.0
Japan	6.6	8.0	12.1	13.6	14.3
Peripherals:					
EC	7.3	7.5	10.7	13.0	13.0
USA	81.2	77.8	66.5	60.5	57.8
Japan	11.5	14.7	22.8	26.5	29.2
Software:					
EC	6.6	7.9	9.8	12.9	12.0
USA	83.3	82.0	79.0	75.6	75.3
Japan	10.1	10.1	11.2	11.5	12.7

Note: Until 1989, *Datamation* listed the largest 15 firms in each of the above segments, on which the above percentages are based.

Source: Malerba and Torrisi, 1989, from *Datamation Top 100*, 1985-89

Some 73 per cent of systems sold are estimated to have been those of IBM or are IBM-compatible (*Datamation,* 1988). Apart from IBM itself, one group of 'follower' firms, which includes Siemens and ICL in Europe, has a strategy of compatibility with IBM, while a second group, which includes Bull, have concentrated on developing proprietary operating systems.

As Table 6.1 implies, the European firms have a very modest share of this mainframe market. Moreover, they are heavily dependent on imported non-European technology (for example, Bull on Honeywell and NEC, Siemens and ICL on Fujitsu and Comparex and Olivetti on Hitachi). It is not a segment that many firms have striven to enter, despite its historical precedence, partly because of huge entry barriers, and in recent years, partly because of the threat from medium-sized systems.

Medium systems

Total turnover of the top-15 minicomputer firms was $19 billion in 1988. Growth in the 1980s at about 12 per cent p.a. was noticeably slower than in the 'takeoff' years of the 1970s but still sufficient to bite deeply into large-systems markets. Between 1984 and 1988 the share of American firms in the *Datamation* top 15 declined somewhat (from 82 per cent to 75 per cent, cf. Table 6.1), but it was the Japanese rather than the European contributors (Nixdorf and Olivetti) who gained.

The *workstation* market was created in the early 1980s by two new firms, Apollo (acquired by Hewlett-Packard in 1989) and Sun (the current market leader). It is expected to continue growing by 25–30 per cent annually for the next five years or so. The large firms (DEC, Hewlett-Packard and now IBM with the RS/6000) have subsequently entered this market, though having to overcome the hurdle of proprietary operating systems. Similar problems, allied with the lack of new entrants, have hindered any major European contribution.

Small systems

About 150 firms marketed a broadly recognised brand of PC in 1988 (*Dataquest*, 1989). Sales of the more important firms totalled $28.4bn in that year (*Datamation*, 1989), although the multitude of very small firms may have raised that figure by close to one-fifth (*Financial Times*, 10 October 1988). Among the *Datamation* top fifteen, the share of American firms fell from 88.3 per cent in 1984 to 74 per cent in 1988, while the sole European (Olivetti) rose from 4.2 per cent to 6.1 per cent. Olivetti was the world's third largest producer in 1985, after IBM and Apple, but has since been well passed by Compaq and is being caught up by a number of others, including the American firm Zenith, now acquired by Bull, and some Japanese firms.

Apart from the two leaders, IBM and Apple, firms can be divided into those producing IBM-compatible machines and those with other operating systems. The former can in turn be divided into producers of quality-orientated compatibles (including Compaq, Zenith and Olivetti), which have used some customised components and differentiated hardware (for example, drives, VDUs and memories); and the clone-makers, assembling standard components purchased in the market. The British firm Amstrad essentially used the latter strategy to produce compatibles rather than clones, but the clone-makers have suffered in the late 1980s from cost pressures (for example, higher component prices) and a shift of demand up-market.

Portable computers are subdivided into various sub-categories according to their bulkiness. Laptops have grown particularly rapidly from a small base – their share of the total PC market is predicted to rise to 16 per cent by 1992 (IDC, 1989). The Japanese share has risen rapidly, and Toshiba is now regarded as the market leader. There is some European interest arising out of niche products like personal organisers and also from the takeover of Zenith Data Systems by Bull, but like the American entrants they face daunting competition from Japanese producers. In technology, the role of energy-saving components such as Liquid Crystal Display (LCD) screens and miniaturised hard discs has been crucial, and this too has tended to favour the Far Eastern producers.

Despite moves towards industry standardisation and systems integration – indeed, to some extent because of these factors – the product range offered by the industry has continued to diversify, for reasons on both the supply side (for example, new components) and the demand side (new needs). Technological boundaries between large and medium, and medium and small, systems are becoming increasingly fuzzy (cf. Figure 6.4). It now seems more appropriate to define the size of the system from the use to which it is put,

Table 6.2 Turnover and ranking of European informatics firms. 1989

	World rank	IS turnover $bn	IS as % of turnover	RD Intensity (%)	Return on sales(%)
Groupe Bull	8	6,465	100.0	9.0	-0.6
Siemens	9	6,011	18.5	11.2	2.6
Olivetti	10	5,573	84.7	5.3	2.2
Philips	18	2,815	10.4	8.0	2.4
Nixdorf	19	2,793	100.0	9.7	0.0
ICL	22	2,643	62.0	10.4	6.8
Alcatel	32	1,800	12.8	10.5	3.7
Nokia	46	1,192	22.4	5.1	0.7
CapGemini	48	1,103	100.0	6.7	0.0
Mannesmann	54	0,819	83.0*	n.a.	n.a.
Amstrad	59	0,717	70.0	0.0	12.2
Brit. Telecom	62	0,693	3.6	1.9	14.2
Finsiel	64	0,663	100.0	4.5	1.9
Racal	68	0,574	19.8	7.0	n.a.
Comparex	69	0,566	100.0	n.a.	3.0
SD-Scicon	83	0,432	93.1	0.0	1.1
Sligos	89	0,401	100.0	4.9	5.2
Sema Group	95	0,379	79.0	0.0	3.5
Norsk Data	99	0,358	100.0	10.1	-16.9

Notes: * = Mannesmann Kienzle only
n.a. = not available (not distinguished in accounts)

Source: Datamation. 15 June 1990

even though this may imply the same type of machine being regarded as a medium system in one application and a small system in another.

III European technological and commercial performance

Statistics published by governments and international organisations on R & D, trade and production in the area of computers are problematic. Classification schemes often do not allow a clear separation of computers from other types of information technology. More seriously for studies of competitiveness, the extensive activities of multinational companies outside their home countries mean that statistics are open to misinterpretation.

In recent years, comparative statistics on company performance have become more widely available. Consultancy firms such as Datamation regularly publish data on production, R & D and profits; and patent statistics held in the SPRU–OTAF data base allow detailed assessments of companies' and countries' technological performance.

There are five points we would like to make from analyses drawing on these two sources. *First,* the three main European firms – Siemens, Bull and Olivetti – are on the lower edge of the top league of world computer firms (Tables 6.2 & 6.3). Following its absorption of the German firm Nixdorf, Siemens is now Europe's largest computer manufacturer, and the company

Table 6.3 Regional sales and patenting of computer hardware firms, ranked by European sales, 1989

	Sales 1989				Patents,1969-86			
	Europe	N. America	Other	Total $bn*	Europe	N. America	Other	Total no.
IBM	35%	43%	22%	21.28	8%	91%	0%	9,746
Siemens	90	9	1	5.41	94	5	0	8,385
DEC	38	46	16	4.92	0	100	0	168
Olivetti	81	9	10	4.51	89	11	0	805
Groupe Bull	63	30	7	4.07	98	2	0	422
Hewlett-P	37	48	15	2.89	5	93	2	1,044
Unisys	29	51	20	2.72	6	92	2	5,419
Nixdorf	93	5	2	2.60	93	7	0	89
STC/ICL	82	10	8	2.17	94	6	0	287
NV Philips	73	21	6	2.05	84	13	3	8,939
NCR	32	42	26	1.70	4	94	2	1,260
Apple	23	64	13	1.24	0	100	0	17
Nokia Data	99	0	1	1.18	79	21	0	112
Compaq	41	54	5	1.18	0	100	0	1
Xerox	32	54	14	0.89	3	87	10	5,949
Wang Labs	31	51	18	0.84	0	99	1	69
Mann.Kienzle	92	7	1	0.75	100	0	0	133
Amdahl	34	58	8	0.71	0	100	0	43
Amstrad	84	9	7	0.60	0	0	0	0
Commodore	69	24	7	0.60	20	80	20	10

Note: Total sales figures in $bn are for Europe only

Source: Sales from *Datamation*,15 June 1990; patents from von Tunzelmann, *Electronics* data base

with the broadest range of products. The second tier of European computer companies (Philips, Nixdorf and ICL) have recently succumbed. In addition to Nixdorf's and ICL's respective takeovers by Siemens and Fujitsu, Philips has announced plans to sell or close down its computer operations. A large gap has therefore opened up between the big three firms and remaining smaller firms such as Amstrad and Norsk Data, which also find themselves in a period of painful restructuring.

Second, the most important international trend in this industry in the past decade has been the rise of Japanese firms in world markets and the parallel relative decline of American firms. This shows up both in movements in trade and production shares (see Tables 2.8 and 2.9 in Chapter 2), and in Japan's rapidly increasing share of patents taken out in the United States (Table 6.4), the latter being especially ominous for the longer-term standing of American and European producers. Events in the European industry have from this perspective been tangential to the main commercial struggle between American and Japanese producers.

Third, European firms' shares of world computer markets have increased, although less markedly than the Japanese shares. In 1988, Japanese firms accounted for 21 per cent of world computer markets (compared with 6.5 per cent in 1975), while the European share in 1988 was 16 per cent (up from

Table 6.4 Patenting in computing in the United States 1963-68 to 1981–86, by country of patent origin

Computing	1963-81	1969-74	1975-80	1981-86
Total patents:				
Macro region:				
Europe	595	1,031	1,195	1,446
N. America	3,191	4,680	4,416	5,648
Japan	48	368	741	2,166
Rest of world	36	58	74	95
Totals	3,870	6,137	6,426	9,355
Europe:				
FR Germany	205	324	363	541
France	85	166	237	264
United Kingdom	159	268	222	248
Belgium–Lux.	9	12	10	16
Netherlands	35	55	60	83
Switzerland	34	50	80	94
Austria	5	12	5	8
Sweden	24	41	67	61
Other Scand.	5	10	16	12
Italy	33	91	130	116
Other Medit.	1	2	5	3
Total	595	1,031	1,195	1,446
Shares:				
MacroRegion:				
Europe	15.4%	16.8%	18.6%	15.5%
N. America	82.5	76.3	68.7	60.4
Japan	1.2	6.0	11.5	23.2
Rest of world	0.9	0.9	1.2	1.0

Source: von Tunzelmann, *Electronics* data base (from SPRU-OTAF data base)

4 per cent in 1975). However, Table 6.3 shows that the European computer industry still makes the great majority of its sales to national and other European markets. The growth of world markets has thus been mainly the result of the expansion within European markets where firms have been able to exploit local advantages or benefit from protection.

The performance of European computer industries is less flattering still when the huge and increasing trade deficits are taken into account (see Chapter 2). The deficit has become very substantial in relation to both the United States and Japan, and in the former case has only partly been mitigated by the growth of multinational production in Europe (what is more, one of its consequences has been to provide more effective channels for goods produced elsewhere in the world to flow into the European market).

Fourth, patent data broken down by geographical area show a rather high level of specialisation among European firms in informatics (computing plus information storage). Table 6.5 shows that the European revealed technological advantage (RTA), a measure of specialisation in computers set against the broad pattern of industrial specialisation, has risen from 0.5 to 0.7, while

Table 6.5 Informatics patents as % of all patenting by electronics companies

	1969–71	1972–4	1975–7	1978–80	1981–3	1984–6	Total
(1) Core companies							
Europe	0.54	0.66	0.85	0.76	0.74	0.67	0.71
USA	1.09	1.06	1.05	1.06	1.04	0.98	1.02
Japan	0.91	1.15	1.00	1.09	1.17	1.24	1.19
(2) Total, including subsidiaries etc.							
Europe	0.67	0.68	0.80	0.70	0.67	0.63	0.70
USA	1.08	1.07	1.08	1.09	1.05	0.97	1.03
Japan	1.07	1.18	1.01	1.11	1.23	1.32	1.29

Note: All figures greater than unity indicate a positive RTA for that region relative to the criterion chosen

Source: Calculated from von Tunzelmann, 1988: Table 1G

it has risen from 0.9 to 1.2 for Japan and declined to around 1.0 for the United States (a figure greater than one indicates above average specialisation).

Fifth, where performance across European countries is concerned, the German industry emerges now as the strongest in Europe, in terms of scale of output, trade balance and patent shares. The United Kingdom, France and Italy have all suffered significant deteriorations in their trade balances, although the British industry has alone recorded a decline in patenting in the United States. All other European countries are substantial net importers of computers.

These statistical measures therefore demonstrate that while some advances have been made, the European computer industry remains in a rather precarious state. Its share of world markets is small, it is still to a significant degree national in its market orientation (the French and German industries relying particularly on their home markets) and it has yet to establish itself as a major player in American, Japanese or other non-European markets.

IV Firm structures and management

General trends

The turbulence experienced by the European industry in the 1980s looks likely to continue well into the 1990s. Hardware production continued to be dominated by the old 'national champions' like Bull, Siemens, and STC/ICL down to the end of the 1980s, but with the takeover of ICL by Fujitsu in 1990 and of Nixdorf and the French firm IN2 by Siemens, a new round of rationalisation is under way. Since they are now the main European players, we shall consider developments in these companies later in this section. Firms like Nixdorf, Amstrad (United Kingdom) and Norsk Data (Norway) experienced meteoric growth for a number of years but then encountered serious problems showing that scale economies and systems still matter in design and marketing. Amstrad is diversifying away from computers and Norsk Data is attempting to restructure after heavy losses and declining sales. There are a number of small hardware firms, especially in the United Kingdom, but few have shown sustained growth after the profusion of new entries in the late 1970s and early 1980s, and some such as the British firm Apricot have sold their hardware divisons. In continental Europe, the more significant SMFs are now mostly owned or controlled by multinational companies.

The large hardware firms are also among the biggest suppliers of software – which was estimated to constitute 11 per cent of Siemens's IS sales in 1989, 8 per cent of Bull's and 9 per cent of Olivetti's, making these the three largest European software producers (*Datamation,* 1990). However, specialist software and services firms like Cap Gemini, Finsiel, SD-Scicon and Sligos are also important (See Table 7.2 in Chapter 7). European hardware producers have faced two kinds of challenges. On the supply side, they face stronger competition by foreign suppliers, based either on higher efficiency from advanced production techniques, scale economies, or cheaper labour. On the demand side, users have sought standardisation along the lines discussed above (compatibility, modularity etc.) which has cramped the ability of producers to exploit traditionally captive markets by differentiating their products.

Together with the technological changes surveyed earlier, notably the blurring of boundaries and the move to open systems, these factors underlay the problems encountered by the smaller European firms. Most conspicuous was the halting of the growth of minicomputer producers like Nixdorf and Norsk Data. The problem was not confined to Europe, as equivalent American firms like Wang Laboratories and Prime Computer suffered equal setbacks, but the overall effect on the American industry has been less serious because of its greater size, strength and diversity. Similarly, European firms were not alone in being slow to move towards open systems, but the tradition of recognised or *de facto* national champions, and the typically small size of national markets in Europe, again exacerbated the difficulties.

European firms are now aiming to put this past behind them, and are seeking salvation in 'systems integration', building on their greater familiarity with indigenous consumer needs and on the customer-led orientation of the software and services companies. A number of SMFs, squeezed between the supply factors of greater competition and short production runs, have tried to move downstream, in search of the greater margins currently on offer in systems provision. But without an adequate range of products that the firms themselves can supply, this may prove even more hazardous.

There is, however, increasing acceptance that individual firms cannot do everything. The rambling product ranges of the European electronics firms of the 1960s and 1970s have in most cases given way to greater specialisation in the 1980s. This has not usually entailed complete product specialisation but 'restricted diversification' involving greater focus on a few core competences. Some firms like Nokia are still in the painful throes of discarding fringe activities. Only Siemens has conducted a policy of attempting to remain present in most branches of electronics.

The managerial problems thrown up by these changes have been substantial and perhaps the decisive factor in determining whether or not they could be met. 'National champion' policies placed an emphasis on the supply side and on developing new technologies, often at too leisurely a pace. The proliferation of user demands and the increasing sophistication of consumer tastes in the face of greater competition undermined these presumptions. Large firms were often too rigid and bureaucratic, while newer entrants like Nixdorf and Norsk Data suffered the opposite failings. Neither category of companies found it easy to respond adequately to the emerging consumer demands. Faced by a rapidly broadening range of product applications and requirements for each particular product, a responsive and streamlined management was crucial, and choices had to be made.

Internally, firms have sought to alter their organisations to be more responsive, in partial imitation of perceived Japanese organisational behaviour. How much the company culture can be transformed by reorganisation, for instance in Siemens, where a major effort has been made to de-centralise and achieve greater flexibility, remains to be seen. A frequently observed deficiency of European firms, both large and small, has been in marketing, and its weak connections with R & D and production.

The most evident changes have, however, been external. If individual firms are unable to achieve competence in a sufficient range of activities, then alliances with other firms have to be sought, to cross-relate both technology and product ranges. The importance of alliances will become apparent when we turn to discuss the firms individually. Product-led alliances tend to be established at the divisional level and it has been common for firms to cooperate in some markets and compete in others (for example, Olivetti and Digital are collaborators in small systems and competitors in medium systems). Even more interesting are the truly 'strategic' alliances, apparently negotiated at the highest levels among firms. The sequence of links between IBM and Siemens may portend a new oligopolistic phase in world competition. The emerging pattern in a number of areas of electronics is for cross-border alliances. Whether these consortia of rival alliances consolidate into international oligopolies may become a key issue in the 1990s.

However, the takeover of ICL by its erstwhile ally Fujitsu has also demonstrated the fate that can befall junior partners in alliances. Another key issue therefore becomes whether European firms can retain their autonomy in the longer run while engaging in international cooperation with larger and stronger foreign firms.

The principal European firms

Philips's withdrawal from computer markets was probably the unavoidable price that had to be paid for becoming overextended in terms of both product ranges and production base. Computers had never been a core activity and the survival of its semiconductor and consumer electronics divisions was strategically the most important priority for the company and, more broadly, the European electronics industry.

STC's sale of ICL to Fujitsu was a greater shock to the computer industry. ICL's progress in the 1980s had been considerable, especially in moving into 'vertical markets'. In retail automation, in particular, ICL claimed to be first in the British market (just ahead of IBM) and third in the world market (helped by its acquisition of Datachecker from National Semiconductor in

1988). It had also attained significant positions in defence, health administration and other systems markets. Fujitsu has therefore gained access to a wealth of computer systems expertise which it had previously lacked. In principle, it has also bought an entry ticket to cooperative European R & D programmes such as ESPRIT in which ICL was a participant. The French and German computer industries have been sharply critical of STC's decision, and by implication of the British government's decision to allow the sale to proceed. As in automobiles, they see Britain acting as a willing 'Trojan horse' for Japanese companies seeking entry to the European market.

The departures of Philips and ICL have made Siemens, Bull and Olivetti much the most significant remaining European computer suppliers. A brief discussion of their recent activities in this area is therefore in place.

Siemens. Following the take-over of Nixdorf, Siemens now supplies an extensive range of products from PCs up to mainframes and supercomputers, where previously its coverage of the market had been patchy. Possibly even more important, Nixdorf's strength in software and services coupled with Siemens's own already large number of software engineers enhanced Europe's largest software house (a combined software workforce of about 19,000), of potentially great significance for Siemens with its traditional orientation towards supply-led rather than demand-led markets.

Alone among European (or North American) companies, Siemens continues to aim at a significant presence in all core areas of electronics (communications, components and computing). Its nearest equivalent in this respect is NEC in Japan. Though sheer size obviously helps, there is concern that Siemens has spread itself too thinly, and cannot forever buck the trend towards partial specialisation. Siemens' transition from lumbering bureaucracy in the 1970s to leadership in advanced technologies in the 1980s has been told elsewhere (for example, Soete and von Tunzelmann, 1988).

As we have seen in Chapter 5, in an attempt to bypass the bureaucracy, chairman Kaske announced a management shake-up in 1988 with two objectives: to take marketing away from central office and back into the operating divisions and to break up the seven huge operating divisions into about 18 units that could be more coherently focused and – in terms of size – more manageable in an autonomous fashion.

Computers, together with PABXs in the old groupings, account for about 17 per cent of Siemens' 1989 sales of about DM 65bn. Siemens' own computer activities have been rather sluggish in the second half of the 1980s, and it has problems in competing with fleeter-footed rivals. The developing links with IBM are especially important. IBM has sold its telecoms acquisition, Rolm (producing PABXs), to Siemens; Siemens is the largest external chip supplier to IBM Europe; and the two surprised most people by announcing, in early 1990, joint development of next-generation 64Mb D-RAMs. Siemens also belongs to the IBM camp through the Open Software Foundation. However Siemens obtains its mainframe technology largely from Japan – from Fujitsu for its high-end mainframes, and from Hitachi for the joint venture with BASF known as Comparex.

In terms of markets, Siemens has cultivated East European links for many years, and is generally thought to have a head-start (for example, its DM 1.5bn sale of PCs to the Soviet Union). It has been less well represented in northwest Europe, partly because of long opposition to hostile takeovers, although its takeover of the French producer of IBM-compatibles, IN2, was relatively bloodless. Its greatest marketing problems lie in the United States, where despite Siemens's efforts, the American share of the company's sales has been declining in recent years and subject to continuing losses. Nor have minority shareholdings (as in AMD) or joint ventures (as with Intel) really prospered.

Bull. The experience of Bull demonstrates that 'national champion' policies may not be as disadvantageous as often alleged, so long as they can be geared to the needs for 'international champions' in the 1980s and 1990s. The holding company, Machines Bull, is still 90 per cent owned by the French government, and this has permitted considerable latitude in financial and other transactions. Bull's strategy, like its fellow French electronics champions Alcatel (CGE) and Thomson, has been based on large-scale overseas acquisitions in the later 1980s. In 1987 it acquired a 42.5 per cent interest in 'Honeywell Bull', matched by an initially similar holding of the American company (which aimed to withdraw into industrial controls and defence). In 1989 this became Bull HN as Bull increased its stake to 65 per cent and Honeywell dropped away. In October 1989 it also acquired the small-system computing interests of Zenith Electronics. By taking these steps, Bull has moved up from being the world's tenth largest computer manufacturer in the mid-1980s to seventh today, and has a target of being fifth in the mid-1990s. Group revenues were $5.3bn in 1988.

Bull's computing strategy has long been based on distributed architecture and networking. Compatibility problems emerged early through the long association with Honeywell and encouraged an emphasis on inter-operability. Bull has its proprietary operating systems but has endeavoured to bring them into Unix.

Its domestic base in France rose in 1988 as Matra and Thomson decided to quit computer production. But like Siemens, Bull is having trouble in raising its presence in North America, where it carries out 40 per cent of its R & D but only 18 per cent of its sales. The acquisitions strategy aimed to remedy the imbalance, but at the cost of clashing corporate cultures which explains why Zenith Data Systems continues as a separate division of the company.

Bull has also forged links with NEC. Though there have been benefits (to Bull from access to NEC technology and to NEC from European distribution channels), this alliance seems to have brought little so far. While Bull enjoys the protection of the French state, it lacks Siemens' spread and depth in technology, and there are doubts whether it can prosper without again becoming more dependent on foreign, and especially Japanese, technology.

Olivetti. From the late 1970s to the mid-1980s, Olivetti was turned around from a stagnating electro-mechanical office equipment company into a leading electronics company by the managerial wizardry of Carlo de Benedetti. De Benedetti then appeared to lose interest as his financial dealings across Europe proliferated, culminating in his daring but ultimately unsuccessful attempt to take over Société Generale de Belgique. Eventually Vittorio Cassoni was brought back from his secondment as head of AT&T's computer division to revive his former company, but again it was a case of moving too late.

Olivetti never aimed at a technological lead, but from its inception the company set high standards of design as the basis of its competitive advantage. In the later 1980s the markets, however, moved too fast to accommodate such strategies, although their advantages in more settled markets could become more apparent. In the meantime, Olivetti depends – at least in part – on Intel for chips, Microsoft for software, Canon for copiers and laser printers, Sanyo for fax machines, and Y-E Data (Yasukawa Electric) for laptops.

The relationships with AT&T are especially interesting. When AT&T's own production collapsed in the early 1980s, Olivetti stepped in and an estimated 36 per cent of Olivetti PC shipments went to AT&T between 1984 and 1987. In return AT&T had bought about a quarter of Olivetti shares in 1983. Cassoni was despatched from Olivetti to resuscitate AT&T hopes in computing and, paradoxically, was instrumental in forging a link between AT&T and Sun, for it was on the basis of this link that AT&T turned away from Olivetti and towards Sun. Out of the latter alliance arose the Unix battles discussed in Chapter 11. The loss of sales to AT&T has had a serious impact on Olivetti, which is trying to offset this by selling PCs to Digital.

More than the other large European firms, Olivetti has survived by being fleet of foot, prepared to enter alliances with a variety of foreign suppliers and to put its label on their products (or have them put their labels on its own products) where necessary. It remains to be seen whether its progress can be sustained without heavy investments to deepen and widen its own technological capabilities.

IBM-Europe. IBM's long dominance of the computer industry appeared to be drawing in during the 1980s. The innovativeness displayed in the 1950s and 1960s appeared to be giving way to bureaucratisation and conservatism. Its organisation appeared to be imitating its conception of technology, built around the central mainframe supplying 'dumb' terminals. IBM entered the market for PCs and later mid-range systems, but tardily. After several years of low growth and profitability in the mid-1980s, it moved towards major restructuring, aiming to move from a technology-driven to a customer-driven company. The leading prongs of this restructuring included: cutting employment, but without trying to overturn the company's paternalistic job guarantees; reducing bureaucracy; decentralising on a world-wide basis; undertaking collaboration with software and systems houses, for example, for CASE development; and increasing market specialisation (*Financial Times*), 24 April 1990). Although its share of the total computer market has declined in the 1970s and 1980s, it is still large enough to have more time to sort itself out than smaller firms – not only is it still dominant in mainframes, but its sales bulk largest in all major sub-sectors of computing except workstations and computer services (its lead over the next largest in software is actually greater than in mainframes).

Its role in Europe can be seen in similar vein. As can be seen from Table 6.3, its European sales are nearly three times those of the next largest supplier, the new Siemens-Nixdorf Information Systems. Four of its subsidiaries appear separately in the 'Fortune 500' list for 1989 of the world's 500 largest companies by turnover (IBM UK Holdings, IBM Deutschland, IBM France and IBM Espana; IBM Italia should also be included). Its main problem has been integrating those national subsidiaries into a more cohesive European whole – in this respect not dissimilar from some indigenous

European countries. Measures to increase coordination at the European level were announced in April 1990 and have involved parcelling out pan-European responsibilities among the various national subsidiaries, as well as increasing the duties of the Europe headquarters in Paris.

The European connection has been invaluable to IBM in the later 1980s. It has supplied much of the growth and profitability of the company, as indeed with a number of other large American computer firms. It has shown a considerable technological record, exemplified by the two Nobel Prizes won by the Zurich laboratory, and its decentralised structure has formed the model for IBM to introduce more generally in the company. Like the other leading American producers, it has been aiming to increase its European image, partly with an eye to EC support for both markets and technologies. And as in those other companies (Digital, Apple etc.) the very growth of European operations and strength is giving rise to considerable ambiguity over the American-ness of IBM. Perhaps its most important gesture was moving its communications system division from the United States to the United Kingdom in December 1990.

V Conclusions

During the 1980s, the diversification of products and technologies opened up possibilities for survival and growth, particularly where specificity of application, knowledge of the market and the need for close producer–user interactions were important. There were significant new entrants in Europe, a well as the revival of firms such as ICL which had previously been in difficulty.

At the beginning of the 1990s, however, the European computer industry again finds itself on the defensive. There has been a strong trend towards industrial concentration, so that there is now only a handful of European firms with significant computer development and manufacturing capabilities. Moreover, European suppliers have tended to cede ground in hardware, particularly to the Japanese, and to concentrate on the software-intensive provision of computer systems and networking. While it is there that the greatest market growth is occurring, there is danger that too much control over the market will be lost if the withdrawal from hardware proceeds much further.

At present, the indigenous European industry comprises three large firms – Siemens, Bull and Olivetti – and a scattering of smaller ones. Among these, Siemens appears to be in the strongest position, as it is in other areas of electronics. However, it would be regrettable if Siemens became the only substantial European computer manufacturer. Plurality in supply is important for reasons of dynamic efficiency and supply security and to ensure that there is some spread of computer manufacturing capabilities across EC countries. In our view, the stage is being reached where further concentration in this area could represent a serious threat to Europe's industrial prospects.

During the 1980s, the foreign multinational presence in European computer markets also increased substantially. IBM has been followed by Digital, Apple and other American firms in setting up R & D and production facilities, various technology agreements have been made between European and foreign firms and now Japanese firms are beginning to build up stakes in computer production in Europe. The European activities of many American firms are now so great (Table 6.3) that their European management is often found seeking a more powerful role within the company – in other words, firms like Digital, Compaq, Apple and even IBM are becoming more 'European'. This has all been to the considerable benefit of the European consumer and has forced indigenous suppliers to become more competitive. Nevertheless, there are legitimate grounds for concern that the foreign presence is beginning to become excessive. Particularly if Europe becomes a theatre for confrontation between the American and Japanese giants, local suppliers, including some of the American subsidiaries which have sunk deep roots in Europe, may find themselves being elbowed out of the way.

This is not an argument for 'Fortress Europe'. Such an attempt at protectionism could make matters worse for domestic industries by encouraging American and Japanese firms to accelerate the expansion of their productive capacities inside Europe. It is also important that European firms should remain open to cooperation with American and Japanese firms in the field of technology. However, this is one area where the EC can ill afford to give up the struggle to retain a significant industrial presence. These problems are discussed further in the concluding chapter of this book.

7 Computer software: the IT constraint

Tim Brady and Paul Quintas

I Introduction

The vital role of computer software in IT systems is now well known but there is a widespread perception that software has failed to keep pace with hardware developments. There have been problems with quality, reliability and maintainability of software. Systems failure in fly-by-wire aircraft and automated stock-market trading systems inevitably result in the finger of blame being pointed at this nebulous commodity. The increase in size and complexity of computer programs leads to doubts about human ability to ensure that fatal flaws are not in-built. Skill shortages have become widespread. Backlogs in the implementation of new applications have become commonplace. Demand for software so outstrips supply that an American study predicts (with inexorable logic and not a little irony) that by 2040 every man, woman and child in the United States will be writing software. As Chapter 6 has shown, many new developments in computer networking depend on new software.

With the software 'crisis' now in its fourth decade, this chapter examines trends in technology, in supply and demand, the prospects for solutions to the problems and the opportunities these may create. The development of software is a highly skills-intensive process which has resisted the transition from craft to science and from human labour to automation. At heart, key elements of software development remain cerebral and inter-personal activities. As we will argue below, the extent to which the decade ahead may witness the consolidation of a software 'engineering' (as opposed to 'craft') approach may be conditioned by social and institutional factors as much as by technological developments.

Discussion of software is made difficult by the esoteric nature of what is, essentially, the information component of IT systems. For example, laws concerned with intellectual property rights have widely regarded software as equivalent to literature and other copyright material, rather than patentable technology. Furthermore, software is not a homogeneous product – there are many different types of software. The main distinction is between systems software and applications software. The former includes the operating system, which supervises the control function of the hardware, compilers (assemblers, translators), which transform programmes from one language to another and a range of application tools which allow end-users to retrieve, organise and manipulate data, data bases and programmes. The second category includes general purpose applications such as word processing and payroll operations and user-specific applications such as management information systems or machine-tool control software.

Software may be delivered in a number of different modes or media and it may be written in any one of numerous languages. The different modes of software include standard packages (a piece of software for generalised use); modified packages (a standard package tailored to the requirements of a particular customer); and bespoke or customised software (written from scratch specifically for an individual customer). The media on which software is stored include floppy disc, tape, cartridge or firmware (this is a special example of software whereby a program is stored on electronic circuits within the hardware). Inevitably, as in the previous two chapters, some technical jargon is unavoidable.

II Software markets and producers

Software was originally produced in-house by the computer users themselves and/or provided by manufacturers when they supplied the computer. A large proportion of software developed today is still produced in-house, either for internal data processing and information systems, or for incorporation in products (for example, by electronic capital-goods manufacturers) and as such is not 'tradeable' in its own right. However, since the famous IBM 'unbundling' decision in 1969, in which the United States anti-trust authorities obliged IBM to separate hardware and software, we have witnessed the growth of independent software suppliers in many countries.

The 'software industry' has evolved to produce and market the tradeable element of software. It consists of software houses; systems houses (who supply complete systems including both hardware and software); hardware manufacturers; distributors; and sometimes subsidiaries of user organisations who 'spin-off' software products they have developed initially for internal use.

There are severe problems in trying to derive statistical information about the software industry. Most of what is available relates to the output of the computer services sector which includes bureau, training and consultancy operations as well as sales of software products and services. There is little compatibility across countries or organisations in terms of what is measured, and even definitions of 'software' vary. Official statistics are at best mediocre and at worst non-existent (for example, in Italy there are no published statistics on the software sector). In the absence of official statistics we have to turn to private-sector intelligence providers such as IDC (International Data Corporation) who provide the software market information for the European Computer Services Association *Yearbooks* and are widely used in OECD and other reports.

The next section presents consultants' estimates of the 'tradeable' element of software for world markets, the United States, Japan, Western Europe and the United Kingdom. For data on internally produced software we report results of a survey of patterns of software use and sourcing in British computer-using organisations.

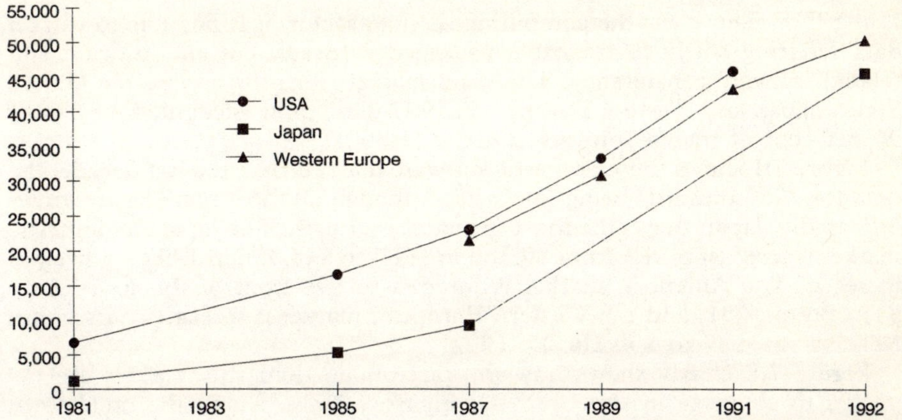

Figure 7.1 World markets for tradeable software, 1981–92
$ millions
Source: IDC/Coopers and Lybrand

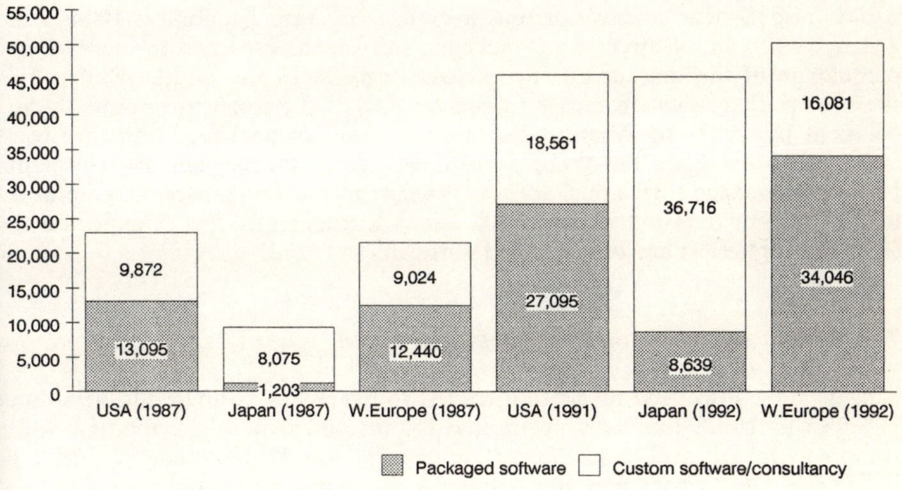

Figure 7.2 World markets for tradeable software, 1987, 1991–92
$ millions
Source: IDC/Coopers and Lybrand

Major world 'tradeable' software markets

Using IDC figures for the computing services sector, it is possible to extract data referring solely to tradeable packaged software, custom software and related software consultancy. The main markets for software are the United States, Japan and Western Europe – in 1987 these three accounted for almost 90 per cent of traded software (OECD, 1987).

Figure 7.1 shows these domestic software markets over the last decade, the data for 1989 onwards being forecasts. Although all three markets are growing rapidly, Japan shows the most spectacular growth. The Japanese domestic market is forecast to rise from $9.3bn in 1987 to $45.3bn in 1992 – a fivefold increase. The American market is forecast to rise from $23bn in 1987 to $45.6bn in 1991, and the Western European market is forecast to rise from $21.5bn in 1987 to $50.1bn by 1992.

Figure 7.2 clearly shows how the three main domestic markets are expected to converge in size by 1992. Figure 7.2 also shows the breakdown between packaged software and custom software and related consultancy services. In 1987, packages accounted for 57 per cent and 58 per cent of software sales in the United States and Western Europe respectively, whereas in Japan only 13 per cent of software sales were packages. One reason for this is that Japanese hardware manufacturers did not unbundle software until 1977 which has inhibited the growth of an independent software sector. Those software houses which have emerged are usually closely affiliated to the hardware manufacturers for whom they undertake very specific contract work and provide pools of programmers. Major Japanese users, such as banks, also appear to have strong in-house software capability (Baba and Takai, 1990). In all three cases packaged software is expected to increase its percentage of the market but by varying degrees. In the United States the share of packages was forecast to rise by only two percentage points to 59 per cent by 1991. In Western Europe the share of packaged software was expected to exhibit a ten-point percentage rise to 68 per cent by 1992. In Japan the package software share was forecast to rise seven percentage points to 20 per cent by 1992. Thus the Western European market is expected to have the highest share of packaged software by 1992.

The Western European 'tradeable' software market

The Western European market for traded software was estimated to be about $21.5bn in 1987. Packages accounted for 58 per cent of the market, with custom software and consultancy accounting for 42 per cent. Table 7.1 shows the sales of packages and custom software and consultancy in Western Europe in 1987.

The domestic markets of Germany, France, the United Kingdom and Italy account for almost 70 per cent of revenues for software and software consultancy in Western Europe. Figure 7.3 shows the former West German market is forecast to overtake France to become the largest in Western Europe by 1991 and to increase its dominance in 1992. (If the former East Germany is added the total would be even greater).

Table 7.1 Western European software market, 1987 ($m)

	Packaged software (%)	Custom software/ consultancy (%)	All
Former West Germany	3,182 (66.3)	1,617 (33.7)	4,799
France	1,981 (43.5)	2,578 (56.5)	4,559
United Kingdom	2,194 (63.2)	1,275 (36.8)	3,469
Italy	1,391 (57.2)	1,042 (42.8)	2,433
Netherlands	795 (55.0)	650 (45.0)	1,445
Switzerland	532 (61.1)	339 (38.9)	871
Belgium	452 (61.7)	281 (38.3)	733
Sweden	446 (61.0)	285 (39.0)	731
Spain	453 (66.3)	230 (33.7)	683
Denmark	297 (58.9)	207 (41.1)	504
Austria	302 (63.8)	171 (36.2)	473
Finland	203 (50.1)	196 (49.9)	399
Norway	212 (58.1)	153 (41.9)	365
Totals	12,440 (58.0)	9,024 (42.0)	21,464

Source: IDC Eurocast, 1988

Figure 7.4 shows the relative shares of packaged software and custom/consultancy services in the four largest European domestic markets in 1987 and 1992. Italy, the former West Germany and the United Kingdom all exhibit high sales revenue for software packages. The French market shows by far the highest proportion of software sales from custom software or related software consultancy (57 per cent in 1987). This is related to language problems – few English-language packages have been translated into French, and packages developed, marketed and distributed in France are relatively expensive because the economies of scale allowed by English-language packages do not exist. These factors encourage custom solutions. In all four countries the share of packaged software is forecast to rise by 1992 – from 66 per cent to 75 per cent in Germany; from 43 per cent to 54 per cent in France; from 63 per cent to 71 per cent in the United Kingdom; and from 57 per cent to 66 per cent in Italy.

The growth in the package market is primarily being driven by sales of application tools – for example, data base management systems, and development aids including CASE (Computer Aided Software Engineering) tools (Lowe, 1989). By 1993 the size of the packaged market is expected to reach $40 billion (more than treble the market in 1987). Application tools will account for 40 per cent of this, systems/utilities 24 per cent and application solutions 36 per cent.

Figure 7.5 shows the source of packaged and custom software in the United Kingdom, the former West Germany, France and Italy. As we saw in Chapter 6, the European computer hardware manufacturers are also the biggest suppliers of software. In 1987, hardware manufacturers supplied 53 per cent of package sales, independent software houses 31 per cent, and systems houses 16 per cent. Much of the Western European software market is made up of the systems and utilities software packages supplied by the American hardware vendors with IBM dominating. IBM produce 43 per cent of all packaged software supplied by hardware manufacturers. The recent

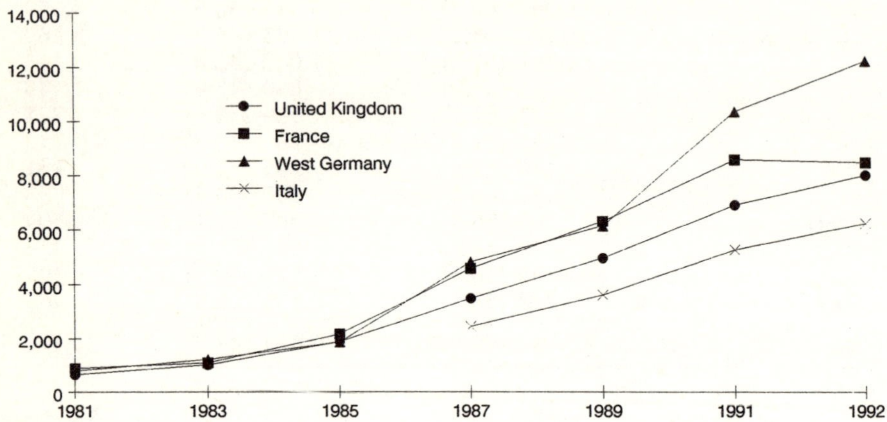

Figure 7.3 Western European markets for tradeable software, 1981–92
$ millions
Source: IDC Eurocast, 1988

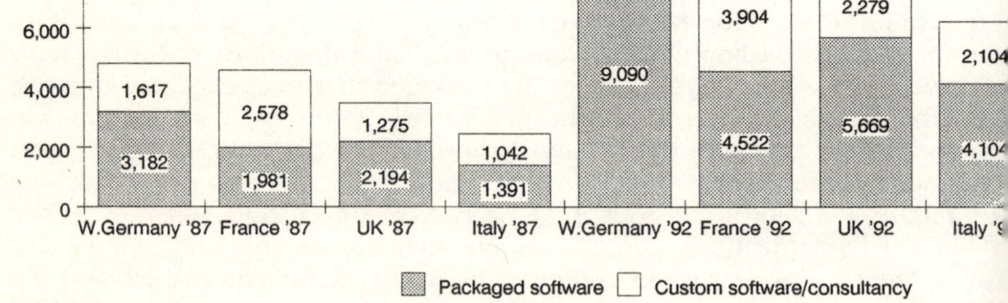

Figure 7.4 Western European markets for tradeable software, 1987 and 1992
$ millions
Source: IDC Eurocast, 1988

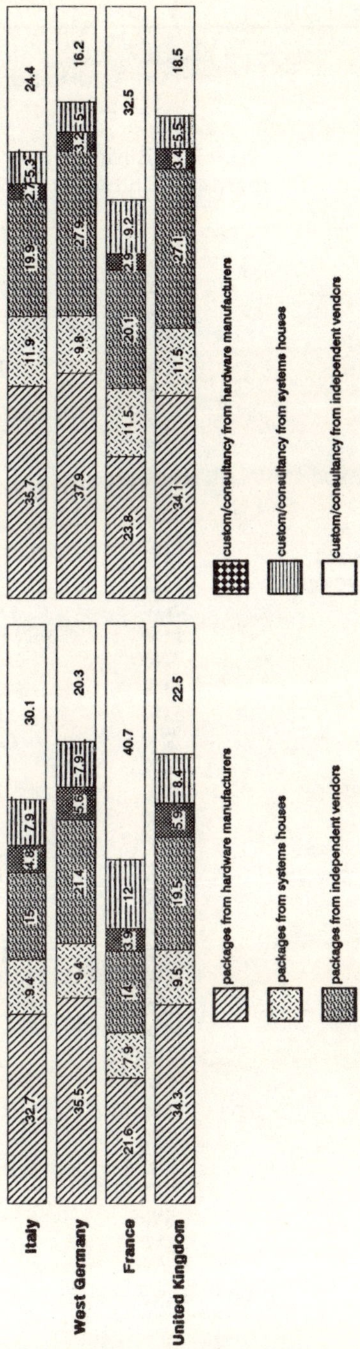

Figure 7.5 Sources of packaged and custom software: Italy, the former West Germany, France and the United Kingdom, 1987 and 1993 (%)
Source: IDC Eurocast, 1988

Table 7.2 Top-40 packaged software suppliers to the Western European market, 1989

Rank	Company	Principal Business (hardware/software)	Base	1989 European revenue (ECU m) Package software revenue
1	IBM	HW	US	2,659
2	Siemens	HW	Germany	500
3	Nixdorf	HW	Germany	470
4	ICL	HW	UK	400
5	Bull	HW	France	395
6	DEC	HW	US	350
7	Olivetti	HW	Italy	296
8	Unisys	HW	US	230
9	Computer Associates	SW	US	191
	Microsoft	SW	US	191
11	Oracle	SW	US	182
12	Hewlett-Packard	HW	US	165
13	McDonnel Douglas	HW	US	150
14	Software AG	SW	Germany	140
15	Philips	HW	Netherlands	126
16	Lotus	SW	US	114
17	Apple	HW	US	100
18	Alcatel	HW	France	91
19	SAP	SW	Germany	75
	Sema	SW	UK	75
21	SG2	SW	France	71
22	Ashton-Tate	SW	US	70
	Dun & Bradstreet	SW	US	70
24	Cap Gemini Sogeti	SW	France	68
25	Cincom	SW	US	63
26	Wang	HW	US	59
27	Nokia	HW	Finland	55
28	GSI	SW	France	53
29	Prime	HW	US	50
30	NCR	HW	US	42
31	CGI	SW	France	40
32	Pansophic	SW	US	39
33	SAS Institute	SW	US	37
	SD-Scicon	SW	UK	37
35	Informix	SW	US	35
36	Compaq	HW	US	30
37	Cognos	SW	Canada	28
	Softlab	SW	Germany	28
39	Tandem	HW	US	26
40	Ingres	SW	US	25
	Total			7,826

Source: OVUM 1990

takeover of Nixdorf by Siemens means that the German company follows IBM, with 7.5 per cent of this market. Bull, Olivetti, ICL and Philips respectively accounted for 3.6 per cent, 3.3 per cent, 2.9 per cent and 1.1 per cent of the packaged market in 1987. Furthermore, European software houses often supply software packages (including systems, utilities, tools and applications) developed in the United States.

Table 7.2 shows recent figures from OVUM, the market research consultancy, listing the top-40 package suppliers to the Western European market. These suggest that 64 per cent of packaged software sales come from American firms.

Internal software development

The shortage of statistics on internally developed software is even more severe than that for tradeable software. An OECD report quotes an estimate by Infomatics, one of the largest software houses in the world, claiming that in 1976, that is in the days before PCs, only between 10 and 15 per cent of all instructions executed on American computers were software purchased from external suppliers. Datamation estimated that in 1978 only 2 per cent of the computer budget of American firms went on external software, whereas in 1982 this had risen to 10 per cent (OECD, 1985:52). Although there is no systematic collection of such data across OECD countries, the OECD report went on to say that 'there is increasing evidence, in most OECD countries, that in-house data-processing departments now tend to focus on the maintenance and enhancement of existing software, an increasing share of which is purchased outside in packaged form' (OECD, 1985:51).

Recent research at SPRU which examined internalisation and externalisation of software development in the United Kingdom tends to confirm this view (Brady, 1990). A survey asked respondents to provide details of their most important systems, utilities, application tools, general purpose application and industry specific application software. Two-thirds of the 1,352 firms responding to the survey said they did some software development in-house, but overall, less than a quarter of the major software items they mentioned were developed in-house. Furthermore, over three-quarters of the items were either standard packages or modified packages. This suggests that much of the internal software development is related to the on-going maintenance and refinement of systems rather than the creation of new systems in-house. It is the firms with larger data-processing departments which do proportionately more of the in-house development. Figure 7.6 shows that the propensity to carry out internal software development also varies across industrial sectors but with some capability in every sector.

Figure 7.7 shows how the proportion of internally developed software varies depending on the type of hardware as well as the type of software. A lower proportion of microcomputer software (15 per cent) was developed internally than for software for minicomputers (27 per cent) or mainframes (24 per cent). This difference was particularly marked for applications software (28 per cent of micro software was produced internally compared to 48 per cent for both mini and mainframe software).

Supplementary interviews (carried out in 41 establishments in non-service

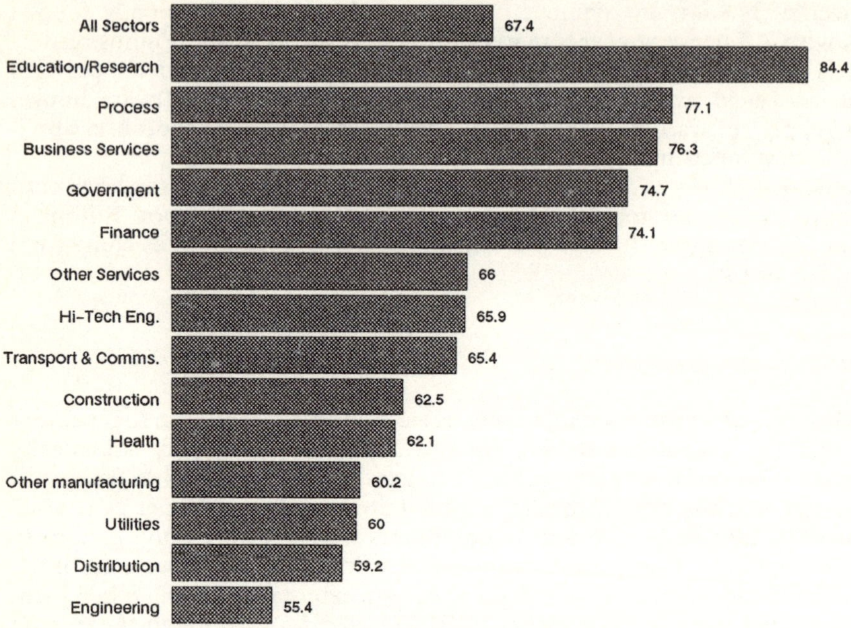

Figure 7.6 In-house software development by broad sector
(% of respondents performing)
Source: SPRU Software Survey

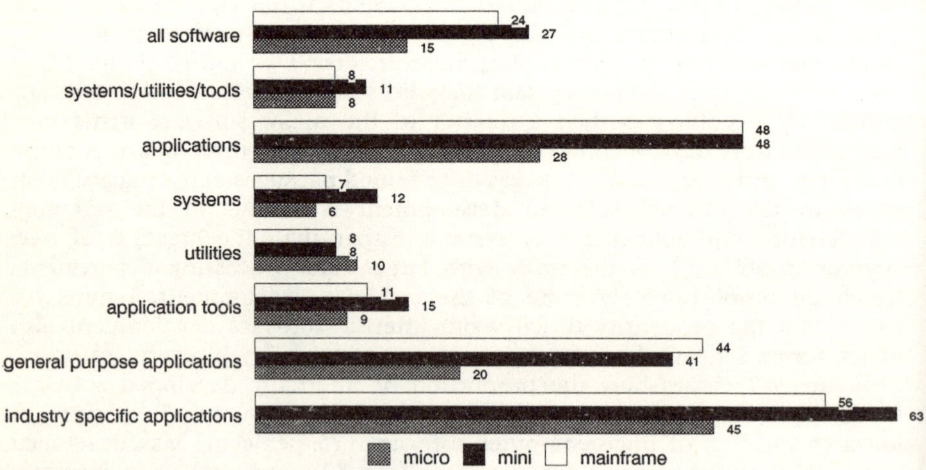

Figure 7.7 Software items developed internally by type of hardware and category of software (%)
Source: SPRU Software Survey

sectors) confirmed that the main area of internal development was applications software. While much of the software in use had been developed internally, the majority of establishments (26 of the 41) always sought a packaged solution as the preferred first option, and this was an increasing trend. The reasons included: increased demand for systems combined with a lack of in-house resources (both manpower and money) to meet demand; the increased availability and reliability of packages; packages are more entrenched in the market with corresponding better support available; problems with maintenance of in-house systems; the need to standardise software across large groups.

Firms opted for in-house solutions when packages were not available at all or did not have sufficient functionality. Additionally, an internal solution was often preferred when a system was thought likely to bestow competitive advantage (a custom solution could also be sought from a software house). Existing investment in internally developed systems discouraged the introduction of packages provided the in-house systems were working well. If there were problems – for example, lack of documentation, high maintenance costs, failure to meet business requirements – then a package would often be sought as the quickest replacement.

These results tie in with the estimates for tradeable software – hardware manufacturers are the dominant suppliers in the markets for systems/utilities/tools with software/systems houses being more important in the application software markets.

The relative position and strategy of the European software industry

Some of the findings of recent reports on the competitiveness of the British software industry in an increasingly international marketplace are relevant to the European software industry as a whole. One of the main concerns is that the domestic markets of individual European countries are too small to sustain a world-class software industry. All the reports recommend a restructuring of the software industry by the creation of larger firms via mergers to enable them to compete in world markets more effectively.

The last few years have indeed been characterised by a spate of mergers and acquisitions as firms have attempted to gain 'mass'. For example, in 1988 the British firms Systems Designers and Scicon merged to form SD-Scicon, and British-based CAP merged with the French company Sema-Metra to form Semacap. Cap Gemini Sogeti (CGS), the largest computer services company in Europe, took a 20 per cent stake in CAP prior to the merger with Sema-Metra and retains a hostile 23 per cent stake in Semacap. CGS, which has expanded via an aggressive acquisition policy, has 14 autonomously managed subsidiaries in France plus 22 more in 12 other countries. In July 1990, CGS bought SD-Scicon's loss-making German subsidiary Scientific Control Systems (SCS) and two days later acquired a 69.5 per cent stake in Hoskyns, the British computer services company previously owned by GEC-Siemens since their takeover of Plessey.

Some commentators (for example ACARD 1986) have urged a strategy of competing with the Americans in packaged software. But while there are some exceptions, the European software houses have not had spectacular success in developing packages, and Grindley (1988) rejects this strategy as being not viable for British firms. Instead he emphasises the need to build on the relative strength of British firms as producers of customised solutions. This would also apply to other European software firms – especially in France where the emphasis is already on systems integration and professional services. The best opportunities for the independent software industry in Europe may revolve around the systems integrator role – providing a total solution bridging both hardware and software. Such a strategy would bring the software houses into competition with hardware manufacturers who themselves are taking a greater interest in software beyond the operating systems with which they have traditionally been concerned. Grindley suggests a policy of collaboration with hardware manufacturers rather than competing with them, with the independent software houses taking the lead as systems integrators. In 1988, the French software house SESA signed an agreement with IBM to set up a joint systems integration company, while CGS and Bull launched a new venture company in the Value-added Telecommunications Networks market. American telecommunications groups have also bought into British computer service companies – BIS was bought by Nynex, while Istel was purchased by AT&T.

We turn now to the key issue of software as a constraint on IT systems development. Software is frequently referred to as the 'bottleneck' in IT applications, and the 'software crisis' is ever present. The following section identifies some of the characteristics of the software problem.

III Problems and solutions in software development

Software development continues to be faced by problems related to the productivity of software developers and the quality of software products. Advances in software have failed to keep pace with the performance and cost improvements observed for hardware. Collectively these problems are referred to as 'the software crisis'. The crisis manifests itself in a number of ways, of which the failure of safety critical systems is the most dramatic. More mundane effects include the widespread backlogs in the implementation of new applications experienced by most user organisations. When installed, many systems fail to meet requirements or perform adequately.

At the core of the software crisis lie endemic skills shortages, recognised throughout the OECD countries. Japan, for example, has an estimated shortage of 420,000 programmers and 550,000 systems engineers (Quintas, 1990). A recent British survey suggested a shortfall of between 4,000 and 7,000 software developers in the British user community alone (Buckroyd and Cornford, 1988). Like many of the problems associated with the 'software crisis' skills shortages are not a new phenomenon, having been recognised in the United Kingdom many years ago (Anderson and Hersleb, 1980).

The crux of the problem lies in the skills-intensive nature of software development, the pervasive nature of IT and the fact that not enough people are being trained. Consequently, skilled staff are in great demand. Firms are reluctant to train people because they fear they will be poached by other firms who have not had to bear the costs of training; most seek to recruit staff with at least two years' experience. This results in an increasing shortage with many user sectors (particularly the public sector) finding that their skilled practitioners are lured away by higher salaries elsewhere. In many countries, policies have been directed towards encouraging more training, including retraining of graduates from other disciplines. The use of alternative labour sources is another approach – developing countries, for example, are becoming a significant source of software labour (Quintas, 1990; Graham and Mody, 1990).

Software technology

Solutions to the twin software problems of productivity and quality are continually sought. The growth in the use of packages shows one response, but packages themselves are not problem-free, often provide only partial solutions and cannot confer competitive advantage on the user organisation. Since the roots of the problem lie in the software development process itself, much attention focuses on technologies and approaches which assist and might ultimately automate this process.

Fourth-generation languages

Fourth-generation languages (4GLs), which first emerged in the 1970s, are high-level procedural languages which are relatively easy to learn and enable a system developer to specify what is to be done, rather than how it is to be achieved. A range of software tools are currently provided by 4GLs which support the software development process and include end-user orientated languages which allow (theoretically) end-users to produce their own applications. SQL (Structured Query Language) is an industry standard 4GL which supports data-base search and analysis. Proponents claim impressive productivity improvements, but critics claim that code produced by them is inefficient, that large systems using them are difficult to maintain and that 4GLs are not much easier to use than standard languages.

Software engineering

'Software engineering' denotes a range of methods and tools which aim to move software development from being an 'arcane and error-prone' craft activity (Hoare, 1982) to the domain of an engineering discipline. 'Software engineering', chosen as the title of a seminal 1968 NATO conference, was intended to imply '...the need for software manufacture to be based on the types of theoretical foundations and practical discipline that are traditional in the established branches of engineering' (Naur and Randell, 1969). However, the term 'software engineering' covers two distinct traditions: one holds that engineering methodology must be founded on a science and mathematics base; the other says that engineering is essentially a practical activity which needs to develop better methods and tools to solve practical problems in the real world (Pelaez, 1988). Crudely, the dichotomy may be stated as existing between the formalists and the pragmatists.

The formalists' approach contains three main strands: (1) structured programming, which seeks to rid software of ambiguity and error-prone bad practices; (2) the use of rigorous methodologies for the specification and design of systems; and (3) the use of mathematically-based formal methods (described below). The pragmatic approach is now often referred to as CASE (Computer Aided Software Engineering), which aims at the provision of software tools (broadly analogous to machine tools in manufacturing) to assist the software development process.

Opponents of software engineering maintain that software production is essentially a 'craft' process impossible to automate fully. In the short to medium term, use of established software engineering approaches (particularly those concerned with managing the development process) would greatly improve the quality and transparency of software. Rigorous methodologies can minimise errors, and tool support will greatly assist, particularly with routine functions such as code generation from (increasingly higher-level) design, and subsequently as support for formal methods. However, the process will remain dependent on (constantly updated) human skills for some time to come.

CASE tools and integrated software development environments

Software tools which support discrete parts of the software development process emerged in the 1970s, and by the late 1980s CASE tools were available to aid or automate most of the software lifecycle, including requirements analysis, design, cost estimation, project management, documentation, code generation, testing, verification and validation. CASE tools for prototyping and to support re-use of existing code have also emerged. Very high productivity increases are often claimed for CASE, but caution is required in interpreting such claims since software development productivity is notoriously difficult to measure. Assessment of the extent of current CASE usage is problematic because of the diversity of the software developer community.

The use of discrete CASE tools presents problems in interfacing with other parts of the software development cycle. The major current push is thus to provide integrated sets of tools and methods. CASE tools integrated within a development environment such as Toshiba's Workbench are claimed to increase greatly both productivity and quality (Coopers and Lybrand, 1989). Full integration of CASE tools within a project management system is provided by an IPSE (Integrated Project Support Environment). IPSEs developed in the United States as Ada Programing Support Environments (APSEs) for military applications.

Commercial IPSE sales began in the mid-1980s in many OECD countries, with installations in large organisations such as utilities, financial services sector companies, and the public sector (Jones, 1988; Lowe, 1989). Smaller users are reluctant to make a commitment in a rapidly evolving market, and tool providers are waiting for standards to emerge (especially for interfaces). While Europe and the United States are each backing different standards for CASE environments, and Japan is promoting Sigma (see next section), IBM's recently announced AD/Cycle, based on a data repository approach, has the potential to become a *de facto* industry standard.

Formal methods

Formal methods are generally mathematics-based approaches to software development. Early practical use of formal methods focuses on the specification of software. Formally expressed specifications may be checked for logical consistency and concurrence with user requirements. Paper-based graphical techniques for system design and formal notations for specifications are beginning to be supported by software tools which increases their practical utility. In the longer term, formal methods may provide software which is formally proven to be internally correct. Formal proof is difficult and time-consuming, and at present limited to small programmes. The use of formal methods is severely constrained by shortages of available skills. However, there are a small but growing number of industrial formal methods projects in both the United States and the United Kingdom. Possibly the largest commercial use of formal methods to date is that by IBM (UK) with the Oxford University Programming Research Group (Collins, 1989). In general, American R & D has focused on formal approaches to verification, in contrast to the British focus on formal specification.

Object-orientated programming

Object-orientated programming (OOP) is built around the re-use of software elements, and allows flexibility, fast development and the ability to model complex systems (Bobrow, 1989). Emerging from AI (artificial intelligence) research, OOP is forecast to become a major part of mainstream software development in the 1990s (Ovum, 1989). OOP languages include Xerox Parc's Smalltalk, and C++, developed by AT&T.

Software metrics

A continuing lack of adequate metrics for the quality and reliability of software products and for software process productivity makes system predictions, investment justification and the comparative analysis of different approaches difficult. The United Kingdom is at the forefront of metrics research, but researchers warn that American and Japanese industry is taking metrics more seriously than the United Kingdom and Europe (Kitchenham, 1989).

Expert Systems

Expert systems (ES) and other AI techniques are beginning to be used within the software development process. Initially used for passive functions, ES are now being used in more active roles, for example, to guide developers in re-engineering, or as pro-active support to system design. IBM is a key player, with AI-based tools forming a central part of the AD/Cycle architecture. The British Alvey programme included research on the use of ES in software specification, prototyping and maintenance. Since the requirements analysis and specification phases of software development are the least amenable to automation, it is likely that these will be the areas where a knowledge-based systems approach may be most valuable.

Standards

A major area of interest is the development of standards of which there are two principal types: quality standards (for example, criteria for performance; compliance with specified development process methods); and compatibility standards (for example, agreed definitions for interfaces, languages, operating systems). Quality standards for performance depend on adequate metrics (see above). Compliance with process methods is currently being introduced for life-critical systems (for example, the British Ministry of Defence's Defence Standard 00-55, which mandates the use of formal methods). Compatibility standards are commercially and politically sensitive; moves to Open Systems and interface standards such as the European Portable Common Tool Environment (PCTE) will serve to increase competition. While the absence of software standards inhibits some developments, their premature imposition in a fast-changing technology may side-track further advance. In a standards vacuum, powerful players may enforce *de facto* standards (for example, IBM and MS-DOS and (potentially) AD/Cycle; see Chapter 11).

Unix

In the move to open standards, as we have seen in Chapter 6, operating systems have been particularly acute. Prior to Unix, operating systems were proprietary systems which locked users into specific vendors. Originally developed more than 20 years ago, Unix emerged from AT&T in the 1980s as the leading contender for a universally standard (that is, open) operating system. Initially backed by American government procurement, Unix has been promoted as a standard environment for the European ESPRIT, British Alvey and Japanese Sigma R & D programmes. It has advanced file handling, multi-tasking and distributed computing capability. In 1986 seven European and two American computer firms formed the X/Open group to promote a Unix standard and provide a common base from which to compete with IBM (Flamm, 1987). In 1988 the Unix world polarised around two camps: the Open Software Foundation (OSF, centred on IBM) and Unix International (UI, centred on AT&T). Hopes in 1989 of reconciliation have proven premature, but Unix continues to advance nevertheless. Most computer manufacturers now offer Unix for their products, and the telecommunications market is opening up. Growth should continue through the 1990s as Unix increasingly encroaches on proprietary operating systems (Coopers and Lybrand, 1989). A fuller account of the problems will be found in Chapter 11.

R & D programmes in software engineering

In a continuing effort to find solutions to the prolonged software crisis, governments have supported R & D in software technology for over 30 years, and the 1980s saw major new programmes in Japan, the United States and Europe. These latter programmes were sparked off by Japan's announcement of its fifth-generation programme which stimulated a rapid response from European and the American governments. The major software-related programmes are shown in Table 7.3. Many have been aimed at tackling productivity and quality issues.

In Japan, the Sigma programme supported a nationally networked software development environment which enabled technology transfer and encouraged standardisation. Sigma is likely to be followed by a $200m 'Super Sigma' programme intended to tackle 'usability' problems (CEC, 1989). In addition to R & D programmes, the Japanese government subsidises software firms via low-interest loans and tax concessions which allow up to 50 per cent of profits tax free and available for reinvestment.

The United States, while not supporting a nationally coordinated software R & D programme, has initiated a range of (often overlapping) programmes mostly supported by the department of defense (DOD) which lists software tools and methods as 'militarily critical'. The Software Engineering Institute (SEI) at Carnegie-Mellon University is the first federally funded laboratory to be established since the nuclear labs of the 1950s. The institute is intended to provide a focus for software R & D, education and training and to promote the diffusion of SE tools. Following the 1984 relaxation of the anti-trust laws, American private industry has established many R & D consortia.

Three large European programmes supported software R & D: the European Community's ESPRIT and RACE programmes and the pan-European EUREKA programme. In ESPRIT I, 20 per cent of the total proposed budget was devoted to software R & D. At the national level, the British Alvey programme supported software engineering R & D in four enabling technology areas: IPSEs and tools, formal methods, reliability and metrics, and knowledge-based software tools. The programme objective was to widen the software engineering community by forging links between academia and industry; and to develop generic software technologies (tools and methods) which were expected to support the development of a multiplicity of software applications in the post-Alvey period, thereby improving the competitiveness of British software suppliers.

IV Conclusions and implications for policy

The development of the software industry has been (and remains) inextricably linked to the hardware industry. It also depends crucially on the behaviour of user organisations. For years proprietary operating systems tied users to hardware manufacturers. Users have since become more sophisticated in their knowledge of IT and, as the relative cost of hardware has declined, software has become the key determinant in many firms' IT strategies. Users have become less willing to accept proprietary systems with the result that manufacturers are slowly adopting an 'open systems' approach.

Changes in the cost and performance trajectories of IT cause organisations to review their systems periodically. Thus, in recent years, we have seen a growing emphasis on relational data bases, fourth-generation languages, CASE tools, local area and wide area networking, end-user computing, the

Table 7.3 Major government-sponsored software-related R & D programmes

Country	Programme	Duration	Software Emphasis	Major R & D funding
United States	Software Technology for Adaptable and Reliable Systems (STARS) – DoD	1984–90	Integrated and automated software design tools	$85m in 1984. Software Engineering Institute $109m between 1984 and 1989
	Strategic computing – DoD	1984–94	Software for multi-parallel processing, artificial intelligence (speech and natural language understanding; vision and expert systems)	$750m to $lbn over entire project life for hardware and software
	Space Station – NASA	1984–94	Network operating systems and software tools	$8bn over entire project life for space station hardware and software
Japan	Very high-speed scientific computing systems	1981–88	Operating systems and programming languages for supercomputers	$200m from MITI over entire project life for hardware and software
	TRON (real time)	1981	To develop first all-Japanese operating system, targeted at PCs, workstations, consumer electronics and office and industrial automation equipment	
	Fifth generation computing system	1982–91	Software engineering and artificial intelligence are critical parts of the project; objective is a Knowledge Information Processing (KIP) machine	$450m from MITI over entire project life for hardware and software

	Intap	1985–92	Facilitating interchange of information from wide variety of sources	$100m over project life
	Sigma	1985–90	Software engineering	$100m (project self-financing but strongly supported by MITI with another $100m)
France	Five-year electronics plan/software programme	1983–87	Software engineering	$685m from French government for computing services and software
	Marisis	1983–86	Supercomputer software	$15m from French government for hardware and software
United Kingdom	Alvey Programme	1983–88	Software engineering	$50m from British government for software over entire life plus $33m from industry
European Community	European Strategic Programme for Research and Development in Information Technology (ESPRIT)	1984–88	Software technology	$100m for software technology programme over entire life; total EC commitment to ESPRIT is $645m
	ESPRIT II	1988–92	Software technology	$2000m for whole programme

use of lap-top computers by sales people and the use of electronic data interchange. Developments in hardware have made it possible to migrate applications from mainframes to minis and micros. This 'downsizing' has the effect of increasing the use of packages and thus stimulates the growth of the market-place.

Above we have seen evidence of the shift by end-users to greater use of packages wherever possible. Forecasts for 'tradeable software' also reveal an increasing share for packaged software. At the same time, the development of bespoke industry/firm-specific systems designed to achieve competitive advantage is more likely to be done in-house or by software houses. Although firm-specific systems may be developed in-house, software tools and languages needed to develop them are more likely to be supplied by software suppliers (compare the forecast growth in the application tools market). The continuing growth in demand for IT systems coupled with IT skills shortages suggests that many smaller systems will be developed using packages.

Whether European software firms can benefit from these changes remains an open question. Despite the restructuring that has taken place in recent years, there has been little change in the overall position of the European software industry – American hardware and software companies still dominate world markets. European software suppliers remain heavily dependent on sales to their domestic markets even after the mergers and acquisitions involving overseas companies. Europe is in a relatively weak position *vis-à-vis* the growing packaged software market with many packages they supply being of American origin. The larger European (particularly British and French) software firms' main strength still lies in providing custom solutions rather than as package developers. As companies increasingly look to their IT systems to provide them with competitive advantage and look for bespoke solutions this could provide opportunities for European systems integrators. But satisfactory bespoke solutions necessarily involve highly skilled people both in user firms and in consultants.

This type of intimate collaboration between hardware suppliers, consultants and user software groups appears to be particularly characteristic of the Japanese software sector (see Baba and Takai, 1990 on the Japanese Banks). Many analysts believe that Japan could become the major challenger to the United States in software because of its (Japan's) emphasis on implementation of software engineering techniques which may eventually deliver a competitive edge in producing low-cost, error-free programmes (Coopers and Lybrand, 1989). Philip Hughes, former Chairman of Logica, has expressed the view that 'Japan may soon become the main world player in computer software' because the Japanese industry has 'developed the vigorous quality and quantity control that software now needs' (*The Independent*, 1 October 1990). Japan has a growing strength in systems for manufacturing process control, and Japanese overseas investment takes with it IT systems incorporating Japanese software (for example, imbedded in manufacturing systems in new plant). The McKinsey Report (1983) identified the 'shortage of leading edge customers' as a major factor in European IT decline.

While Europe has strong capability in software R & D, and is internationally recognised for its contributions to theory, the European Commission has noted that European achievements in software engineering R & D have, as yet, failed to translate into industrial software products (CEC, 1989). European ideas, theories and prototype tools have formed the basis for American CASE products, with the result that, in spite of supporting world-class R & D, Europe is losing out on the lucrative CASE product market to the United States.

European programmes in the 1980s have concentrated on pre-competitive R & D to provide a range of enabling technologies. Evidence from the United Kingdom (Quintas *et al.*, 1989) suggests that the take-up of such R & D currently faces formidable barriers focusing on institutional and social factors within the software community. Europe has lacked programmes which support industrial innovation, technology transfer and training in the way that the Japanese Sigma and American SEI initiatives have done.

However, future programmes are calculated to remedy some of these deficiencies. EC plans for the 1990s include the European Systems and Software Initiative (ESSI), a proposed five-year programme focused on the transfer and industrialisation of results from R & D programmes such as Alvey and ESPRIT. The programme would involve both producers and users, and would support education, training and pioneer applications (Tully, 1989). ESPRIT II has also moved distinctly closer to users and applications.

There is more scope for 'enlightened procurement' policies which specify the use of advanced software engineering technologies for public-sector applications which could act as demonstrators. Users need to be involved from the earliest stages in demonstrator projects to ensure relevance to real needs. Mandatory software quality standards for civil safety-critical systems would promote the adoption of software engineering, increase systems reliability and demonstrate the advantages to other application areas such as financial services in which quality and reliability are vital but not safety critical. However, the enforcement of such standards without adequate provision of trained personnel may well set back the cause of formal methods, since formal methods can only be effective if used by skilled staff (Ince, 1989).

It remains unlikely that software development will be fully transformed from a craft activity to an engineering discipline, its outputs proven and validated to conform to the user's requirements and with guaranteed reliability. Since user requirements must inevitably change, software must be constructed in such a way as to allow for and support such change. Software development is thus inherently linked to the structure, organisation and development of user organisations. Its design is thus a function of human interactions.

At the firm level, skills deficiencies are the major barrier to the take-up of software engineering. Raising the number of new graduates is not an ade-quate strategy since thousands of existing software developers must also have their skills continuously upgraded, and the demographic shift in the 1990s will mean that new recruits will become even more scarce. While some companies continue to neglect the training arena, competitive advantage is likely to accrue to software developers with the most highly skilled workforce. Training and retraining policies are thus an area where major national and EC initiatives are a *sine qua non* for European success.

8 Evolving telecommunication infrastructures: organising the new European Community market-place

Robin Mansell and Kevin Morgan

I Introduction

At the heart of the diffusion and application of a growing variety of business and consumer services lies the telecommunication infrastructure. Forecasters suggest that telecommunications, including equipment, network and service provision could account for as much as 7 per cent of the European Community's gross domestic product by the year 2000. Regardless of the accuracy of such forecasts, there is virtually no argument about the widespread impact of innovations in the telecommunication field over the past several decades. Of particular significance is the arrival of 'intelligent' service applications that rely on an increasingly complex telecommunication infrastructure. These will be reviewed in the following Chapter (9). This chapter concentrates on the telecommunication infrastructure.

Since the invention of the telegraph in the 1840s, telecommunications have become a central factor in the restructuring and expansion of markets and the transformation of social and cultural patterns of communication. For the business community in the 1990s, services based on the telecommunication infrastructure are now major weapons in the arsenal of competitive strategies. They permit information processing and exchange in ways that extend the geographical reach of management decision-making and control and increase the speed of response to changes in world markets. Increasingly, too, it is becoming essential for firms to be linked into a 'network' of intra-firm and inter-firm communication services.

For the consumer, the telecommunication infrastructure holds the promise of important two-way linkages through tele-shopping and banking services, public health and other information services, the delivery of diverse entertainment products and interactive participation in electronically mediated social and economic networks.

The key to the emergence of a more flexible and complex telecommunication infrastructure is the introduction of computer-based digital switching and transmission techniques. These have transformed relatively homogeneous telecommunication services of the past, which really only provided voice telephony and telex, into an array of information-processing services. But this transformation has been uneven. Some new applications have become widely diffused, while others have provided enhanced capabilities to only a relatively small number of larger firms and scientific organisations.

The historical tension in telecommunications between the development of services which serve public and social policy objectives and the development of innovative business services has been exacerbated by these technical developments. At the centre of this tension lies the history of the development of a monopolistic structure for the provision of telecommunications. Services which depend on a physical delivery link to the consumer, for example electricity and gas as well as the old-style telephone, have been characterised by many as 'natural' monopolies. Nineteenth-century experience suggested that it was neither sensible (too much digging up of roads) nor economically feasible to provide the dedicated link to the customer's premises competitively, given the technologies available at the time.

But the rise of private and public monopolies in telecommunications at regional and national levels is also attributable to the refusal of service operators to interconnect with municipal or regional telephone systems and other economic and political factors. In this sense, the monopoly national PTOs which emerged in the 19th century were anything but natural. In some countries telephone service was provided by interconnecting regional operators. In others municipal operators continued to thrive. As technologies changed the underlying cost characteristics of the industry, the 'natural' justification for the monopoly in telecommunications came to be challenged with increasing frequency. New suppliers sought to compete using innovative technical systems and on the basis of quality and price.

Another justification given for the rise of national PTOs was the added complication of what economists call 'externalities' – the wider the network, the greater the overall benefit gained by subscribers even though these benefits were difficult to capture by the price mechanism. There is no conclusive evidence as to whether these 'externality' benefits could have been captured through the interconnection and interoperability of separate networks. Nevertheless, national publicly owned utilities and regulated private utilities emerged with a monopoly which extended to all components of the telephone network. These utilities were encouraged to meet a variety of public or universal service objectives. These included the extension of services throughout their territory and the maintenance of relatively low prices for local access to their networks.

Recent developments in telecommunication technology have made it increasingly difficult to sustain the so-called 'natural' monopoly in telecommunications. Cellular telephony, satellites and other innovations mean that there is no longer any need to be physically linked into the network at the customer's premises. Long-distance services also can be provided using competing technologies. The burgeoning of new services offers some customers a range of choice which makes the concept of monopoly seem absurd. Yet elements of monopoly control of service provision do remain. Today, and for the foreseeable future, most residential and smaller business customers in the EC will continue to rely on fixed connections to the telephone network. These customers have little choice but to purchase telephone service from their national PTO, which retains a monopoly in the provision of voice telephone service.

When competition in the supply of data communication and other advanced services is permitted, it is tempting for new entrants to concentrate on the lucrative business market and neglect their potential application for the more dispersed domestic users. In these circumstances the challenge for public policy is to create an environment which reduces distortions in service access, while at the same time stimulating the diffusion of new services (and efficiency gains). When competition and monopoly provision of different services exist within the PTO, there is a need for effective regulation to ensure that domestic consumers are not disadvantaged. For new services such as public health and medical information or home education innovative policies are needed to ensure that PTOs are responsive to social policy rather than mainly to business needs.

The transformation of telecommunications is taking place on a global scale. In the past, when telecommunications meant largely voice telephony, public monopolies dominated investment patterns, equipment design parameters and service characteristics with only loose government oversight. As voice telephony gave way to the wider range of services, these public utilities were perceived to have failed to produce the quality of services and diversity of choice demanded by the user community. The combined pressures of technical change, growing private-sector expertise in the operation and management of their own 'captive' or leased telecommunication networks and competition have focused attention on the need for new policies and institutions.

It is this task that confronts the CEC as it seeks to restructure the European telecommunication equipment and service markets. Its aims are to 'develop the conditions for the market to provide European users with a greater variety of telecommunications services of better quality and at lower cost, affording Europe the full internal and external benefits of a strong telecommunications sector' (CEC, 1987c). As it pursues these aims, it confronts disparate strategies among the major actors in the member states. The Public Telecommunication Operators (PTOs), government policy-makers, equipment manufacturers, service suppliers and countless users are all developing new ways of organising the supply and use of the telecommunication infrastructure. Currently this leaves the level of technical development and investment in the network extremely uneven among the various member countries.

The technical promise of telecommunications is relatively easy to see. But the combination of government policies and private-sector strategies that can respond to the new market realities is more difficult to forecast. In this chapter, we examine the tensions underlying efforts to achieve greater efficiency in telecommunication supply and to develop services that will support public participation in the 'information economy'. As with the previous chapters in Part II, some technical jargon is unavoidable. The reader is referred to the glossary and list of acronyms etc.

II The evolution of the telecommunication utilities

In the early 1980s, European telecommunications epitomised the economic problems of 'non-Europe'. Each member state had its separate regimes and there was no Community-wide dimension to policy-making. This fragmentation of markets made the introduction of new products and services especially difficult since they had to be certified and tested separately in each of the 12 member states. Most countries also had differing standards, which meant substantial costs for equipment modification in each national market.

Table 8.1 Telecommunication trade with non-EC, 1988 (ECUm)

	Total	USA	Japan	EFTA	SE Asia	Rest of World
EC exports	3782	412	51	1038	186	2095
EC imports	3686	819	1220	950	342	355
Net exports	+96	-407	-1169	+88	-156	+1740

Source: CEC, 1990b.

Nevertheless, overall the member states had traditionally recorded a trade surplus in telecommunication equipment, and it had been regarded as an area of comparative strength. But during the 1980s this balance dwindled with rapidly rising imports from Japan and the United States (Table 8.1).

In 1983 the CEC established a task force to develop policies for telecommunications, and under their auspices the development of a Europe-wide Integrated Broadband Network (IBN) has been encouraged. Simultaneously there have been pressures to open telecommunications markets to competition, and in 1987 the directorate general responsible for Telecommunications, Information Technology and Innovation, (DG-XIII) published proposals for a radical liberalisation of the sector. In effect, the proposals meant the end of the structures, organisation and procedures which until then had sustained Europe's 'natural' monopolies.

The changes that are now being implemented as part of the 1992 agenda in the EC market attack the heart of the monopoly utility concept in so far as they acknowledge that most of the innovative developments in advanced services are in no sense 'public' goods and can be provided competitively in the market place. Hence, the PTOs need not provide them under public service obligations to extend services as widely as possible, a change which in itself requires a substantial reform of the 'culture' of the PTOs. The PTOs of the 1990s must establish close links with their corporate customers to provide customised services and at the same time devise new ways of expanding service offerings to their smaller business and residential subscribers.

The challenge to the CEC is to gain consensus on the best way forward. This is a formidable task. Each of the national telecommunication networks in the EC has achieved different levels of network modernisation, and governments and the PTOs have varying views on the proper balance between competitive and monopoly supply. The PTOs have devised new organisational strategies to strengthen their responsiveness to corporate demand, but varying national political priorities have meant different weights are being given to efficiency gains, competition and public-service mandates.

DG-XIII has been given the responsibility for opening telecommunication markets to competition, and it has sought to act through political consensus among the PTOs, third-party service suppliers, equipment manufacturers and the major corporate users. The push to complete the internal market, however, has come from DGIV (the directorate general responsible for competition and the internal market). Their wish to act quickly to liberalise markets has led the Commission to exercise its powers under the Treaty of Rome. The 1987 'Green Paper on Telecommunications' (CEC, 1987c)

Table 8.2 Subscriber lines, digitalisation and ISDN by EC member countries, 1987 and 1990 (estimated)

Country	Subscriber lines per 100 inhabitants	Digitalisation* rate (%)		ISDN commercial launch date
		1987	1990 (estimated)	
Belgium	34.5	1.6	31.5	1990
Denmark	52.0	2.0	25.0	1991
Germany (former West)	45.6	2.0	9.0	1988
Greece	35.0	0	5.0	1993
Spain	26.2	7.0	22.0	1991
France	44.6	55.0	70.0	1988
Ireland	22.3	42.0	65.0	1991
Italy	33.3	13.0	30.7	1992
Luxemburg	45.9	12.0	23.0	1991
Netherlands	42.7	7.4	35.0	1989
Portugal	15.8	0	20.0	1993
United Kingdom	40.8	9.6	42.0	1989
EC	38.1	17.2	35.0	—

* Percentage of user lines linked to digital local switches at end of year

Source: CEC, 1990b

declared that the CEC would if necessary use its powers under Articles 85 and 86 (which deal with anti-competitive collusive agreements between firms and with abuse of dominant position by monopolies) and noted its powers under Article 90 of the Treaty which enable the passage of legislation 'necessary' to secure compliance without reference to the European Council of Ministers. Since 1984 there have been a series of Council recommendations aimed at the transformation of the telecommunications sector. Recommendations, resolutions and directives have addressed policy with regard to public procurement, R & D, equipment approval, financial assistance for telecommunication development, standards, ISDN narrowband and broadband introduction, mobile and paging services and the interconnection of public and private networks. It was not until 1989 that the Commission used its powers under Article 90 to issue a directive on competition in the terminal equipment market.

Most of the problems encountered in this sector are associated with the high rate of technical change in switching, transmission, software and services. At the heart of these changes has been the digitalisation of the networks which we discuss in the following section.

Table 8.3 European telecommunication market 1987 and 1993 (estimated) (1987 US$ rate)

Equipment Category	Market value (US$m) 1987	1993 (estimate)
Switching:		
Public (trunk/local)	6,000	8,080
Private (PABX)	1,400	2,000
Transmission:		
Fibre-optic Systems	2,489	5,511
Digital microwave	2,017	3,841
Sat. earth stations	971	2,003
Optical systems	23	45
Mobile communications:		
Cellular radio/paging/telepoint	1,800	6,570
Customer premises equipment:	3,300	6,180
Sub-assemblies:	600	1,600
Total:	18,600	35,770

Note: Includes EC 12 plus Austria, Finland, Norway, Sweden, Switzerland

Source: Frost and Sullivan in *Communications International*, January 1990

III The digital era in telecommunication and the supply of equipment

Structural changes and the introduction of competition in telecommunications began to emerge in the United States in the late 1950s, long before the widespread diffusion of digital technologies. The progression of switching technology from electromechanical to analogue electronic and then to digital has led to the introduction of complex software-controlled switches which receive, direct and monitor traffic, perform new service functions (for example, call forwarding, holding etc) and process digitally encoded information. Substantial gains in the efficiency of network operation are expected as networks become fully digital and services are tailored to meet a wider spectrum of user requirements.

Advanced digital switching is being combined with improved transmission. Digital microwave and fibre-optic links are replacing the old analogue public switched telephone network (PSTN). The cost per minute for the use of transmission cables has declined very rapidly. For example, the cost per minute of the transatlantic TAT-1 fell from $2.53 in 1956 to $O.O4 in 1988 (OECD, 1989a), and data transmission speeds have improved substantially over public networks. Whereas the old PSTN allows for the transfer of a 1,200 kilobyte file in 44 minutes and 27 seconds, the same amount of data can be transferred using the Integrated Services Digital Network (ISDN) in 2 minutes, 30 seconds (Stewart, 1989.)

The rate of diffusion of digital technologies and the introduction of advanced service applications varies regionally within countries and across the member states of the European Community. Table 8.2 shows the estimated rate of digitalisation in EC member countries for 1987 and 1990. France is in the lead. By 1990 70 per cent of its local switching exchanges and over 70 per cent of its long-distance exchanges were capable of handling digital signals.

Progress toward the full commercial implementation of ISDN, which considerably improves the data handling capabilities of public networks, has also been uneven (Table 8.2). The ISDN enables integration of voice and data applications, improves the quality of voice services and supports improved services such as high-speed facsimile, videotex and electronic mail. Again France was in the lead in early introduction and penetration of ISDN, but by the end of 1989, ISDN commercial service had been introduced to varying degrees in the former West Germany, the United States, Japan, the Netherlands and the United Kingdom.

The 'intelligent' capabilities of the present generation of switching systems are complemented by equivalent functions located in terminal equipment. A Private Automated Branch Exchange (PABX) can be used to support corporate networks and to augment the services supplied by PTOs. The multiple ways of configuring public and private networks have created tension between suppliers of public-switching equipment and those of 'intelligent' terminals. Both seek to respond to corporate networking requirements and to expand their share of equipment markets.

The cost of terminals should begin to decline as competitors enter the market-place, but restrictions on equipment certification are still in use which curtail competition. As Table 8.3 shows, the market for public switching and transmission equipment is still very substantial. It has proved more difficult to open up than terminal equipment because national PTO procurement policies have tended to favour 'national champions'. Pressures for opening markets are coming from the equipment manufacturers themselves since they recognise that a secure domestic market is no guarantee for survival, and they are anxious to secure their position in the wider European context. No single European country accounts for more than 6 per cent of the world market, and the pressure is substantial to break the links that have existed between national PTOs and their suppliers.

The United Kingdom's experience attests to the fact that equipment market liberalisation can bring about extremely rapid changes in the fortunes of domestic suppliers. Since liberalisation in 1981, the United Kingdom has moved from a position of modest surplus in telecommunication equipment to become the main trade deficit country in the Community (amounting to ECU564 m in 1988). Between 1976 and 1986 employment in the telecommunication equipment industry in the United Kingdom fell from 75,291 to 38,837 – a drop of 48 per cent, compared to a fall of 16 per cent in France and a gain of 11.4 per cent in West Germany.

Table 8.4 gives details of the sales of the major world suppliers of telecommuncation equipment. The upfront costs of research, development and design of the new digital switching systems are increasingly leading to concentration. The national champions survived into the early 1980s, mainly thanks to the continued protection of their home markets through public procurement. In the United Kingdom, three companies, GEC, Plessey and STC, provided all (the then nationalised) British Telecom's (BT) equipment

Table 8.4 World telecommunication equipment revenue of top-15 companies, 1987 (ECUm)

Company	Country	Telecom turnover	Total turnover
1. AT&T	USA	7,900	29,114
2. Alcatel(CGE)	EC	6,800	11,200
3. Siemens	EC	5,200	24,800
4. Northern Telecom	Canada	3,980	4,206
5. NEC	Japan	3,700	16,340
6. Motorola	USA	3,000	5,812
7. Ericsson	Sweden	2,600	4,430
8. IBM	USA	1,850	46,982
9. Fujitsu	Japan	1,810	12,324
10. GPT	EC	1,740	1,740
11. Bosch	EC	1,650	12,245
12. Philips	EC	1,300	22,540
13. Matsushita	Japan	1,100	20,030
14. Italtel(STET)	EC	1,000	1,000
15. GTE	USA	880	13,363

Source: CEC, 1990b

needs; in France the market was shared between Thomson and Alcatel; in West Germany Siemens and SEL, a subsidiary of ITT, met supply needs; in Italy there were STET and Italtel, even at that time exploring links with AT&T. In addition, Ericsson from Sweden and the American multinational, ITT, had established market positions, and Philips, which originally had interests only on the transmission side of network provision, linked with AT&T in 1983 with the aim of strengthening its product offering in telecommunications. Hence there were 11 equipment suppliers in a market which was widely expected to be able to support only three or four. At R & D and investment levels required to support the first generation of digital equipment (approximately $600m at then price levels), it was reckoned that annual sales of approximately $11bn were required (Dang Nguyen, 1985).

The shakeout came in the later years of the 1980s, sparked by the demise of ITT, and Alcatel's purchase in 1987 of its European telecommunication interests. Within France, Alcatel had already (1983) taken over Thomson's telecommunication interests. This 1987 move made Alcatel Europe's biggest telecommunication equipment supplier, in world terms second only to AT&T (Table 8.4). In the same year (1987) GEC and Plessey amalgamated their telecommunication interests to form GPT. In 1989 this was subsequently amalgamated with Siemens telecommunication interests in the Siemens/GEC takeover of Plessey. Ericsson, the Swedish company, remains the third major player in the European market. Philips withdrew from their joint venture into switching systems with AT&T, but retain a major interest in transmission equipment. Italtel and STET in Italy have firmed up links with AT&T. Hence where there were 11, there are now only six – Alcatel, Siemens (and GPT), Ericsson, Bosch and Italtel (STET) with the first two accounting for about half the total European market.

A major objective of the Commission's policy in creating an open internal market for telecommunications by 1992 has been to retain competence in the supply of telecommunication equipment. Their policy has been to encourage concentration in order to capture scale economies in the EC internal market. DG-XIII favours the creation of a 'space' for European equipment producers to rationalise production before allowing large-scale entry by non-EC competitors. DGIV, by contrast, is worried about concentration in this market and is anxious to open markets to competition both within the Community and with other countries. Meanwhile, EC producers are entering alliances, frequently with non-European firms, in an attempt to gain a stronger footing in the restructured market, as in the example of Italtel. A CEC directive on public procurement treads carefully in advocating an increase in intra-Community trade in public switching equipment from present relatively low levels. The new provisions favour suppliers of EC origin and insist on conformance to EC standards.

IV Trajectories in network evolution

The tension between the major players in equipment markets is mirrored in the development of service markets. The PTOs believe that their ability to meet public-service objectives could be threatened by the erosion of their revenue base by new competitive suppliers. The latter include private corporate networks and third-party suppliers who lease capacity from the PTOs.

As we shall see in the next chapter, the telecommunication infrastructure supports a variety of services which combine information production, manipulation, storage and/or distribution with the use of telecommunication facilities and software functions (OECD, 1989b). Advanced applications such as financial services, electronic data interchange and electronic mail are developing rapidly with annual growth rates as high as 25 per cent. These services differ from traditional voice and data services in their reliance on software which provides the support for functions and applications. The estimated relative growth rates of the service market are shown in Table 8.5 for 1988–93.

The CEC must create conditions whereby equipment manufacturers, PTOs and competitive service suppliers can provide flexible, high-quality services at competitive prices. But proposals for the development of a more competitive and efficient market must deliver these services to smaller firms and consumers as well as to the larger users. They must also create incentives for the extension of services to the less-favoured regions of the Community.

Thus the 1990s will see the further development of three parallel trajectories in the evolution of the telecommunication infrastructure. The first is the upgrading of public networks to support ISDN and 'intelligent' service applications. The second is the growth of in-house corporate networks and third-party services. The third is the hybrid network which includes public and private services in a complex interconnected web of standardised and proprietary applications.

The public switched telephone network, engineered originally to support voice traffic, has been augmented by public data overlay networks during the 1970s and 1980s. In the late 1980s, it became clear that the integration of

Table 8.5 European telecommunication service markets: estimated growth rates, 1988–93

Type of service	Annual growth rate (%)
Local telephone	7.9
Basic ISDN	135.1
Cellular	19.3
Domestic telephone	
Long distance	8.5
Primary ISDN	70.4
800 service/green no.	27.4
Leased lines	6.3
Virtual private nets	58.1
International Telephone	4.8
Data	6.3
Facsimile	21.9
Telex	7.6
Information services	22.6
Value-added services	28.2
Total services	7.3

Note: Includes EC 12 plus Austria, Finland, Norway, Sweden, Switzerland

Source: Northern Business Information, 1990

voice, data and image transmission through ISDN was the favoured approach of virtually all the European PTOs in spite of the slow take-up of applications. However, progress toward an integrated infrastructure has required the early agreement of standards. The narrowband ISDN, the dominant design of the 1970s, reflects the design requirements of voice telephony. Limitations on its transmission speeds render this type of ISDN inappropriate for some high-speed applications such as CAD/CAM. In addition, bandwidth requirements for applications such as high-definition television are difficult to meet using narrowband ISDN configurations.

The CEC is hence investing in R & D programmes to accelerate the introduction of integrated broadband networks (IBN). The RACE (Research in Advanced Communications in Europe) programme promoted by the CEC under its Framework Programme (see Chapter 5) is experimenting with optical fibre and switching techniques that could bring interactive services and television signals into the home and the workplace. But end-to-end optical links will be needed if IBNs are to achieve the universal reach of the existing public switched telephone network, and this would entail a new domestic distribution system. Many services such as home information, meter reading and security monitoring can be carried over the existing public switched telephone network with narrowband ISDN, and it is not clear that a 'universal' broadband infrastructure is really necessary. Problems arise in justifying so major an investment when the time horizon for achieving reasonable returns could extend far into the future. Its advocates maintain, however, that in the longer term, plans for IBN-based services represent an inevitable and necessary step in the evolution of public network infrastructure.

The development of public networks needs to be considered in the light of the parallel development of private corporate networks and services provided by third party suppliers. Since the 1970s, this second trajectory in the evolution of the telecommunication infrastructure has seen the larger corporate users leasing capacity from PTOs and configuring their own applications. The growth of these networks in Europe has been slowed by differing national regulations which affect private and public network use for data and voice service applications. In most countries private network operators have been required to limit their services to 'closed user groups', for example, banking, automobile sectors etc.

Public switched networks are being upgraded to introduce services for both the corporate and the residential community. A major incentive for investment in public networks stems from the PTOs' attempt to persuade the larger corporate users to mount their service applications and route their traffic through the public switching systems rather than through private networks. This strategy is also reflected in the development of 'intelligent network' services that aim to simplify network management and operation by combining digital signalling with computerised data bases. Virtual private networks, Freephone and alternate billing services could be supported by these network configurations. The PTOs' plans for the introduction of 'intelligent' networks aim to offer services that large corporate users require without the disadvantage of having to maintain or operate private corporate networks.

Since the early 1980s, the United Kingdom has retained its status as the most liberalised market for telecommunications within the European Community. Liberalisation, privatisation and the removal of restrictions on the use of networks have accelerated the growth of private networks. In 1986, there were approximately 11,060 private data networks in Western Europe with the British market representing some 28 per cent of all data networks and 45 per cent of installed equipment (See Chapter 9).

The third main trajectory in telecommunication infrastructure development is the 'hybrid' or mixed solution where larger users opt for a combination of public and private networks to meet their needs. As regulatory restrictions on network use in the EC are relaxed, this alternative is likely to become the predominant pattern.

The hybrid alternative has implications for the investment strategies adopted by PTOs. In their efforts to capture the most advanced customers, they have moved to develop costly business applications, based on their public switching equipment, sometimes at the expense of allocating resources to service development for domestic consumers and smaller businesses who represent a less profitable and sometimes more risky market.

The development of hybrid networks also raises the difficult issue of standardisation. In order for the components of the varying public and private networks to work together a minimum level of interface compatibility has to be achieved. Although 'intelligent' switch and terminal manufacturers increasingly support 'open' international and regional standards, they are far from developing fully standardised applications. Many continue to opt for proprietary implementations that will make communication across the boundaries of the EC more difficult and costly. In addition, the regional and international standards-setting process itself has often failed to produce standards in the time required. By 1989, for example, full standards for ISDN equipment had not been delivered, and the process of reaching agreement on 'intelligent network' standards had barely begun (see Chapter 11 for more detailed discussion of the problems in standardisation).

The difficulty in allocating investment priorities between large corporate user services and mass consumer services is aggravated by difficulties in establishing demand profiles. If business users find it difficult to forecast future demand for their new services, the mass consumer market exhibits only the most vaguely defined preferences. Demand forecasting is even more difficult in countries where voice telephone service has failed to reach substantial penetration. As PTOs face competition in basic voice telephony markets, they become understandably resistant to the high-risk service development strategies that are needed to encourage consumers and smaller businesses to experiment with new advanced services.

V Regulating the new hybrid networks

The issue of competition in service supply has proved difficult to resolve. There have been major disagreements as to how to draw a line between the voice telephone service which the CEC believes should be 'reserved' as part of the public service monopoly, and more advanced services which involve processing and conversion of voice or data signals. France, Italy, Belgium and Greece had sought to include data transmission using the public packet switched data networks within the 'reserved' category. The United Kingdom and Germany have pushed for the inclusion of data transmisson services in the competitive domain. These data transmission services support transaction and information service applications which are targets for entry by both foreign and domestic third party-service suppliers.

A services directive, under the powers of Article 90, and a directive on Open Network Provision (ONP) under other provisions, have been used by the CEC in their attempt to harmonise the Community market. The latter is intended to ensure the right of interconnectability between public and private networks and that PTOs do not place anti-competitive restrictions on the use of the network infrastructure which they continue to control. The agreement reached in December 1989 was that the services directive would reserve voice telephony and telex services for monopoly supply; but the deadline for compliance was extended to 1993 (with a further extension to 1996 for countries giving priority to the modernisation of the infrastructure, for example, Spain, Portugal and Greece). To date, competitive entry by service suppliers has been limited by PTO prohibitions on resale of leased circuit capacity and the interconnection of private corporate networks with the public network. In addition, many of the EC countries have set the prices for the lease of private circuits at a level which encourages use of the public switched networks, even where switched capacity is not required. The resale of such capacity to third-party service suppliers has been perceived as a threat to the economic viability of the PTOs because of the unbalanced structure of PTO prices for the switched telephone network. In general, local usage prices have been kept down and revenues from long distance have contributed to the costs of providing other PTO services in line with what are perceived to be public service obligations. 'Cream skimming' of the more lucrative markets has been expected to undermine the PTOs' ability to subsidise loss-making or low-profit elements of their service.

As with terminal equipment, the United Kingdom has taken the lead in liberalising circuit resale. The first steps in this direction were codified in a General Licence for VANS in October 1982. The subsequent relaxation of restrictions on the interconnection of private and public networks resulted in the United Kingdom becoming the European leader in the establishment of private wide-area networks in the United Kingdom.

In several countries restrictions on resale and the availability of leased circuits still remain in place. Until recently West Germany was regarded as the laggard in the 'liberalisation' trend. However, new legislation passed in 1989 is bringing competition in advanced service provision, eliminating restrictions on the use of leased circuits and, according to some observers, will make this country the most liberalised market in Europe. The new policy continues to protect the reorganised DBP Telekom's monopoly over transmission and voice telephone service.

The Commission's challenge to the monopolistic PTO structure will have little effect unless it can bring about substantial change in pricing practices. The CEC has called for greater 'cost-orientation' of tariffs, but the differences between member states in prices for public switched and leased services remain enormous. Nearly all PTOs and national governments acknowledge the need to restructure their tariffs. The problem is one of degree and timing. The administered tariffs of the past reflected historical 'rules of thumb' and broad economic and political objectives. Prices in the future will have to respond more to the market, and charges for access and use of switched and leased capacity by the PTOs will crucially affect investment in network infrastructure well into the next century.

The CEC has called for a '... fair trade-off between cost-orientation and the aim of universal service on reasonably the same terms for all' (CEC, 1988e). Operating as utilities, PTOs have been free to average their prices both geographically and between different types of services. Cross-subsidies have been used (if rarely empirically documented) to meet social and economic objectives. In the face of new entry and technical change, PTOs would in any case have been forced to reconsider their price structures – the CEC directives are, in this respect, merely reinforcing a trend. It is unlikely that prevailing price structures in domestic markets can be maintained. For example, prices for leased circuits in Germany, Spain and Italy are often more than four times their estimated costs. Only in the United Kingdom and Sweden are prices roughly in line with costs. This situation does not bode well for a truly integrated EC internal market.

The Commission has sought to influence change by issuing directives which require compliance by the member states and by gaining consensus for recommendations which variously urge and cajole the European tele- communication industry into greater responsiveness to the imperative of creating an advanced and efficient market for services and equipment production. But the member states have their own conception of the speed and shape of change, and the process of restructuring has itself introduced problems such as new incompatibilities in standards, greater price distortions and the continuing segmentation of service and equipment markets. Nor have all member states been altogether happy with the high-profile position adopted by the CEC.

VI Business and private use of the networks

The influence of larger business users on the evolution of the telecommunication infrastructure is considerably greater than their numerical significance suggests. In the United States, some 3 per cent of business users account for around 50 per cent of telephone revenues, and a similar pattern occurs in Europe. These users represent the cutting-edge of new applications. Their needs extend beyond the present capabilities of public networks and are likely to continue to do so even in the face of rapid investment.

The direct costs of private corporate networks arise from the development and implementation of 'intelligent' software-driven applications and from the costs of the underlying transmission facilities. As we have seen in Section V, in the EC with the exception of the United Kingdom, leased circuits, the basic building blocks of private networks, have traditionally been priced higher than the costs of switched network capacity. The option of purchasing bulk leased capacity in the United Kingdom stimulated the development of private networks and the United Kingdom became the home of a powerful lobby favouring increased flexibility and competition in network and service use. In recent years, the British regulatory authority has required British Telecom to increase the price of some leased circuits which may make this option less attractive for large users.

However, price advantages for the large corporate user as between public and private networks can be outweighed by a variety of non-price factors. Among these are the need to maximise control, reliability, security and international connectability, and to ensure that network management and training fit within the firm's overall strategy. The corporate user's evaluation of the costs and benefits associated with public or private networks has to be considered alongside the longer-term improvements taking place in public networks, and likely changes in the firm's pattern of internal and external communication needs.

There are also indirect costs associated with the private-network option. These relate to costs of exiting from established systems and pressures stemming from technical complexity, including the growing demands on managerial time. Corporate users in the past have opted for private networks in order to gain maximum control over their telecommunication activities, but the advent of 'intelligent' public network services, (that is, virtual private networks with managed data services) makes it possible for users to achieve a significant degree of control without the direct costs and management problems of running their own networks.

Telecommunication networks support the flow and use of information within and between firms. They are now crucial to the organisation of the global marketplace. Larger firms are using their networks to 'lock-in' suppliers and customers and thereby effectively to 'lock-out' their competitors. Examples of such network applications now abound particularly in travel, the airlines and the financial services sectors but also in such industries as automobiles and clothing where they have revolutionised stock control methods and introduced a new flexibility to production.

Case studies suggest that these new service applications have not weakened the dominant position of larger firms or substantially changed economic power relations (Morgan and Davies, 1990). Many applications simply speed up information flows at various stages of the production and marketing chain. Typically, smaller firms dependent on orders from major firms remain so. Indeed, the need to respond more rapidly to requests for changes in design or batch size can actually add to costs.

A large number of these new applications are compatible with both public and private networks. Only a very small number will require bandwidth and data-transmission speeds that will outstrip the capacity of public networks. Some are likely to be better suited to switched configurations, others to dedicated private circuit links, depending on geographical location, traffic volume and security requirements. For many, narrowband ISDN may be perfectly acceptable for the foreseeable future; for others, it may offer too much or too little bandwidth. In general the large users will need a variety of intricately linked public and private, switched and unswitched networks. The appropriate mix will depend not just on price but on how the range of other factors is assessed.

The range of choice for smaller firms and consumers is much more restricted. Even in the United Kingdom where competition in the provision of local exchange services is being encouraged, few subscribers perceive that they have alternatives (DTI, 1990a). Mercury, British Telecom's major competitor, has yet to target the smaller business market effectively. Cellular radio services are currently not price competitive with the PSTN, and even the newly announced Personal Communication Networks are expected to levy charges for usage at least 20 per cent higher than a normal voice telephone call. Thus, in spite of widespread market liberalisation and competition, for the vast majority of telephone subscribers in the EC, there is unlikely to be realistic competition for traditional services for some time.

This is not true of the more advanced service markets. The Minitel videotex system developed by France Telecom (see Chapter 9) provides a clear demonstration that services with widespread consumer appeal can be developed, but this requires substantial up-front investment and pricing strategies that encourage use. The Minitel service is a rare example of successful diffusion where advanced services are targeted at *both* the ordinary consumer *and* the business user. Nevertheless, the pattern of usage is gradually shifting toward the professional applications rather than the building-up of public information services.

VII Conclusions

Competition in telecommunications is gradually achieving a more pervasive presence in most of the EC market, but the strategies being adopted in each of the member states differ considerably. The larger users, represented by organisations such as the International Telecommunication User Group (INTUG), have mounted an effective campaign to promote greater competition. They argue that the main barrier to an innovative telecommunication infrastructure stems from the remnants of the monopolistic telecommunication environment that thwarted their ability to capture earlier productivity gains from digital technologies and software-based service applications.

Their lobbying has had some major successes. The CEC stands at a crossroad in the transition from monopoly to competitive supply. The steps it has taken to implement its goal of achieving a common market in telecommunications have sought, in general, to weigh the development of sophisticated applications for the corporate users against the need to establish a universal infrastructure which will enable PTOs to meet their public-service obligations.

At the national level, the adjustment process is characterised by a variety of initiatives. At one extreme is the British Government's attempt to leave the development of the telecommunication infrastructure to market forces. Beginning with the decision to license Mercury to build a new nation-wide terrestrial network, the net has been cast wider in an effort to induce competition in local exchange services. However, for the market to work effectively, the British regulatory authority (Oftel) and the Department of Trade and Industry have been obliged to intervene in many areas, including standardisation, market entry, pricing and rights to network interconnection. While the experiment has encouraged a flowering of supply-side services for the business user, it has had only limited success for the ordinary domestic consumer and continues to pose difficult regulatory problems.

Policies in the other EC countries have been more restrained. In some countries, competition in the provision of advanced services is being encouraged, but through the public switched network. Other countries support competition in nearly all 'value-added' services (electronic mail, electronic data interchange etc) regardless of whether public or private networks are used.

The best example of a successful infrastructure strategy which provides a hybrid solution is France Telecom. It intends to continue to promote the use of its public data network, Transpac, which provides the infrastructure for its successful videotex network but seeks also to encourage private corporate networking and is upgrading its national leased line service, Transmic. Such a mixed solution may be the key to longer term success in promoting the dual social and economic objectives of telecommunication policy.

The promotion of a hybrid strategy such as that of France Telecom can encourage firms at the leading edge of network development to engage in experimentation and learning without imposing major costs on the public network. This experience can then be incorporated in more advanced public network applications, and the risk of a mismatch between public services and corporate requirements can be reduced. At the same time, resources can be directed to the enhancement of the public network which must support a broader spectrum of users, many of whom do not need access to advanced applications and are unlikely to wish to pay the related increased charges.

As alliances among actors in the EC ebb and flow, the Commission is divided as to how fast to move towards open markets. While the competition directorate (DG-IV) is concerned to ensure that greater efficiencies are achieved in the supply and use of telecommunications, the directorate responsible for telecommunications, information industries and innovation (DG-XIII) is also concerned that a wholesale opening of markets might jeopardise the ability of European equipment firms to compete with foreign entrants. This also could endanger the maintenance and extension of public services. The problem of organising telecommunication supply in a mixed monopoly and competitive environment is one which will have to be addressed if the results of the transformation of the telecommunication infrastructure are not to be skewed to the benefit of the larger corporate users.

Given the structure of the market-place, there is a need for policies that stimulate investment for the benefit not just of big business but also look to the interests of the domestic consumer and smaller business. From experience to date, it seems likely that the new market environment will not bring widespread benefits unless coordinated national and European institutions play a role in planning for the balanced development of the new infrastructure. Yet it is clear that there can be no return to the traditional protected national telecommunication markets of the past. Competitive entry across the spectrum of telecommunication applications is a reflection of fundamental changes in technology and a growing perception that competition is viable in major segments of the telecommunication infrastructure and service market. Yet elements of monopoly survive as long as the provision of services to domestic customers and segments of the infrastructure are protected from competition. As long as these elements of monopoly survive, vital issues such as interconnectability and the price to be paid for it, as well as services prices, require continued regulation and control. The issue for policy is whether the CEC can develop effective strategies to hold the balance between competing demands and simultaneously help to promote its industrial and social goals.

9 New Telematic Services in Europe*

Nick Jagger and Ian Miles

I Introduction

New Telematic Services (NTS) are made possible by the interfacing of telecommunications and computing: they exchange information between computer systems through telecommunications networks. They are widely seen as being among the key activities which can unleash the power of new Information Technology (IT). The United Kingdom and France represent alternative strategies for gaining effective use and leadership in these services. This chapter examines the current position of the United Kingdom and its European partners regarding New Telematic Services and the challenges and policy options that lie ahead.

II New Telematic Services

Computers' abilities to acquire, process and present information used to be limited to data from their local environment. This limitation is rapidly eroding as computers can communicate with other computers via telecommunications networks. This greatly increases the ability of the technology to integrate and present information from spatially distant locations.

This chapter examines NTS offered to third parties and does not consider the enormous intra-organisational non-traded exchange of information through 'Wide Area Networks' (WANs) or 'Local Area Networks' (LANs). (However, many NTS have grown out of such in-house systems and corporate learning from their use is important for NTS development.) Roughly speaking, NTS are what are typically described as VANS or VADS (Value-Added Network/Data Services). We focus here on computer-to-computer communications, excluding the emerging 'hybrid' technologies such as audiotext, voice messaging, video-conferencing and videophones. We focus also on telecommunications as the delivery medium and thus do not examine either data broadcasting (including satellite business services) or the rapidly growing optical publishing industry (CD-ROM databases etc.).

NTS can be classified by primary function. Elaborating on the categorisation of Thomas and Miles (1989), the four main types of NTS are information services, communications services, transaction services and Managed Data

*The research underlying this chapter was partly funded by the ESRC as part of their Programme on Information and Communication Technology, under the Measuring and Mapping the Information Economy Project.

Network Services (MDNS). This classification system is set out below. Many services – especially MDNS suppliers – in practice offer a combination of these functions. However, each function has different market and diffusion characteristics and generates differing policy debates and strategies which is why it is convenient to deal with them separately.

(i) Information services, such as accessing databases, Until recently, such online services were the main form of electronic publishing and NTS. Examples include Dialog, which began as a service covering material from scientific and technical publishers; Lexis, a service providing access to legal material operated by Mead Data Central; and the Textline service, provided by Reuters, which provides material on companies and industries from news sources.
(ii) Communication services, involving text messaging of various kinds such as electronic mail (e-mail), bulletin boards, computer-conferences and analogous services. The main British example is the e-mail service run by BT as Telecom Gold; computer-conferencing facilities are less commonly used, but the CIX system is one example in the United Kingdom aimed at computer specialists.
(iii) Transactional services, such as Electronic Data Interchange (EDI), which is employed as a means of companies' sending invoices and orders, and consumer teleshopping, which also permits ordering and paying for deliveries online. The latter services are available on public videotex systems, and in the Keyline initiative to be launched soon in the United Kingdom. Some sectors have been fertile ground for the development of transactional services (that necessarily involve some information service component): examples include travel agencies, insurance brokers and the automobile trade.
(iv) MDNS provide basic digital communications and network management functions. Usually, such a simple transport function underlies other NTS such as information services. However, the transport services are often sold separately and can be used by customers to create their own in-house or third-party services. Examples include Tymnet in the United States, Istel's Infotrac in the United Kingdom etc.

As can be seen from the examples, NTS contribute to the responsiveness and flexibility of firms working in the 'Information Economy'. Their strategic importance to users is bound up with the complex structure of national computer and telecommunications industries, and the whole area has been subject to numerous policy statements and initiatives. The importance attached to telecommunications by the Commission was reflected in the 1987 Green Paper which argued that:

> A technically advanced Europe-wide and low cost telecommunications network will provide an essential infrastructure for improving the competitivity of the European economy, achieving the Internal Market and strengthening Community cohesion, which constitute priority Community goals reaffirmed in the Single European Act [Green Paper, CEC, 1987c].

Despite this perceived importance, official statistics are very poor at documenting NTS. Even the massive CEC statistical publication on the *Panorama of EC Industry* (1990), which is innovative with respect to many difficult statistical areas, does not attempt to provide tables on this topic. In part this reflects the statistics' novelty, in part commercial factors and in part the dispersion of NTS actors across a wide spectrum of industries – telecommunications, computer hardware and software suppliers, scientific publishers, financial services and a wide range of other service providers. NTS activities may be picked up as parts of these widely divergent activities rather than as a distinct activity.

III Dynamics of NTS diffusion

New Telematic Services are not only made possible by the interfacing of computing and telecommunications, but they also reflect the often conflicting traditions of these industries. Although the two component industries – telecommunications and computing – are themselves relatively young, the traditions and institutional structures which determine the dynamics of innovation and diffusion of each of them are different.

Telecommunications has a tradition of relatively stable and upwardly compatible standards combined with the concept of universal service and interconnectivity. This stems from a combination of the basic economics underlying telecommunications and the tradition of state control and regulation. Computing on the other hand has a history of using conflicting standards as a competitive weapon, which has created a situation where there are real problems with interconnectivity. Early data communications tended to share such characteristics with the computing industry, although (as in computing) *de facto* standards and international standards efforts are increasingly important (see Chapter 11).

These different traditions have tended to generate different patterns of use by companies for voice and data communications. The SPRU Data Communications Survey found voice communications were used for both internal and external functions, while data communications were restricted primarily to internal functions (Aksoy *et al.*, 1990).

As we shall see later the United Kingdom and France have similar-sized NTS industries, but the United Kingdom's industry has developed out of the computing tradition, while the French developed from the telecommunications tradition. This means that the two industries have different dynamics and approaches, with the United Kingdom being characterised by application-specific solutions and niche markets, while France has a standardised and near universal service.

This difference is perhaps best illustrated by the comparative fate of videotex developments in France and the United Kingdom. In France, in the universal service tradition of telecommunications, Minitel terminals have been distributed at a subsidised cost by France Telecom to over five million households, with Minitel videotex terminals capable of accessing their Teletel service over the Transpac network (Allen, 1988; Thomas, 1990). The terminals were primarily distributed as an electronic alternative to printed telephone directories, the savings in production, printing and distributing the printed directories supposedly covering the costs of the terminals. The actual economics of the operation is in dispute (Roussel, 1989), but the project has succeeded in stimulating an enormous market for consumer-orientated NTS in France and in so doing achieved one of its secondary purposes.

In the United Kingdom the main videotex developments have been outside the pioneering public (now BT's) system (Prestel), and each project has tended to have its own, application-specific, standards and independent networks (Jagger, 1989). For instance, the London Stock Exchange's TOPIC (Teletext Output of Price Information by Computer) was launched as early as 1980 to provide price information. TOPIC uses the videotex display standard, but, owing to requirements for updating, does not use BT's Prestel standard (Currie, 1983), the value of the information provided outweighing the extra costs of separate standards.

User acceptance, especially among a relatively small and definable user group, can be increased by providing training. The package-holiday company Thompson, on introducing its videotex-based tour availability and booking system, trained 10,000 travel agents in a period of three months. This commitment to training was one of the critical factors in the success of the system (Feeny and Brownlee, 1987). Other successes have been in sectors where existing experience in computing has circumvented the barriers to use caused by user inexperience. These successes include financial information systems in the City and scientific and technical bibliographic services for the R & D community.

IV National positions: overall perspectives on NTS

The pattern of NTS developments has been determined as much by the absence of direct policy initiatives (or their mismatching) as by their presence. The main impact of policy on the NTS market has been through measures to liberalise and re-regulate the whole telecoms market (see Chapter 8). In this section we survey briefly the position of the various EC member states.

The United Kingdom has seen very rapid growth in NTS. The Business Monitor series (SDQ9) provides statistics on various subsectors of computer services – including the 'Value Added Network Services' (VANS) and Database Services (DTI, 1990). There are problems of definition and composition with this data set (they have been changed four times in the last ten years) which makes reliable time series impossible to construct. Since 1988, reporting has been mandatory for organisations with a turnover of over £4.5m but is on a voluntary basis for other firms. This sample data set (estimated at about 40 per cent of the total turnover), although not being a reliable indicator of market size, does provide a useful indicator of growth rates. Table 9.1 is based on the combined values of data for VADS and Database services, which fall within our definition of NTS. (They are sometimes combined for confidentiality purposes.) It covers a recent period during which there have not been definitional changes.

Assuming the data to reflect 40 per cent of the market, they suggest that the British NTS industry generated a total turnover of only about £800m in 1990, although the industry is now over five times the size of three years ago. This rate of growth, together with the strategic nature of the industry, explains the interest in NTS. The recent plateau in activity, especially in exports, is probably explained by the role of Reuters, which industry sources suggest generates as much as 50 per cent of British data-base revenues. Its revenue is largely dependent upon the level of trading in the financial markets, which have dropped off sharply in the late 1980s.

Given the absence of compatible European data, it is necessary to rely on consultancy estimates and tangential official statistics to get any overview of

Table 9.1 British value-added network services and data-base services Billings (£1,000)

Period (quarter)		To foreign clients	Parents/ associates	To other UK clients	Total billings	(%) Data-bases
90	3rd	13,801	*	*	83,916	57.4
90	2nd	9.585	*	*	81,577	56.9
90	1st	10,441	*	*	81,665	56.2
89	4th	7,392	*	*	75,753	57.9
89	3rd	8,308	*	*	72,517	59.3
89	2nd	8,298	*	*	72,182	59.9
89	1st	10,568	21,427	40,695	72,973	65.1
88	4th	9,359	20,133	40,293	69,785	65.7
88	3rd	5,547	10,757	44,030	58,334	62.6
88	2nd	5,243	8,898	40,777	54,914	68.8
88	1st	2,986	5,932	46,057	54,991	63.5

*The headings 'Billings to Parents/associates' and 'Billings to other UK clients' discontinued as a result of the Review of DTI statistics in the 2nd quarter of 1989

Source: DTI SDQ9 Business Monitor Series

relative national activity in other countries. One useful source is a recent report on the skills needs in the NTS sector, commissioned by the European Commission, which has generated estimates of past, present and future national employment figures (*Eurostrategies,* 1989). Even taking the estimates of future growth with a large pinch of salt, the figures probably do reflect the relative strengths of national NTS industries within Europe.

The data presented in Table 9.2 confirm the high rate of growth in the British NTS activities from 1980 to 1988 and the sharp slow-down in 1989–90. They also indicate that the United Kingdom represents roughly 40 per cent of the Euopean industry. (Again this is largely explained by the international importance of British financial services.) France has the second largest NTS sector in terms of employment. The Federal Republic of Germany, despite a large and powerful economy, has only 17 per cent of NTS employment.

As a background to these differences, Table 9.3 presents a range of relevant telecommunications statistics and characterisations of regulatory regimes, derived from the 1986 International Telecommunications Union (ITU) statistical yearbook and an 1988 OECD survey. The pace of change renders some of these data of historical interest rather than a guide to current circumstances, while the gaps reflect the difficulty in obtaining comparable statistics in this field.

The data in Table 9.3 makes clear that the telecommunications infrastructure in each of the EC countries is very heterogeneous, reflecting differing national histories and regulatory policies. Indeed the pattern of NTS employment (Table 9.2) can, in part, be explained by this heterogeneity of regulatory regimes. The United Kingdom's early lead – 65.8 per cent of EC NTS employment in 1980 – reflects the relatively low price of leased lines and the early relaxation of the regulatory regime in the United Kingdom. But it also reflects factors such as the size of the British financial services sector;

Table 9.2 Employment as a measure of NTS developments, (ranked by 1987 NTS employment)

	1980	1987	Annual average growth (%)	1992 (estimated)	Annual average growth estimated (%)
			1980–87		1987–92
United Kingdom	1,350	16,700	43.2	22,200	5.8
France	350	8,050	56.5	11,400	7.2
Germany	300	7,000	56.8	10,600	8.6
Italy	–	2,700	–	5,700	16.1
Belgium	50	1,500	62.6	2,300	8.9
Netherlands	–	1,450	–	2,800	14.1
Spain	–	1,300	–	3,200	19.7
Denmark	–	800	–	1,300	10.2
Greece	–	500	–	1,300	21.0
Ireland	–	300	–	450	8.4
Portugal	–	300	–	500	10.8
Luxemburg	–	150	–	250	10.8
EC	2,050	40,750	53.3	62,000	8.7

Source: Derived from *Eurostrategies*, 1989

the presence of Japanese and American multinationals, (major users of NTS); the historical strength of British scientific publishers, (who have been developers of online S & T information sources); and the importance of the English language in the international NTS market.

France's strong growth is in large part based on infrastructure investment, especially around the public use of videotex terminals. A series of conscious policy decisions were signalled early in the Nora–Minc Report (Nora and Minc, 1981) which stressed the importance of telematics (at a time when most comparable national studies were focused on computerisation).

Given the relative size and strength of the German economy, its relatively low position in EC NTS employment is surprising. The PTO has been taking an active role in upgrading the infrastructure, but it is widely argued that despite their relaxed rules on third-party resale and the recent revoking of the monopoly of terminal attachments, the Federal Republic of Germany's pattern of regulation has had a negative impact on NTS growth.

The Netherlands, Belgium, Luxemburg and Denmark have small domestic markets, but specific national considerations have also been important. Denmark has an advanced and relatively liberalised telecoms infrastructure and an orientation to the Scandinavian market; the Netherlands has a strong publishing tradition, especially in scientific publishing, which is expected to lead to high future growth; Belgium has had a rigid regulatory regime inhibiting NTS developments, though its central position in the EC means that the SWIFT interbank settlement system is partially based there.

Greece, Spain and Italy are late arrivers – at least if the forecast of high growth between 1987 to 1992 is realised. These countries have had relatively poor infrastructure (indicated by their relatively low levels of telephones per capita), but are now investing heavily. The regionally concentrated nature of

Table 9.3 NTS Relevant telecommunications indicators (European Countries ranked by main telephone lines per 1,000 inhabitants)

	Main lines per 1,000 inhabitants	Mins of PSTN calls to pay for 250km internal M1040 leased line	No. of public videotex subscribers, 1989	Value-added services
Denmark	529	48	2,000	Competition
Germany	440	1,720	130,000	Competition
Luxemburg	437	*	200	Competition
France	436	1,513	4,500,000	Competition
Netherlands	424	338	30,000	Competition
United Kingdom	390	–	90,000	Competition
Greece	347	–	–	Public Monopoly
Belgium	345	1,566	3,000	Sig. Add.[1]
Italy	333	1,610	25,000	Defacto Competition
Spain	262	329	300	Competition
Ireland	225	3,085	1,000	Fut. Lic.[2]
Portugal	161	3,300	–	Comp. Lic.[3]
EC	374	–		

Notes: * Luxemburg is too small to have 250km leased lines
[1] Sig. Add.: Significant Value Added necessary
[2] Fut. Lic.: To be licensed in near future
[3] Comp. Lic.: Competition subject to license

Sources: OECD *Telecommunication Based Services:* ICCP Report No. 18 and various issues of *'Communication Week'*, and LINK cited in White, 1990

their industries may make installation of advanced services relatively easy (for example, in Athens). Ireland and Portugal have lower levels of telephones per capita and high costs for leased lines (though these data need interpreting with caution, since the stated tariffs may not reflect the negotiations between PTOs and large users). The absence of important finance sectors and large industrial sectors has meant little pressure for NTS development, though some countertendencies are visible – Ireland for example using its English-language facilities to supply offshore office services and attempting to play a hub role in international communications.

V Specific NTS developments

Online information

According to the statistics in Table 9.1, data bases represent approximately 60 per cent of the British NTS market. Reflecting their size and early development, these are the best documented services although even here comparative data are limited. A useful cross-national historical perspective is provided in a report from the Germany ministry for research and technology (Table 9.4).

Table 9.4 Comparative turnover for online information services, 1987 (in US$m)

Country	Scientific & Technical	Economic Information	Totals
USA	2,794.0	3,878.8	6,672.8
Europe	44.6	851.7	896.3
United Kingdom	7.7	315.2	322.9
France	13.6	108.3	121.9
Former West Germany	10.5	148.7	159.2
Italy	4.1	84.7	88.8
Rest of Europe	8.7	194.8	203.5
Japan	67.1	522.0	589.1
Totals	2,905.7	5,252.5	8,158.2

Source: Specialised Information, 1988

Within Europe, the United Kingdom generated the bulk of online database turnover: according to Table 9.4 36 per cent of European online information turnover. (Compare to Table 9.1 which shows the United Kingdom with 41 per cent of the NTS employment.) Within Europe, France shows a relatively high usage of Science & Technology (S & T) services while the United Kingdom has the largest turnover in the economic information category. However, Table 9.4 demonstrates most strikingly the dominance of the United States which had 81.8 per cent of world turnover. The high turnover of S & T services in the United States (96.1 per cent of total) is in part a matter of higher usage, mostly because the world's major scientific and technical databases such as Chemical Abstracts, Medline and the Science Citation Index are United States-produced. The Japanese market where there has been rapid growth stimulated by the Database Promotion Center is now also larger than the European market for online information, 30 per cent of available world data-bases are produced in Japan (IMO, 1990).

There is relatively little intra-EC trade in online services. National state support for data-base production tends to create a market fragmented by language. This is partly explained by the lack of an integrated, cheap and reliable data communication network for the EC as a whole. For the last three years EUSIDIC, the European online users organisation, has conducted a survey of data communication problems experienced by its members. Some of the results of the most recent survey are presented in Table 9.5. The problem of differing national networks is compounded by varying international call charges between the EC members which discourage direct access to other national networks. When combined with differing standards for terminal equipment and the difficulty in obtaining inter-network leased lines, this means that intra-EC NTS traffic is much lower than it might be. Improvements could stimulate NTS activity considerably.

The EC's 'Information Market Observatory' (IMO) aims to improve the coverage and consistency of statistics about NTS and other information services market in Europe. This has involved liaison with Eurostat on the development of definitions and market taxonomies and the generation of new data. The IMO has conducted surveys of European and world-wide data-base

Table 9.5 1989 International Data call (IPSS) failure rates

Country* of call origin	Failure (%)	Total calls
Germany	10.9	175
France	13.9	777
Netherlands	16.0	929
Luxemburg	22.8	57
Denmark	22.8	123
Belgium	23.4	265
United Kingdom	29.6	1,427
Greece	33.3	15
Italy	36.6	588
Ireland	36.9	65
Spain	40.0	959
Portugal	62.9	35

*Ranked by percentage of failed data calls

Sources: EUSDIC Survey cited by J. Walko, 1989 in *Communications Systems Worldwide*

production and European host operations and set up a representative panel of European users of information services who are questioned regularly on a range of topics. The IMO's 'Guidelines for improving the synergy between the public and private sectors in the information market' (CEC, 1989), largely based on those drawn up by the United Kingdom's Confederation of Information Communication Industries, seek to ensure that state agencies do not exploit privileged access to data for commercial purposes. These guidelines currently have no legal force, but are likely to do so in future.

The future of information services is rather uncertain. Some areas of growth seem likely to be:

(i) *real-time financial information* as the European financial markets become more integrated and market information becomes more important;
(ii) *European-wide business intelligence*, credit-checking, markets, contracts competitors etc.;
(iii) *scientific and technical online data bases*, especially with full text retrieval;
(iv) *new audiotext and fax data-base services*

Already some innovations in each area have had considerable success in consumer markets (often illicitly using office equipment). It is hard to forecast the development of such activities, especially given the paucity of research attention they have received.

Transactional services

There are very few statistics which cover transaction, messaging or other simple 'carrier' services, although it is widely (and plausibly) asserted that the potential economic importance of these services is greater than that of information services. We consider below available data in terms of distinct service categories.

Videotex

Videotex was designed as an information dissemination technology, to enable mass access to data bases. The original 1,200 bps downline data transfer rate compared with 75 bps upline, and the crude keypad for data entry reflected the philosophy that, overwhelmingly, information was to flow *to* users, not from or between them. There has, however, been limited public demand for such services (unless they are not readily available by other means – the French strategy of putting the telephone directory online has succeeded in bringing many people into using this service).

In contrast, transactional services have become an important area of videotex activity. The SPRU survey of British videotex host systems (Jagger, 1989) illustrates this point. It found that the number of systems dedicated to transaction-type applications, as distinct from information-type applications, had increased with time. The British use of videotex for transaction services has, however, been confined to a few sectors – travel, insurance, automobile distribution and some retail sectors such as white goods and office consumables. It has played the role of an easy-to-learn entry system for NTS, and its use became near-universal in British travel agencies in a matter of a very few years in the mid-eighties, (owing largely to pressure from a few large companies – the tour operator Thompsons, for example, simply pulled the plug on its traditional telephone booking service).

However, the position of videotex, even in this area of greatest success, is now being challenged by competing airline booking and reservation systems, which serve the business travel market (more profitable than the holiday trade) without using the videotex format. The current videotex terminal suffers from speed limitations on information delivery. Our survey found that most British videotex operators, especially those using it for transaction-type services, found the slow speed of the up-line connection on the older systems a severe disadvantage. Transaction services require more than a one-way flow: they need upline communication and preferably integration with other office systems and printers. Given the current (perhaps temporary) decline in profitability of packaged holidays, it seems likely that videotex terminals will be gradually displaced by the more sophisticated technology offered by information systems such as Sabre, Amadeus and Galileo.

The future of videotex is in doubt in other applications, too. As computer systems, telecommunications and user sophistication continue to develop, the need for so basic a format diminishes. It would be possible to improve videotex and innovate within its framework – there has already been much of this, for example, better keyboards, gateways to other NTS, keyword searching, rapid two-way communication flows etc., but a question mark must be placed over strategies like that of France (and now being adopted by a number of other countries), where videotex has been seen as the lead into the new telematics era and consequently attracted much public investment (for example, free terminals).

Table 9.6 World EDI users, 1989

Country/Region	Share (%)
USA/Canada	70.4
United Kingdom	15.5
EC1	3.6
EC2	1.4
EFTA	2.1
Others	7.0

Notes: EC1 Denmark, the former West Germany, Netherlands
EC2 France, Italy, Spain, Portugal, Belgium, Ireland, Greece
Others New Zealand, Australia, Pacific Basin

Source: Sarich 'An Update on EDI in 1989' in CEC, *EDI–1992 and Beyond*

Electronic Data Interchange

The need for – and potential success of – Electronic Data Interchange (EDI) is typically explained by estimates that 4 per cent of the value of domestic trade and as much as 15 per cent of international trade is attributable to paperwork. The aim of EDI is to eliminate this paperwork with orders and invoices transmitted directly between the trading partners' computer systems (Ungerer and Costello, 1988). There has been considerable publicity for EDI and astounding predictions of its market growth.

At present, however, there is little diffusion: in the United Kingdom only 2,000-2,500 companies were using EDI in 1989 (Guest, 1990). Table 9.6 sets out estimates which suggest that the United Kingdom accounts for 79 per cent of EDI users in the European Community and 15.5 per cent of the world users (Sarich, 1989). An unpublished EC survey shows, not surprisingly, higher awareness and use of EDI in Northern Europe than in Southern Europe, but even among the larger firms in Northern Europe there is much ignorance and low levels of current use, (although there are also high expectations of future use especially in transport).

One stimulus to greater use would be the use of the UN's Edifact standards by European customs and a great deal of attention is currently being paid to this. It is even more necessary, however, to reconcile conflicting EDI standards across Europe (see Chapter 11). A number of special systems and standards have been developed between trading partners in vertical trading situations. For example, Tradacom is used by the British chemicals and retail trade; ANSI X.12 by many American companies; Edicon by the British construction industry; Cefic by the European chemicals industry; and Limnet and Rinet by the insurance industry (Vanguard, 1989). In the automobile manufacturing sector, (where incompatibilities and 'standards wars' between different suppliers have been intense) some component firms have to maintain several distinct terminals on the same desk in order to communicate with different manufacturers. Much effort has been put into bringing these standards into alignment with the United Nations and Commission approved Edifact standard. In the long run Edifact itself needs to be integrated with the X.400 messaging standard, so that EDI can break away from single dedicated networks (Vanguard, 1988).

Even if these standards problems could be reduced, there will remain organisational problems in EDI. The main issue seems to be integrating EDI information into existing trading, accounting and production systems. Although the use of nonstandard message formats could make this more difficult, in most respects British companies probably have a head start in systems integration, the main reason for this being the wider availability of leased lines and managed data network systems (MDNS) in the United Kingdom. It is estimated that the United Kingdom has 25 per cent of Europe's leased-line-based 'Wide Area Networks' (WANS) and accordingly more experience with interconnecting different computer systems.

As for other transactional services, EFTPOS (electronic funds transfer at point of sale) is experiencing rapid growth in retail and other outlets, and the outlook for growth must be rosy. The more futuristic teleshopping initiatives have so far failed to take-off: they have perhaps been most successful as subsidised services for the housebound. However, innovators are continuing to seek to improve technical (interface) and social (delivery) features of such systems, and an upturn in fortunes is conceivable (Miles, 1989).

Messaging Systems

The link between EDI and MDNS mentioned above is mainly due to the latter offering store-and-forward E-mail systems. With E-mail, messages can be sent into the network and stored in a 'postbox' until the recipient 'logs-on' to the network to collect the next batch of messages.

E-mail has many other business uses over and above its link with EDI, and despite its apparent eclipse by fax, is continuing to experience steady growth. There were approximately 200,000 users of public E-mail services registered in the United Kingdom in 1990, although growth in the latter part of the 1980s was slower than earlier. A major obstacle to expansion of E-mail (in addition to its complexity compared to using fax) is the difficulty of communicating across systems. Moves towards creating simpler directory structures and interconnectivity are underway, but it looks as if it will remain a facility whose use is limited to special needs. For example, users who need to transmit text directly to each others' computers (for example, journalists, researchers) are likely to intensify their use but elsewhere its growth seems likely to remain slow.

This forecast may be upset by innovations which bring E-mail closer to fax. The 1990s may well see the development of an electronic notepad/communicator which will be able to receive and store text messages as well as serve as a voice telephone. Such an innovation – being discussed currently by several suppliers – would open the way to a range of new transactional and information services. Messaging has been a success story in the French Minitel experience (and has been one of the more dynamic areas of Britain's Prestel, mainly catering to computer hobbyists). While it is easy to overstate the extent of messaging (only 5 per cent of users of Minitel actually use the facility), to date it has proved one of the most profitable NTS activities on videotex, owing in part to the proliferation of chatline-time systems. This indicates scope for electronic messaging services to substitute both for conventional post and to provide new types of social interaction. But whether such services will take off and what will be the scope for international trade remains anyone's guess.

VI Network issues

The development of a European MDNS network

There have been various efforts to implement a Europe-wide MDNS, most notably by the CEPT (European Association of Telecommunication Administrations). Agreement was not forthcoming from all countries, but some are proceeding with joint plans. Given the problems that the administrations have in working together, the most likely organisation(s) to implement a European MDNS could well be a third-party operator leasing lines from the PTOs and/or linking up existing national EDI networks.

With standardisation through Edifact and X.400 there will be less need for a single network. ISDN networks (see Chapter 8) should also make basic data communications easier across national borders. Despite this, some companies may still require, for security reasons, the high levels of service which only MDNS can provide. There is strong competition in Europe over creating such international networks with AT&T-Istel and BT-Tymnet posed to compete with France Telecom, INS (ICL-GEISCO) and IBM. Infonet, with its public telephone authority shareholders, is also poised to attack the same market. Critical to all these developments will be the attitude of the International Telecommunications Union to international leased circuits, since the problems with using international leased lines for third-party use has meant that few bodies outside the PTOs have the necessary (international) network management experience. Thus the takeovers of Tymnet by BT and Istel by AT&T are interesting developments, combining MDNS and PTO skills.

Commission intervention

The European Commission has long viewed telecommunications and NTS as an important policy area and there has been a wide range of Commission interventions and policy initiatives aimed at improving the basic telecommunications infrastructure, especially in the peripheral regions, and at stimulating advanced telecommunications, especially broadband and information services.

Over the next few years the most important question facing NTS development in Europe is whether the European standardisation effort can keep pace with de-regulation which will allow the rapid development of non-standard applications. The standardising role of the PTOs has historically been extremely important. In the United States, when AT&T operated the whole network, Bell Labs were obliged to put all their results into the public domain and had a tremendous impact on standards; de-regulation has led to a decline of this standard-setting role (Harris, 1990). ETSI, the European Telecommunications Standards Institute, has been set up to take the lead here, although there is considerable industry scepticism about its capability and clout. Standardisation may be more critical for SME and domestic use of

NTS than for large corporations. (The skills necessary to operate in an non-standard environment are more likely to be available to the large corporations.)

VII Conclusions

Within Europe, France and the United Kingdom have the two largest NTS sectors, but these have been nurtured under very different policy regimes and feature very different composition. The British NTS sector has largely developed on the basis of its financial services sector and its 'de-regulated' telecommunications environment. France's NTS sector has depended on state intervention to alleviate critical mass problems through the distribution of Minitel videotex terminals. The size of the French domestic and SME NTS market shows the importance of producing this standardised environment, while the British sector is orientated around the needs of large corporate users often with non-standard application-specific solutions.

Critically the future of NTS in Europe depends on the relative balance between the standardisation forces versus the 'de-regulation' forces. This will be especially important for the development of any mass or domestic NTS market. But in any case the rate of growth is likely to be high. The two major trends in the future of European telecommunications – ISDN and further de-regulation, especially of new services (for example, mobile communications) – will impact on these developments profoundly. E-mail and EDI are both likely to be very important for meeting the requirements of the internal market. There are considerable opportunities for well-positioned actors, but there are also growing risks, which are leading to a variety of new alliances being forged.

10 The new consumer electronics

Alan Cawson and Peter Holmes

I Introduction: major international trends

In its early days, the electronics industry was almost synonymous with the radio industry. While capital goods were always important for transmitters and communication equipment, the bulk of production from the 1920s through the 1950s was for consumer radio and later TV. However, during and since World War II demand for radar, computers, instruments, electronic medical equipment and other capital goods has grown even more rapidly than demand for electronic consumer goods so that capital goods now account for the greater part of production in most leading industrial countries. However, both capital and consumer goods share a common basic technology and benefit equally from the fantastic rate of technical advance in components (Chapter 5). Many firms, especially in Japan, have a strong presence in both capital and consumer goods as well as in components and there are some synergies, especially in R & D.

Hitherto in Part II we have considered mainly computer-related industries, telecommunications and the new information services based on telecommunications. In this chapter we turn to consumer goods which continue to account for about a quarter of the total output of electronics and remain an extremely dynamic sector of the industry.

The European consumer electronics industry has experienced far-reaching changes in the past ten years, in terms of the pattern of international trade, the composition of its products, the structure of its firms and the pattern of intervention by national governments and the European Commission. The industry of the 1960s and early 1970s was a mosaic of relatively independent national markets served predominantly by small-scale production within national boundaries. The major product in the 1960s was colour television (CTV), which was afforded a degree of protection against imports by the adoption of unique television transmission standards in PAL and SECAM. As shown in Tables 10.1 and 10.2, the dominant position of the European industry has been maintained for CTVs, with European firms increasing output and their share of total production. But for VCRs, audio products, compact disc players and camcorders they have lost ground to Japanese and other Far Eastern producers.

After a costly format war fought between the Philips-Grundig V2000 machine and two rival Japanese VCR formats (VHS and Betamax), Philips was forced to concede defeat and license VHS technology in 1985. The second biggest firm, Thomson, had declined to cooperate with Philips and developed an alliance with JVC through its acquisition of the German firm Telefunken, which had formed a joint venture with Thorn EMI and JVC. In

Table 10.1 European consumer electronics consumption: share of imports by product category, 1989

	Apparent Consumption (ECUbn)	Share of imports (%)
Colour television	6.5	20
Video cassette recorder	2.9	46
Video camcorders	0.8	99
Audio products	3.2	89
Compact disc players	0.8	74
In-car entertainment	1.4	58

Source: Data supplied by the European Commission

1987 Thomson acquired Thorn EMI's consumer electronics interests and thus a 50 per cent stake in the joint venture, which is now Europe's largest VCR producer.

Although European production of CTVs remains relatively stable, the level of imports is rising as products from South Korea and China begin to penetrate the European market. In addition the composition of European production is rapidly changing as an increasing proportion of European output is produced by the European subsidiaries of Japanese firms. These plants produced over 16 per cent of European CTVs in 1986 and almost 25 per cent in 1987. The position of the Japanese firms in VCR production is much stronger, having steadily replaced imports from Japan by local production which amounted to over 40 per cent of European production in 1986.

The response of many European CE producers (especially in Britain) to the increasing competition from imports and Japanese production in Europe was to leave the industry, and those left behind became concentrated within three large groupings: Philips (which controls Grundig), Thomson (which took over Telefunken, Nordmende and Ferguson) and Nokia (which bought the ITT subsidiary in Germany and Oceanic in France). These firms now account for three-quarters of European CTV production and half of VCR production. Both Philips and Thomson have undertaken a series of plant closures in order to concentrate European CTV production into larger and more efficient plants and have shifted the production of small-screen CTVs to new plants in the Far East which now account for about 50 per cent of their world-wide CTV production. In 1987, Thomson shed 2,500 consumer electronics jobs in France and created 1,200 in Malaysia and Singapore. In 1990, after registering the first losses in its 99-year history, Philips announced a programme to shed 45,000 jobs world-wide. Unlike Philips and Thomson, however, Nokia has almost all of its consumer electronics manufacturing in Europe.

During the 1980s the NICs have been rapidly increasing their CE production and exports to Europe, especially in VCRs. The three largest Korean producers (Samsung, Goldstar and Daewoo) had by 1986 developed the capacity to produce 4.7 million VCRs, significantly bigger than the entire VCR production in Europe (OSI, 1988:359). In a single year from 1986 to 1987 the Korean firms increased their VCR market share in Europe from 6 per cent to 16 per cent. By 1988, the annual output of the Korean VCR

Table 10.2 Growth in European CTV production capacity, (000s) 1984-8

Firm	1984	1985	1986	1987	1988
Philips	3,050	3,150	3,100	2,370	3,400
Matsushita	230	250	300	360	400
Sony	310	355	535	670	850
Thomson	2,100	1,850	1,050	1,900	2,650
Sanyo	—	150	150	570	700
Sharp	—	—	5	50	100
Bang & Olufsen	130	140	145	150	150
Hitachi	200	300	370	380	500
Mitsubishi	95	110	125	150	200
Nokia	500	600	680	2,130	2,250

Source: Data supplied by the European Commission

industry had reached 8 million, and that of the Taiwanese 1.5 million. It is not surprising that the European firms viewed this trend with alarm and filed allegations of dumping against three Korean and two smaller Japanese firms in 1988. As the Japanese firms did a decade before, all three Korean firms have now established manufacturing plants within the EC.

All the European firms are experiencing increasingly fierce competitive pressure in consumer electronics products, which is intensifying as the proportion of value accounted for by hardware products tends to diminish as markets mature. Sales of software products have been increasing much faster than hardware in all the major markets; for example, in Japan software is estimated to account for 72 per cent of total video sales, and in Britain the comparable figure is 64 per cent. This factor helps to account for the moves by leading Japanese firms such as Sony and JVC to acquire stakes in the music industry and film production.

In hardware the Japanese firms have moved the production of 'mature' items such as CTVs and low-cost VCRs offshore to South-east Asia, and have invested heavily in product and process innovation for higher value-added products. Competition in the industry will increasingly depend on successful innovation, and we can learn something of the prospects for the European firms by examining their record of patenting in comparison with Japanese firms. While the R & D performance of firms in this industry is hard to measure directly, it can be assessed indirectly by analysing data on patents.

If we take patents registered in the United States by foreign firms for 'sound' and 'image' applications, some very clear patterns emerge. Philips's commitment to internal growth can be seen in research activities which dwarf all others: in 'image' patents, for example, out of a grand European total of 1,086 patents over the period 1963–86, it notched up 514, against Thomson's 159 (Table 10.3). The latter firm has invested in growth through acquisition of other firms and has of course seen its turnover rise faster than Philips. It must be stressed that patenting alone does not guarantee success, as the United Kingdom's high patenting figure but poor industrial performance in the 1960s and 1970s shows, and indeed Philips's current difficulties echoes, but it is indicative of the technological dynamism of the sector. However, Japanese sound and image patents dwarf those of other world competitors.

Table 10.3 'Sound' and 'image' patents filed in the United States by European firms, 1969–86

	1969–71	1972–4	1975–7	1978–80	1981–3	1984–6	Total
Image:							
European firms (all)	155	201	190	281	271	345	1,533
Philips	91	75	84	69	77	118	514
Thomson	12	11	24	33	30	49	159
Sound:							
European firms (all)	120	146	125	132	183	192	1,106
Philips	29	27	19	16	31	44	166
Thomson	0	3	1	4	6	3	17

Note: Post-1986 data are not included because in 1986 a change in the method of recording patents by company makes comparison difficult

Source: Data supplied by Nick von Tunzelmann from SPRU patents database

Table 10.4 EC consumer electronics production, trade and employment, 1980–88

	1980	1985	1988
Apparent consumption (ECUm)	11,350	14,304	21,055
Net exports (ECUm)	–3,468	–5,409	–8,149
Production (ECUm)	7,882	8,895	12,906
Employment (000s)	180	134	123

Source: CEC, 1990b

II Protection and inward investment

The flood of Japanese imports in this sector has not gone unanswered. The consumer electronics industry is one of the most highly protected in Europe and in the United States. Along with steel, cars and textiles, it has seen at various times the application of Voluntary Export Restraints (VERs), exceptional increases in tariffs, anti-dumping duties, controls on inward investment, technical barriers to trade, financial support (especially to Thomson in the early 1980s) and industrial policies designed to promote the industry. The United Kingdom has long had unofficial VERs, as has France where there have also been technical barriers (notably the SECAM system). Pressure has been put on the EC Commission to introduce 'anti-dumping' measures which many observers feel act as a deterrent to all price competition, not merely unfair 'predatory' pricing (NCC, 1990). Even so, imports have continued to rise (Table 10.4), and European producers have lost market share, with the British firms largely quitting the industry.

However, protection helps the Japanese even more. This is no paradox: protection may help European producers in the short run, but it helps the Japanese even more by bolstering their profits and enabling them to invest in the EC. Agreed import quotas (Voluntary Export Restraints) and 'price understandings' linked to anti-dumping actions have allowed the industry to keep up prices in the short run, and as we look across Europe we see that the most protected markets have the highest prices, profits being up to 50 per cent higher in France than the United Kingdom. But the high level of prices has encouraged entry of new firms from Asia. Thus in the short run the industry resembles a 'collusive oligopoly', but in the long run it corresponds to 'monopolistic competition', and excess capacity causes the profit potential from high prices to be dissipated. The number of firms in British industry has been roughly constant in recent times, though their identity has altered (Cawson *et al.*, 1990:Chapter 10). European restrictions on imports from Japan and Korea have encouraged both a flow of inward investment from those countries and a spectacular rise of imports from countries not yet controlled, notably Turkey and China whose share of TV imports has risen dramatically recently, and will no doubt go on doing so unless import restrictions bite. They could easily be followed by Malaysia, Thailand or Brazil. To impose quantitative restrictions on all imports from new sources denies the process of international competition as well as being incompatible with GATT.

EC employment in consumer electronics has fallen, by about 3.5 per cent per annum, since the mid-1970s, slightly faster than for manufacturing industry as a whole (Table 10.4). Typically, protected industries have 're-structuring' plans involving replacement of labour by capital, so the industry's reduced labour requirements for assembly are in part a substitution of capital and indirect labour for direct labour – not a pure productivity gain.

Import controls have led Asian firms to develop their EC production, but on the other hand the allocation of import licences to European-owned producers has actually encouraged them to buy cheaply from associated concerns in Singapore or South Korea without allowing free competition from there. The Koreans have been discouraged by threats of anti-dumping actions from using low prices to enter the EC market. Protection has also influenced the pattern of production by EC producers. As we have seen European firms, above all Thomson and Nokia, are disproportionately orientated towards CTVs.

The multinationalisation of Japanese firms corresponds to a very familiar pattern of overseas investment. Since the turn of the century, but above all in the last 25 years, it has become increasingly common for firms to begin establishing production facilities in their major markets rather than seeking to produce everything at home. In this respect what is surprising is that the Japanese do this relatively rarely. It is therefore arguable that Japanese CE firms would have been led down this path eventually anyway. Their advantage clearly comes from something other than the cost of labour in Japan and, unless economies of scale dictate that production be all on one site, they have nothing to lose by relocating: Japanese wage levels are much higher than those in SE Asia, and total CE production costs in Japan are close to European levels.

Location by Asian firms in the EC market-place keeps them close to customers but it also signals their good intentions to oligopolistic rivals. As long as Japanese producers are paying EC wages for a significant fraction of their output they have an incentive to align their prices for local production with EC firms, and to tolerate 'voluntary export restraint' agreements and threats of anti-dumping against the next generation of newcomers. This gives the EC industry a false sense of security: the higher prices gained by the Asian firms enable them to earn extra profits to reinvest, a perverse but predictable consequence of the 1983 EC agreement on prices and quantities of Japanese VCR imports. The current financial difficulties of Philips illustrate vividly the nature of the problems.

The purely national character of EC protection has led to a proliferation of sub-optimal-sized plants by EC producers and to a perpetuation of market fragmentation. However, this will no longer be possible if the 1992 plan succeeds, with a barrier-free internal market and a single external trade policy for the entire EC. With full implementation of Article 9 of the Rome Treaty all imported goods must be given free circulation throughout the EC along with EC-assembled and EC-produced goods, regardless of 'local content'. The open question is of course how restrictive any unified policy is.

III The 'Europeanisation' of the industry

At the beginning of the 1970s the European consumer electronics industry was characterised by national firms producing and selling in national markets. Partly because of incompatible technical standards in CTV, and partly because of differences in national consumer tastes, there was relatively little intra-European trade in CE products. The largest European producer, Philips, maintained plants in nine countries but these were run by national organisations or 'missions' operating relatively autonomously from the headquarters in Eindhoven. Its subsidiaries such as Philips UK or La Radiotechnique in France were well integrated into the national industries of those countries and active members of their trade associations.

This pattern began slowly to change at the end of the 1970s as a consequence of the competitive strength of the Japanese industry which first in audio products, then in colour televisions and finally and most significantly in VCRs, began to erode the market shares of the European producers. The effects were felt most strongly in the two most liberal markets in Europe: the United Kingdom and Germany. As we have seen the response of many of the smaller firms in the United Kingdom was to leave the industry, and this process reached its conclusion in 1987 with the sale of Thorn-EMI-Ferguson to the French firm Thomson and the exit of the last remaining independent British television manufacturer Fidelity. In Germany two of the major firms, Telefunken and Grundig, lurched into financial crisis which raised fears that the German industry would fall into the hands of the Japanese following the lead set by Sony which had acquired Wega in 1975.

However ill-founded these fears might have been, they precipitated a struggle between Philips and Thomson for control of Grundig. Thomson had already acquired three of the ailing German firms, Dual, Saba and Nordmende, and it took control of a fourth, Telefunken, when it became clear that Philips was unwilling to relinquish the quarter share of Grundig that it already possessed. The outcome of this episode was that by 1983 Germany was left with only one independent CE producer of significance, Bosch, and its biggest firms were owned by Dutch, French and American companies. The American-owned firm SEL was eventually bought by the Finnish firm Nokia in 1988.

This 'Europeanisation' of the industry was not, however, the consequence of a deliberate strategy of securing economies of scale to match Japanese production volumes, nor was it part of a push to break out of self-contained national markets. For Thomson, one of the main motives was to acquire market share in what was the biggest European market and to provide an outlet for its Vidéocolor tube subsidiary. In addition the purchase of Telefunken offered the opportunity of an entry into the co-production of VCRs with the Japanese firm JVC – a move which had earlier been rejected by the French government when Thomson wanted itself to enter into a consortium with JVC. For Philips, which was involved in trying to promote its own VCR technology against its Japanese rivals, control of Grundig was a means of maintaining its partner's commitment to the V2000 system, which would certainly have been dropped had Thomson been successful with its bid for Grundig.

The consequence of these takeovers, and subsequent acquisitions, has been to create a European industry dominated by three large groupings centred around Philips, Thomson and Nokia. Each of these firms is faced with considerable problems of integrating disparate parts into a coherent whole at the same time that they manage the challenges posed by the introduction of the new technologies outlined in the following section. So far this trend towards 'Europeanisation' has led to concentration of ownership and control, with some evidence of increasing intra-European flows of components but so far only limited integration of markets for CTVs. For new products such as Compact Disc (CD) and eventually High Definition Television (HDTV) the picture is somewhat different, with firms establishing single production centres to serve the whole European market.

The strategy of the Japanese firms as they began to establish local subsidiaries from the early 1970s was initially tailored to the Europe of separate national markets. Their branch plants, which in all cases started off assembling imported components but later moved to more extensive local sourcing, were relatively self-contained operations tightly controlled from Japan with few intra-European links. But recently there have been increasing signs of a move towards setting up integrated European companies. Sony led the way by centralising its European CTV manufacturing under a headquarters in Stuttgart, and by creating a 'European executive committee' to coordinate the development strategies of the subsidiaries and prevent duplication of effort. Others, including Hitachi, Matsushita and Pioneer, have followed by setting up Europe-wide organisations to co-ordinate their subsidiaries. In this respect the Japanese have some advantage over the European firms: having grown largely organically rather than by acquisition, the problem of integrating plants with different national and corporate cultures is much less acute. Quite apart from the technological lead that the Japanese established in consumer electronics in the 1980s, there is an important organisational dimension which suggests that the Japanese will be more effective at 'Europeanisation' than the Europeans themselves.

Both Philips and Thomson are keenly aware that competition with the Japanese in consumer electronics is not simply a contest over European markets, but is a competition between global production systems. Thomson's purchase of the RCA division of the American General Electric Company, and Philips' integration of the North American Philips Corporation into its reorganised divisional structure, are indications that neither company can afford to restrict its horizon to the European market. In addition both companies have responded to the pressure on production costs by shifting manufacture of audio and small-screen television sets to South-East Asia; in 1986 half of the import licences granted in France for consumer electronics products were to Philips and Thomson. In this context the concept of 'national' protectionism loses its original meaning and becomes the protection of one regional set of plants of the multinational firm against another.

IV New products and new technologies

Consumer electronics technologies have been characterised in recent years by shortening product cyles and the rapid introduction of new products. Even 'mature' products such as colour television have been subject to rapid change through the application of integrated circuits, first in additional features such as teletext, and currently through the enhancement of picture quality by storing and processing television images in microchips.

European firms, and Philips in particular, have played an important part in consumer electronics innovation with the compact audio cassette, PAL and SECAM colour television systems, teletext, 110° picture tubes, stereo TV sound, compact discs and MAC (Multiplexed Analogue Components) television transmission standards. Despite this record of innovation, however, technological capability has seldom been translated into profitable product launches. In particular defeat in the format war between different European and Japanese video technology in the early 1980s led to the retreat of the European firms from this technology and their absence from subsequent video developments such as camcorders and spinoffs from VCR technology such as Digital Audio Tape (DAT). Having perfected the domestic application of video technology by commercialising what was originally American and European fundamental research, the Japanese look set to maintain their lead by further developments such as Super VHS and the fully digital VCR.

In important new areas such as Liquid Crystal Displays (LCDs), digital signal processing and optical recording techniques, the European firms are maintaining a foothold by spending on basic R & D, but they have been slower than the Japanese firms in moving from the laboratory to the marketplace. In LCDs, for example, Japanese firms have gained valuable expertise from the early manufacturing and marketing of miniature displays, and the signs are that they will move more quickly than the Europeans towards the marketing of larger screens. Similarly the European firms' progress towards the fully digital CTV has been slower than that of the Japanese, in part because of greater Japanese strengths in the development of VLSI chips.

The Europeans' innovative efforts are hampered by the considerable uncertainty surrounding the introduction of new television technology. Having pioneered the development of a new system for Direct Broadcast by Satellite (DBS) in the form of MAC, the European countries were divided on the precise specifications because of the need to adapt the signal to diffusion through existing cable networks, particularly in France, Germany and the Benelux countries. This uncertainty has delayed the development of chip sets and increased the cost, given the need to cater for different variants. The European Commission issued a directive making one of the three variants of MAC mandatory for high-power DBS transmissions, but the private Astra satellite launched in 1987 uses a medium-power signal which has allowed the Sky channel to introduce DBS using the existing PAL transmission standard. Despite the relatively low cost of the equipment (dish and decoder) needed to receive Sky, initial market penetration was sluggish, but may pick up. What is evident from the experience with DBS so far is that while the manufacturers have every interest in introducing a new system, and especially one which would act as a form of passive protection against the Japanese, consumers have so far shown relatively little interest. This throws into doubt at least the more optimistic forecasts for the acceptance of HDTV, which is discussed below.

Meanwhile the much discussed convergence between consumer electronics, computing and telecommunications is difficult to detect in the strategies of the major European producers. In contrast to Nokia, which has built by acquisition a range of interests in mobile communications and data systems as well as consumer electronics, Philips and Thomson have in the past few years concentrated their activities in the audio-visual sector and have made themselves dependent on maintaining strengths in core products such as CTV. Philips has shed most of its activities in computers and has retrenched in telecommunications, while spinning off its domestic appliance interests. Thomson sold its telecommunications interests, and exchanged its medical electronics division for part of GE-RCA. The only major Japanese firm so dependent on audio-visual products, Sony, has been diversifying at the same time that the European majors have been concentrating. It has increased its business in professional electronics and has made two major acquisitions, CBS Records and Columbia Pictures. These last two are a clear indication of Sony's assessment of the synergy between hardware and software in consumer electronics. These moves contrast with Philips's vacillation over the future of its PolyGram music publishing subsidiary in the early 1980s. New CE products such as VCR, DAT, and CD and its derivatives are critically dependent for consumer acceptance on software availability.

In other potentially significant product areas the characteristics of the hardware, or software, or both are pushing traditional 'brown goods' manufacturers into new roles and new partnerships. In the case of CD, where new interactive applications (CD-I) are under development, the production of software demands the integration of skills from the traditionally separate areas of publishing, computing and entertainment. Here, unlike audio CD, DAT or VCR, there is no existing software base that can readily be transferred to the new medium, and leading firms such as Philips and Sony have themselves to act as software systems integrators and instigators. In the field of home automation, where discrete functions such as security, energy management, audio-visual entertainment and lighting are linked through a local area network, the development of products using common standards requires the collaboration of domestic appliance manufacturers, telecoms authorities, energy utilities and others. It is perhaps a significant pointer to the future that, led by Hitachi, some Japanese companies are beginning to enter the 'white goods' market in Europe.

The firms which are likely to be able to exploit such new developments are those with a broad range of activities. Compared to Sony or the major integrated Japanese producers such as Hitachi, Matsushita and NEC, whose activities span most sectors of electronics including components, Thomson in particular appears to have a dangerously narrow base. In traditional product areas new entrants from Korea and Taiwan are forcing on the industry ever more rapid cycles of innovation and product development, thus shifting the locus of competition towards the commercialisation of innovation – the precise area where European firms have been weakest in the past.

V Cooperation between firms and HDTV

The creation of new markets for emergent consumer electronics products has been intimately linked to the issue of standards. In product areas with purely consumer applications the setting of standards has been largely a consequence of the outcome of competition in the market-place, where the successful product becomes the *de facto* standard for the industry. In the audio field the compact cassette introduced by Philips eclipsed the 8-track cartridge format, and innovation, especially by Japanese firms, has, despite its technical limitations, centred around the cassette. In the video field Philips faced stiffer competition from a by then stronger Japanese industry, and its failure to establish its own V2000 format proved costly, even leading the firm to consider pulling out of consumer electronics altogether in 1983. Philips was no more successful with its LaserVision videodisc, although in this case its technology was adopted by the Japanese firm Pioneer and established as a successful consumer product in the Japanese market.

In the late 1970s Philips, Sony and others were developing rival technologies for recording audio digitally on optical discs, but fresh from the debacle of V2000, Philips decided to establish common standards with Sony for the compact disc. Sony had also suffered in the VCR format battle, where its own Betamax technology had proved unsuccessful against the JVC VHS system. Without the uncertainty created by rival formats, compact disc technology was licensed by all the major producers, and CD's penetration of the market outstripped the rate of any previous consumer electronic product. The cooper-ation between Philips and Sony extended from audio CD to the application of CD technology to data storage (CD-ROM) and audio, video and text (CD-I). This kind of collaboration in the setting of standards tends to happen when firms perceive themselves to be relatively equally placed in the market, and the risks entailed in being a first mover loom larger than possible monopoly profits.

Broadcasting standards, however, involve governments and public broadcasting organisations as well as manufacturers; in this case standard-setting becomes more of a political process where organisational interests are more powerfully represented than consumer ones. Broadcast television standards are set by an international organisation, the CCIR, comprising the representatives of national governments. In 1965, faced with rival CTV systems proposed by Germany and France in addition to the existing American system, the CCIR decided to permit three incompatible standards to co-exist. The consequence of this decision was that European countries, and especially France, whose system was not widely adopted, were able to establish non-tariff barriers which gave their domestic industries a measure of protection against Japanese competition. As these patents began to expire in the early 1980s, the Europeans were faced with a proposal from the Japanese for an entirely new transmission/reception technology (MUSE) for high-definition television (HDTV) to be accepted as the single world standard.

Fearing that, if accepted, this proposal would ensure the permanent hegemony of the Japanese electronics industry, the European firms assisted by the European Commission succeeded in 1986 in persuading their governments to vote for a postponement of the CCIR decision until 1990. They immediately embarked on a crash programme to develop a European HDTV system, which became the Eureka EU95 project involving subsidies from national governments and the Commission. The consortium, led by Philips, comprises the leading European-owned players in the European industry, backed by the national broadcasting organisations in the major countries. The

European system, which was successfully demonstrated in 1988, is based on the MAC standard and is intended eventually to provide for 1,250 lines instead of the present 625-line standard. However, the CCIR meeting in Dusseldorf in 1990 resolved to put off making a decision until 1994, by which time some observers think the analogue MAC system might be becoming obsolescent in the light of developments in all-digital television technology.

There remains, thus, considerable uncertainty as to whether the EUREKA, or indeed the Japanese MUSE, route to HDTV will be followed. The original goal of a single world standard for HDTV seems unattainable given competing European and Japanese proposals, especially since a ruling from the United States Federal Communications Commission in 1988 that HDTV systems should be compatible with the existing television system (NTSC) in the United States. Since the CCIR decision to postpone a decision in 1986 there has been a flurry of activity in the United States to develop alternative routes to high-definition images, which involve digital processing techniques to enhance the quality of existing television signals. Despite their formal commitment to the EUREKA project and MAC-based HDTV development, the European firms are also heavily involved in alternative routes. Both Thomson's recently acquired American subsidiary and North American Philips have been developing improved definition technologies using semiconductor frame stores, and much of this work is also relevant to Europe.

The advantage of improved definition television (IDTV) over HDTV is that it does not at least initially require the replacement of existing studio and transmission equipment. Despite the claims made by the EUREKA consortium of an 'evolutionary' system compared to the 'revolutionary' Japanese system, any programme material transmitted in MAC or High Definition MAC would require the consumer to invest in a new set or a MAC–PAL converter to receive the pictures at all, let alone in high-definition format. With IDTV, consumers making the investment in a new set receive improved pictures; the remainder of consumers are unaffected. The great unknown in the HDTV debate is whether consumers will be willing to spend heavily to improve the image quality of their television – rather than, for example, spending on additional programmes delivered by satellite. IDTV may prove attractive as an intermediate technology, especially with an oblong 'letterbox' picture, while an all-digital HDTV system is developed.

The HD–MAC system presupposed the diffusion of MAC TV technology through satellite broadcasting, but in November 1990 the two direct broadcast channels in Britain, BSB and Sky, merged, which clouded the future of the MAC system. Initially Philips and Thomson denied this would adversely affect HD-MAC.

Despite the brave words of the EUREKA proponents, the HDTV coalition in Europe may be much more fragile than it appears, and collaboration may dissolve into fierce competition if it appears that ranks might be broken. For this reason the experiments between Grundig and the German broadcasting organisations in transmitting enhanced definition pictures in PAL were played down by both parties. If, for example, Thomson achieved success with its own IDTV system (Advanced Compatible Television – ACTV) in the United States, it would be tempted to transfer this technology to its existing factories in Europe to steal a march on Philips. This would not necessarily be a bad thing if the consumer interest is taken into account in the calculation of the costs and benefits of collaboration. There is a lot to be lost if such collaboration locks Europe into producing a technology whose products fail to find a market. In this respect on the worst scenario the EUREKA HDTV project might be a repeat of the Anglo–French collaborative experience with Concorde.

VI 1992 and the foreclosure of policy options

The 1992 programme and the associated telecommunications and 'audio-visual' strategies of the Community have already foreclosed many of the national policy options once available. The shape of a common post-1992 policy is already beginning to emerge.

Until now, Community policy has relied on a combination of community-level tariffs and national quantitative restrictions and technical barriers to trade. A truly common external trade policy as laid down in the Rome Treaty's Article 113 has become a key goal of the 1992 plan. Pressure will also be put by the competition authorities on Japanese producers to end the price discrimination between markets that they have so far enjoyed in return for agreeing to segment the EC market. Technological change is also going to unify markets. New technology makes the differences between SECAM and PAL far less relevant than before, and it is very clear that new technical developments are going to require pan-European compatibility if not absolute unformity of systems. Even if the pan-European character of the firms were not enough to enforce this, the EC's 1987 Green Paper on telecommunications (CEC, 1987c) makes it very clear that the EC Commission is determined for politico-cultural reasons to create a 'European audiovisual space'. Nor will industrial policy be in future up to individual governments. In a positive and negative way the EC dimension is expanding. The Commission's DG-XIII, responsible for Telecommunications and IT, is now the second biggest spender after agriculture in Brussels. At the same time, DG-IV, responsible for competition policy, is determined to avoid 'unfair' state subsidies to home industries.

In this context one of the demands of the industry is entirely fair. There has been a tendency for EC member states to outbid one another in the campaign for inward investment. This distorts competition and gives unnecessary gifts to the Japanese. The Commission should rigorously enforce the existing competition law in this area, making the rules of the game clear and non-discriminatory.

If the national VERs are indeed removed, as we argue is inevitable, the question is – what will replace them? The EC Commission is at present acting in two somewhat contradictory ways. It is greatly increasing the number of anti-dumping duties that it applies against CE manufacturers, at the same time resolutely denying that it seeks to be protectionist. The Commission has to take initiatives in trade policy in order to establish legitimacy for its claim that Brussels is reponsible under Article 113 of the Rome Treaty, but it is answerable to the Council of Ministers, who in turn are pressed by the industry. The vigorous but selective use of anti-dumping measures is part of an oligopolistic game in which EC is signalling that excessively 'disruptive' price competition will provoke retaliation but that gentlemanly competition is to be tolerated. This means a certain privatisation of trade policy, as producers in Europe and Japan negotiate with each other to create industry-to-industry understandings, though these risk falling foul of the anti-cartel rules of Article 85 of the Rome Treaty.

For political reasons, the Commission may tacitly tolerate agreements on total sales between the European and Japanese producers, as long as the integrity of the internal market is respected. The gradual decline of Japanese imports of finished goods into the Community is more than a mechanical response to existing import controls and even to threat of new ones: it is also a gesture of cooperation which offers local production as a kind of hostage to fortune and a promise of good conduct, exemplified by the Japanese producers' tacit approval of measures against the Koreans. European production does not insulate Japan against EC pressures, but it indicates willingess to see a common interest: in the kind of trade war in which drastic mandatory restrictions could be put on imports, discrimination could easily be introduced against inward investments, as indeed it has been.

The foregoing argument still leaves open a number of options for the Community in this industry. It can be more or less welcoming to inward investment, more or less generous in financial terms, more or less interventionist, more or less discriminatory, selective or discretionary in its aid it can bias its policy towards competition or collusion and it can seek to impose its own priorities or accept those of the firms. Chapter 23 of this book returns to these issues in a wider context.

11 Standardisation in the European Community: the example of ICT

Marta Rosario and Susanne Schmidt

I Introduction

The major reason for the difficulty in achieving agreement on standards is, of course, the existence of incompatible vested interests in diverse existing products and practices – no effort at standardisation starts with a *tabula rasa*. While this is so for all EC regulations (see for example, Chapters 18 and 20) the discussion here exemplifies the general issue by concentrating on Information and Communication Technology (ICT).

As all the previous chapters in Part II have indicated, standardisation has become a central issue for the achievement of compatibility among all the component parts of an ICT system. The coordination or imposition of standards for ICT equipment and services is a major point of discussion within firms, industrial sectors, national, regional and international, private and public committees and other institutions. Compatibility standards are a decisive factor in creating a competitive environment for trade as they make possible the free flow of information and interchange of equipment within and between countries and industrial sectors; decisive but immensely difficult to achieve.

II Standardisation in ICT

Three main factors have recently increased awareness of the importance of technical compatibility standards. The first is the growing convergence of information and communication technologies and their increasing importance for the global economy. The second is the pressure from users of ICT to have a more stable environment in which to make their technology acquisition decisions, given the increasing investment costs. The third is the changing market strategy of suppliers of ICT equipment and services based on both globalisation – to recover R & D costs from a wider market – and specialisation, caused by the diversity of today's information industry in which nobody can cover the whole field (Ergas, 1986).

The increasing digitalisation of telecommunications and the networking of computer systems accounts for the necessity of compatibility of interfaces between them. But as already pointed out in Chapter 8 these technologies have different 'cultures' in relation to standardisation. In the communications field, standardisation was essential to the very existence of systems which had to intercommunicate; standards were developed in regional and international committees or established by the power of monopolies, both private (such as AT&T in the United States) and public (the European PTTs). The computer field, being highly competitive, tried to avoid standardisation; large computer manufacturers used incompatibility as a way to lock-in their customers to a proprietary system in order to maximise profits.

In an economic sense, standards have two key functions: lowering transaction costs (for example, by making it simpler for all the parties to recognise what is being dealt with) and exploiting the external economies of the network (OECD, 1988). The latter can come through a reduction of variety of products and systems, as in the early automobile industry, with a tendency towards convergence of standards. The reduction of variety lowers the costs of information acquisition and yields resource savings in production through economies of scale. The need for interface compatibility may lead on the one hand to concentration of production, if considerations of economies of scale predominate, or, at the other extreme, to vertical disintegration as products become standardised and modularity is achieved.

Another way of achieving compatibility and to avoid lock-in is the development of 'gateway' devices. These permit the interconnection of system components that would otherwise remain isolated. A good example was the development of devices for AC and DC systems in the early days of electricity generation and distribution (David and Bunn, 1988; Arthur, 1988).

Users' evaluations are among the main reasons for firms to look for compatibility. However, the incentives for firms to produce compatible products or services vary. Firms with good reputations, large existing networks and proprietary standards will tend to oppose compatibility in order to maintain high barriers to entry into their market position. Firms with smaller networks, weaker reputations and non-proprietary standards, will in contrast favour compatibility even perhaps where the social costs of compatibility outweigh the benefits (Katz and Shapiro, 1985).

Hence, a firm's incentive to look for compatibility depends on its market position and on the type of its products: proprietary or licensed, basic or complementary. A proprietary system allows suppliers to control the kind of equipment that can be attached to their network and, if there are no rivals, to charge higher prices. Besides, a proprietary system means many years of investment in development which firms are unwilling to give up. Change to compatibility involves costs to producers. This could include the costs of designing a compatible product, the costs of selecting a standard (if any), the costs of testing and monitoring applications and the costs of introducing a new, compatible product. When the costs fall more heavily on some firms than on others, there is a free-rider problem that tends to discourage the firms from attempting to achieve compatibility.

The international/regional agencies

In the field of ICT systems interconnection involves three types of technologies – telecommunications, information processing and electronics – so there are three international organisations involved, each with a defined regulatory framework.

The three institutions are the ITU-CCITT, the ISO and the IEC – International Electrotechnical Commission. The International Telecommunications Union (ITU) is a United Nations treaty organisation and the leading organisation in the telecommunications field. Its telecommunications standards-setting work is undertaken by CCITT members which comprise governments, telephone administrations (public and private) or Recognised Private Operating Agencies but no users. This structure is in transition; for example, as a result of a recent modification in CCITT's procedures: standards are now approved after a 70 per cent consensus, rather than by unanimity as previously, and the time needed for approval has been cut from four years to one.

The International Standards Organisation (ISO) is not a UN agency but undertakes international standardisation in all fields other than telecommunications and electrotechnology. Its main activity in ICT is the OSI model (a seven-layer reference model that covers all aspects of interconnecting different computers, from physical attachment of communication lines to the way in which different applications and common services link together in a network).

The IEC is concerned with standardisation over the whole range of electrical and electronics engineering. It is based on the voluntary participation of scientists and engineers from various countries, constituted as national committees. The increasing overlap of the areas of expertise of the three organisations has fostered cooperation among them to ensure that differing standards are not forthcoming.

At the regional level there are European institutions such as the European Commissions' standardisation bodies, CEN and CENELEC – the European Committee for Standardisation and the European Committee for Electrotechnical Standards – whose voting members are the national standards bodies of EC and EFTA countries; and ETSI – the European Telecommunications Standards Institute. ETSI was established in 1988 by CEPT – the Conference of European Postal and Telecommunication Administrations. It was set up as a private organisation with no formal connection to the CEPT and its activities; resources relating to technical specifications were turned over to ETSI. Its members are telecommunications administrations, public network operators, manufacturers, users including private service providers, research bodies as well as national and European organisations concerned with telecommunications. Other European and non-European organisations can become observers.

Among the private non-governmental organisations the main European ones are ECMA, SPAG and EUCATEL. ECMA – the European Computer Manufacturers Association – has members from some 21 countries, including large American companies based in Europe; their main concern is standardisation in computer hardware. SPAG – the Standards Promotion and Application Group – is an organisation of ICT manufacturers and telecom operators. Its main purpose is to enhance existing but incomplete established standards from major organisations. EUCATEL – the European Conference of Associations of Telecommunications Industries – is not a standards-setting body, but a European trade association of telecommunication equipment manufacturers who advise CEPT in specific matters.

The rising importance of ETSI, TC1 (United States) and TTC (Japan) as regional standards organisations has been seen for some time as a challenge to the predominance of the CCITT in global standards setting. Therefore the first Interregional Telecommunications Standards Conference was held in February 1990 in Fredericksburg (Virginia, USA) where the basis for a new framework of cooperation between the world's leading telecom standard-makers was laid.

Three other important international (private) organisations for standards are X-Open, the Open Software Foundation (OSF) and UNIX International (UI). These are involved in the so-called 'UNIX War', the dispute between the IBM-led OSF and the AT&T-led UNIX International over which version of AT&T's operating system (UNIX) will be the standard in the computer market. (The UNIX case is analysed later).

III European telecommunications policy and standards

The official awakening of the Commission of the European Communities (CEC) to standardisation matters began in 1977, with a directive setting coordination procedures for the award of public supply contracts (CEC DG XIII, May 1988). This directive introduced standards into the public procurement activities of the member states – but not any ICT-related standard because it was then generally accepted that the Treaty of Rome provided no direct competence for action by the Commission in the telecom sector (Schneider and Werle, 1988; Schmidt, 1989).

Telecommunication standardisation began to be tackled in 1979 with the launching of a programme for 1979–83 in the field of data processing (later embracing ICT as a whole). After 1983, the EC began to develop a comprehensive policy framework for ICT. In that year another directive was issued laying down a procedure for the provision of information in the field of technical standards and regulations. In 1984 the first Action Programme in telecommunications was adopted, aimed at the creation of a common European telecommunications policy. In the same year a Council Recommendation was issued, concerning the implementation of harmonisation in the field of telecoms. In 1985 a Council Resolution set a new approach to technical harmonisation and standards which meant the signing of cooperation agreements between the CEPT and CEN/CENELEC to develop common European telecom standards in tune with international rules, incorporating national standards. Member states were required to publish their national technical regulations and standards to avoid investment in incompatible systems.

In 1986 Council resolutions and recommendations were issued regarding the mutual recognition of type approval for telecoms terminal equipment, the adoption of common technical specifications in relation to direct satellite television broadcasting and the coordinated introduction of the Integrated Services Digital Network (ISDN) in the EC, with a common approach on standards, tariffs and services.

In 1987 new decisions and recommendations were issued for standardisation in information technology and telecommunications; for the introduction of public pan-European digital cellular mobile communications and the introduction of a Community communications network programme on Trade Electronic Data Interchange Systems (Tedis); and for developing the electronic exchange of commercial and administrative data, facilitating European firms' activities (CEC, 1988d).

The Green Paper approach

As Part II has demonstrated Europe lags well behind Japan and the United States in data processing and microelectronics. The CEC has tended to see telecommunications and broadcasting equipment as Europe's last chance for catching up in advanced electronics this century – and this has been the rationale behind the development of the CEC's telecoms policy. As Schmidt (1989) puts it, this concern is mirrored in the twofold drive of European policy endeavours: the introduction of R & D programmes to coordinate European Technology research in a more 'MITI-like' fashion in programmes such as the ESPRIT, RACE, STAR etc; and the attempts to harmonise the European market in order to come closer to the economies of scale achieved in the American market.

The CEC's efforts to achieve full regulatory competence in telecommunications culminated in the 'Green Paper' of 1987. In it a general framework of a common European telecommunications policy was outlined with the delineation of the market segments that should be liberalised. The Green Paper's major aim was to harmonise the telecommunications systems of the Community countries as an important precondition for the achievement of the single market. This, it is assumed, will be achieved through the liberalisation of the market and privatisation (undercutting political obligations of networks, such as the provision of universal service) as happened in the United States, Japan and in the United Kingdom at the beginning of the 1980s (for details of these processes, see Hills (1986, 1989); Davidson (1987) and Chapter 8).

Several obstacles to the implementation of the Green Paper have arisen which are rooted in the reluctance of the member states to yield important sovereign rights and in doubts about the appropriateness of the proposed measures. The directive liberalising terminal equipment, issued in 1988, has been challenged at the European Court of Justice by some member states over the use of Article 90 of the Treaty of Rome (which allows the Commission to issue laws to curb public monopolies without consultation with EC governments). Because of the great heterogeneity of national telecoms systems and domestic industries, it is doubtful if a single, common strategy is best suited. There is fear that the richer countries might benefit disproportionately while the poorer countries lose technological capacities and surrender competence over their telecoms development. Furthermore, the CEC's perspective was for years dominated by considerations of trade in equipment. If a common strategy for the introduction of ISDN is initially related to the achievable economies of scale for equipment, the relative neglect of the services side could have grave consequences. The abolition of trade barriers in the Single Market, finally, poses the question whether European firms will be better placed to exploit the situation to broaden their market shares.

The CEC has also been involved in serious disputes with those who see the benefits of liberalisation stifled by too far-reaching re-regulation. Corporate users, private-sector providers and computer manufacturers see much of the CEC's post-Green Paper policy leading to a situation where detailed technical specifications enable European-wide interworking in a style similar to the former PTT monopolies.

Standards in the Green Paper

In the ten propositions of the Green Paper, three directly tackle standards – items D, E and G. Item D sets 'strict requirements regarding standards for the network infrastructure and services provided by the Telecommunications Administrations or service providers of comparable importance'; Item E asks for clear definition by Community directive 'of general requirements imposed by Telecommunications Administrations on providers of competitive services for the use of the network, including definitions regarding network infrastructure provision (...) Consensus must be achieved on standards, frequencies and tariff principles... Details of this Directive on Open Network Provision (ONP) should be prepared in consultation with the member states, the Telecommunications Administrations and the other parties concerned...'; item G determines the 'separation of regulatory and operational activities of Telecommunications Administrations. Regulatory activities concern in particular licensing, control of type approval and interface specifications, allocation of frequencies, and general surveillance of network usage conditions' (CEC 1987c).

Item D is self-evident in its purposes. Item E demands the establishment of ONP, a matter of controversy between the CEC and private network administrations. The main rationale behind the ONP concept is the necessity for rules that allow private service providers in a world of restricted network ownership to offer their services on an equal footing with the PTTs. As such it is an enabling mechanism to achieve the liberalised services market proposed by the Green Paper; it consists of three building blocks: technical interfaces (to be developed by ETSI wherever existing specifications are not sufficient), usage conditions and tariff principles. While all three areas are necessary to rule out discrimination against competitors by the PTTs, serious problems have arisen due to the heterogeneity of telecoms networks in the 12 member states. Rather than driving liberalisation, ONP has thus been in danger of becoming a re-regulatory supplement to the directive liberalising services.

Beyond the initial framework directive establishing the principles of the concept, directives were at first planned for leased lines, public data networks and ISDN which gave rise to fears of too far-reaching harmonisation. Large corporate users and private service providers were concerned that an extension of ONP towards liberalised services could involve their compliance with ONP rules, which they found unacceptable. From the technical side, requirements to meet specific standards and to provide for interfaces with public networks were seen to lead to a considerable limitation of possible service offerings and thus to hamper innovation. At the end of 1989 it was agreed to concentrate work on ONP for restricted telephone and leased lines and to make ONP voluntary rather than mandatory for private service providers.

The Green Paper's item G is partially responsible for the creation of ETSI. The proposal reflects two related policies: complete separation of operational and regulatory roles and regulation by reference to standards developed in institutions in which affected interests can be represented.

However, the rationalisation of the European standarisation process did not happen with the creation of ETSI even with the EC's backing. To handle the 'grey area' between telecommunications and computers a coordinating committee was set up between CEPT and CEN/CENELEC called ITSTC – Information Technology Steering Committee, in which the organisations work together, trading information and holding joint meetings. This does not prevent overlaps in areas such as ISDN. To avoid such problems, ETSI and CEN/CENELEC have been discussing the formation of a formal common standards body to oversee the standards organisations, since the establishment of only a single body to manage standards in Europe does not seem feasible.

ETSI is also likely to be obstructed by its own structure. This was made clear when the institute took a step towards steering the booming European mobile communications market, at the beginning of 1990. In a report, ETSI's strategic review committee recommended that the institution's role should be expanded beyond standards setting to foster innovation and steer marketing and industrial and regulatory policy in mobile communications, given the rapid development of this technology. The ETSI's technical assembly, dominated by Europe's Telecommunications Administrations, rejected the strategic review committee's recommendation that a user-applications inquiry on mobile telecoms should be undertaken. In the view of the technical assembly, the institute should restrict itself to the definition of technical standards (*Communications Week International,* 1990:2). It seems that ETSI is also subject to capture by dominant groups, and the institute is facing the same problem that standards-setting organisations are facing world-wide: overlap of tasks and lack of coordination.

IV Three case studies

The Open Systems Movement

As has been shown earlier in Part II, the Open Systems Movement was gathering strength in the 1980s. It was intended to enable users to run the same program on any manufacturer's computer system which, in turn, communicates freely with any other and with peripheral interfaces. To achieve such a goal users might implement a set of protocols called Open Systems Interconnection (OSI), which covers all aspects of interconnection of different computers and UNIX software, an operating system which can run on a variety of hardware. A truly open system would allow the full integration of tasks in an office or factory and the full development of network-based services such as Electronic Mail (E-Mail) or Electronic Data Interchange (EDI).

IBM had developed its own set of networking standards, the Systems Network Architecture (SNA), which provides compatibility for its own full range of products. First, it sought to make this standard industry-wide by allowing it to remain an open proprietary standard. The other major manufacturers, fearing the increased market power that SNA could give to IBM, united behind the OSI project, as a means to preserve their autonomy.

IBM dragged its feet as much as it could to prevent OSI from taking off, but following the demand pull for open systems, the company joined the OSI project in late 1988. The computer giant uses its proprietary standards within its networks and OSI for linking with the outside, non-SNA world. In 1990 IBM introduced several products that will allow users to mix its products with other vendors' equipment more easily, by supporting OSI protocols.

Other big system suppliers also have mixed feelings about OSI; although it might help to avoid IBM's dominance, it is a double-edged blade, since users can purchase particular machines or parts in this information system from another supplier. The suppliers' commitment to open systems thus varies according to their market position. A proprietary system allows suppliers to influence what kind of equipment can be attached to their networks and, if there are no rivals, to charge higher prices for certain parts of it. Large vendors may have many years of investment in development of their proprietary systems which will be at risk with a fully open system.

It is, however, increasingly difficult to ignore the trend, since big private and public users are pushing hard for OSI. The public nature of OSI protcols make such support easy, but the use of proprietary protocols still overshadows OSI to some degree as was shown in Chapters 6 and 7.

The EC is encouraging its member states to use their procurement power. In February 1990 the Commission signed a contract to create common purchasing guidelines for European governments and asked French, German and British standards bodies to develop an European Procurement Handbook for Open Systems (Ephos) as a technical guide based on OSI standards. Ephos is important because government spending accounts for 20 per cent or more of all IT spending in Europe; it can therefore become a model for private companies in their future purchases.

The UNIX war

UNIX is a potentially universal operating system with enough portability to run on a wide range of hardware. In 1969 it began to be developed in AT&T's Bell Labs initially for internal software development purposes. UNIX remained mainly in academic research for a decade because until 1984, when AT&T was divested, it was not allowed to enter the computer market (other factors were its user-unfriendliness and lack of business-orientated applications). In the early days UNIX was licensed under nominal fees to universities, but in 1980 AT&T introduced a commercial version into the market. The licensing policy created many different versions of UNIX, making its portability difficult.

In 1984 a group of European computer manufacturers joined together with the aim of attacking American vendors' (IBM) dominance by stimulating the market for open systems. The pioneers were initially called the Bison Group – Bull, ICL, Siemens, Olivetti and Nixdorf. Later DEC, Hewlett Packard and Philips joined the group, which had changed its name to X-Open. X-Open does not set standards, but selects and adopts either *de jure* or *de facto* standards for the CAE – Common Application Environment (a core set of portable guidelines). UNIX System 5 was then chosen as the operating system upon which portability would be achieved.

The trouble began when AT&T decided to invest in UNIX as a source of revenues. In October 1987 the company announced plans to develop a high-performance operating system based on UNIX, in a joint venture with Sun Microsystems. In January 1988 AT&T announced plans to take an equity stake in Sun, a highly successful UNIX licensee and a leading maker in the emerging workstation market. This led to an outcry from vendors who feared that AT&T would be able to leverage its ownership of UNIX into advantages for itself and Sun and from rivals of Sun who feared that the company would gain a competitive advantage as a result of their unique position (Saloner, 1989).

X-Open ran into trouble because its members had tied themselves to AT&T's strategy for managing its proprietary standard. AT&T had the option of either progressively restricting the use of UNIX in order to draw maximum profits from licensed sales or promoting the expansion of its market by leaving the standard open. It applied both strategies.

The members of X-Open reacted defensively and in May 1988 most of them (DEC, HP, Apollo, Honeywell-Bull, Siemens and Nixdorf) joined with IBM to found OSF – Open Software Foundation – to create and disseminate a new standard derived from UNIX but independent from AT&T, based on IBM's version of UNIX, called AIX. AT&T's reaction was (in October 1988) to form an alliance with other computer firms (including Amdahl, Control Data, Fujitsu, Gould, Prime, Sun, NCR, Olivetti and Unisys) in a group now called UI – UNIX International. At stake is the American market for computers that now run UNIX, a market that is expected to grow to US$25 billion a year by the early 1990s, according to industry estimates (Financial Times, 1989:26).

X-Open later took on the role of a bridge between OSF and UI, and in May 1989 both OSF and UI joined X-Open. Besides UI and OSF, X-Open supporters include a users' council with the likes of British Airways and General Motors; independent software suppliers such as Lotus Development and Microsoft; and governments of many different countries. The issue in the UNIX war is how best to consolidate the different versions into a single standard. But the prospects for full reconciliation between the groups are poor. Some analysts claim that the disadvantages of having two versions of UNIX are not too serious as users will not see much difference between them. Besides, CAE and other instruments stand as possible bridges between rival or competitive UNIXes.

At present, UI members sell more UNIX systems than their competitors, as they had a first-move advantage. Besides, the full version of OSF's UNIX (called OSFl) had not reached the market at the time of writing, as OSF decided to change from AIX to a version based on Carnegie-Mellon University's Mach software. Some industry analysts argue that this is IBM's deliberate strategy because, as Mach is in an earlier stage of development than AIX, OSFl will take even longer to reach the market, giving IBM more time to sell its proprietary products. However, it runs the risk of being deserted by other members of OSF who fear being too late to market, as for instance Siemens has now done.

The EC first committed itself to using UNIX internally in 1984, but recently a major boost was given towards open systems: in May 1990 the Commission announced its intention to move to specific X-Open standards across all systems – mainframes, departmental systems and workstations.

Electronic Data Interchange (EDI)

The final example concerns network-based services, Value-Added Network Services (VANS) or New Telematic Services (see Chapter 9), that is, the exchange of information between computer systems through telecommunications networks. Such services are heavily dependent on the creation of widely accepted compatibility standards (and, in Europe, also on the harmonisation of the many different regulations of the telecommunications administrations concerning tariffs and the provision of leased lines).

One such service is EDI – Electronic Data Interchange. EDI is the transfer of commercial documents (purchase and payment orders, delivery notes, invoices, debt and credit advices, insurance notes, customs declarations etc) as data through computer networks. Such delivery of data can be done via public telephone networks, via one participant's private telecommunication network or through a network provided by a value-added network dealer (a clearing-house service). A VAN eliminates the need for compatible equipment because it provides any necessary translation of messages to ensure that the second machine can receive the data. The necessity of formatted messages is clear because VANs only accept messages which are designed for EDI (Rosario, 1989:65).

An estimate by Euromatica (1986) puts the cost of paperwork supporting international trade at between US$260bn and 300bn. The electronic transfer of commercial and administrative documents leads to the reduction of paperwork, to a faster and more accurate information flow and closer links with suppliers. In short, EDI raises productivity, profitability and competitive advantage until rivals also link up. But EDI requires considerable initial investment, especially for small firms. Suppliers who do not tie in risk losing their customers, since large firms increasingly demand to do business via EDI.

EDI has been used for more than 15 years, mainly in the United States and the United Kingdom, but is now available in most countries. With the increasing use of EDI and the necessity of firms to trade globally, it has been necessary to develop generally accepted standards for EDI and to harmonise the existing standards. This challenge was met by the United Nations Economic Commission for Europe (UN/ECE), which has developed Edifact (Electronic Data Interchange for Administration, Commerce and Transport). Edifact is arriving now in the market-place, sponsored by users and producers of ICT, by industrial organisations and international bodies. It has to face the challenge of replacing already established sectoral and proprietary standards, which have the advantage of addressing users' particular needs but lack the desirable scale.

In 1985 there were basically two variable format EDI standards in the world: the 'Guidelines for Trade Data Interchange' (GTDI), developed by the UN/ECE, whose members are countries and organisations from North

America, Western and Eastern Europe; and 'X-12', developed by the Transportation Data Coordinating Committee (TDCC/EDIA) and the ANSI Accredited Standards Committee (ANSI/ASC). Edifact has features of both.

The establishment of Edifact as the EDI format standard depends upon the acceptance by old and new users of its features. It is heavily backed by the UN/ECE Working Party 4 which is responsible for the project. WP 4 has the participation of 29 member states, 11 observers, and 23 international organisations such as GATT, ISO, ITU, CEC and SWIFT. Linked with WP 4 are the UN/Edifact Rapporteurs. A trend towards the adoption of Edifact by VANS suppliers and sectoral industry projects is undeniable. Edifact is also being used to provide a gateway between different EDI services networks.

Another important push for EDI/Edifact is the European Commission's Tedis Programme (Trade Electronic Data Interchange Systems). Tedis has as its aim the coordination and promotion of EDI within Europe. The two-year programme was launched in January 1988 (a second phase is now under way). The coordination activities in Tedis embrace the national organisations concerned with the facilitation and simplification of trade procedures and those European sector-specific associations which have a particular interest in EDI. It has gained these associations' support for Edifact. As EDI places new demands on existing networks, Tedis supports the EC's initiatives in the field of open network provision and attempts to ensure that the needs of the EDI community are met within the framework of a European telecommunications policy (Wakeling, 1989).

These developments have placed the EDI industry in a classic standards impasse. In this case, innovative industrial sectors, such as automotive and retail, have defined proprietary standards and gained competitive advantage by doing so but are now trapped by their 'communities of interest': if these players want to trade internationally, they will have to adopt international standards which may differ from those established and working in their own sector. This adoption will require complex and expensive modifications (gateways) to existing systems; many players are reluctant to undertake this investment. Late arrivals benefit since new entrants – countries and industrial sectors – are opting into EDI based on international Edifact standards. The trend seems to be a convergence toward internationally agreed standards.

V Conclusions

The convergence of computers and telecommunications and the pressure of users of these technologies have brought demands for greater compatibility of ICT equipment and services. As a consequence, standardisation in ICT has become a central policy issue in many governments' and firms' agenda. This has caused a global trend towards the creation of a more rapid, efficient and reliable standards-setting process in regional and international ICT-related organisations.

In Europe the Commission of the European Communities awoke to ICT standardisation issues in the beginning of the 1980s; by the middle of the decade a comprehensive policy towards standardisation began to be developed with the publication of the Single European Act, aimed at removing trade barriers between the EC's 12 member states. This policy led to the Green Paper on telecommunications and many collaborative research and development programmes, which have fostered the harmonisation of technical standards.

Since then the CEC has used its own procurement power to promote ICT technical compatibility standards in many cases, such as OSI and UNIX-based specifications, and has backed internationally accepted standards such as Edifact, having urged its member states to do so as well. In the long run this will benefit consumers everywhere. However, the promotion of standardisation is a delicate policy issue given the characteristics of the information and communication technologies (increasing returns to adoption, network externalities, different cultures in computer and telecoms etc). The sheer diversity of telecommunications policies and the uneven development of data communications across Europe make telecommunication standardisation an even more difficult task. This diversity of policies and interests in ICT has to be acknowledged if the CEC is to succeed in setting a new, more liberal telecommunications policy framework for which the development of interface compatibility standards is essential.

The reassurance of the European commitment to international standards and conformance testing organisations is very important. Technical harmonisation implies delegation of responsibility and diffusion of information between the actors involved. For this to be effective the EC should promote coordination of the activities between European standards-setting bodies at the regional and international level so as to avoid duplication of efforts and the issuing of incompatible standards.

Part III
Other Industries and Technologies

Preface

We turn now in Part III to consider other major sectors of the economy – the chemical and drug industries, the engineering industries, the automobile industry and the construction industry – in which European competitive performance has been very much stronger than in electronics. But it must always be remembered that, as we argued in Chapter I, ICT is not just an industry or a set of industries but a technology which pervades all sectors of the economy and all functions within enterprises. This pervasive influence is most clearly apparent in engineering and vehicles where the very expression 'mechatronics' indicates the type of technical change which is now characteristic of these very large and important industries.

To keep pace with Japanese firms in robotics and flexible manufacturing systems will be by no means an easy task for EC industry, as Chapters 15 and 16 make absolutely clear. Japanese success in the automobile industry has been spectacular. In his chapter, Andrew Graves demonstrates that the strength of R & D and design activities in Japanese auto firms and their capacity to reduce lead times and introduce new features in their models, such as new materials and electronic control systems, will present an even stronger challenge in the 1990s.

Even in such areas of strong European comparative advantage as drugs and chemicals, Japanese competition (and new sources of supply in petrochemicals), should not be underestimated. The demand for new speciality products is closely related to the growth of electronics as well as to aerospace. Intimate technical cooperation between users and producers is a feature of these newer materials, once again conferring some transfer of competitive advantage to United States and Japan. Information technology is now critical in materials R & D, in product and process design, in process control and in marketing throughout the chemical industry. Finally, in a traditional area such as construction, which many might have supposed to be still remote from the influence of information technology, David Gann shows in Chapter 14 that 'intelligent buildings' have left the realm of space fiction and are at the cutting edge of international competition in what has become a global contracting industry. Here too, the use of robotics and the scale of R & D in the leading Japanese firms must give cause for reflection in Europe.

12 The chemical industry

P. Lesley Cook and Margaret Sharp

I Introduction

As already indicated in Chapters 2 and 3, the chemical industry is a sector where the European performance is strong by comparison with that of both the United States and Japan. Table 12.1 shows that the trade balance of Western Europe was greater than for any other trading bloc, and its market larger.

The chemical industry is notoriously difficult to define. At one extreme are product areas such as salt or titanium dioxide akin to mineral extraction where selling prices are low; at the other the production of highly sophisticated chemicals used as ingredients for pharmaceuticals or agrochemicals, requiring many manufacturing steps and selling for thousands of dollars a gramme. A good part of the industry provides bulk chemicals and intermediates for other parts of the industry, creating a clear distinction between the 'upstream' industry and the 'downstream' sector, making products sold to consumers and/or other industrial sectors. Table 12.2 illustrates the breakdown of output and markets for the Western European industry. A broad distinction can be made between the (upstream) 'bulk' or 'commodity' chemicals – petrochemicals, plastics and inorganics – and the (downstream) speciality chemicals, pharmaceuticals and agrochemicals. In this chapter we discuss the bulk-chemical sectors in Section II and speciality chemicals in Section III. Pharmaceuticals and developments in biotechnology are discussed in the next chapter.

Another important feature of the industry is the extent to which it is already a 'global' industry. Among the hundreds of firms in the industry there are some 20 very large multinationals. They account for about a quarter of the total sales (Table 12.3). Of these 20, 12 are European, (including three Swiss, which however have many activities in EC countries). The leading American and European firms have in the main held their position – the pre-war giants, BASF, Hoechst, Bayer (together forming IG Farben before 1945), ICI and Ciba Geigy still dominate the industry. They have, however, been joined in the post-war period by some of the major international oil companies, a number of government-dominated firms (such as Elf Aquitaine in France and Norsk Hydro in Norway) and recently by the leading Japanese companies.

The position of the largest firms has been reinforced recently by a wave of mergers. Sometimes they involve whole companies, sometimes just parts of companies. Table 12.4 sets out the latest data on acquisitions and mergers world-wide and illustrates the continuing importance of Western Europe; 41 per cent of acquisitions originate from firms in Western Europe (whereas only 31 per cent of acquisitions were in Western Europe). This gives some indi-

Table 12.1 World chemicals market, 1989 ($bn)

Region	Output	Exports	Imports	Home demand	Net trade
W Europe[1]	340	52	25	313	+27
N America[2]	275	33	22	264	+11
Japan	190	20	15	185	+5
E Europe	170	15	19	174	–4
Central/S America	54	6	15	63	–9
Far East	39	14	32	57	–18
India/Pakistan	25	1	5	29	–4
Africa	15	2	9	22	–7
Middle East	12	5	10	17	–5
Australasia	10	1	5	14	–4
Total	1,130	149	157	1,138	

[1] Excludes W. European intra-trade
[2] Excludes N. American intra-trade

Source: United Kingdom Chemical Industries Association, quoted in Allen, 1990

Table 12.2 Breakdown of output (by value) and markets for the Western European chemical industry (%)

Breakdown of output		End-use markets	
Organics	19	Construction	5.7
Inorganics	8	Services (hospitals, research centres, transplants, etc)	19.2
Fibres	4		
Fertilisers	7		
Plastics	13	Consumer goods	27.4
Pharmaceuticals	15	Agriculture	10.3
Dyes	2	Food industry	3.1
Paints	4	Metal industry	6.5
Detergents	4	Electrical/electronic	3.8
Perfumes and cosmetics	5	Mechanical engineering	3.8
Others	19	Automotive	3.6
		Textiles	6.6
		Paper/printing	3.0
		Other	7.5
Total	100	Total	100

Source: European Chemical Industry Association, quoted in the *Financial Times*, 13 July 1989 (Supplement)

cation of the degree to which Western European firms were acquiring firms in North America – an issue which is beginning to concern American industrial experts. In 1988 the most active firms in terms of acquisitions were all European – Elf Aquitaine, Rhône Poulenc and ICI each participated in 13 acquisitions in 1988. As Table 12.4c indicates, the most active sectors were speciality chemicals, pharmaceuticals (see Chapter 13) and health care.

Table 12.3 The World's largest chemical companies

	Rank		Sales Revenue ($m)	
	1984	1988	1984	1988
Europe				
BASF (Germany)	2	2	14,230	24,329
Bayer (Germany)	3	3	13,873	23,011
Hoechst (Germany)	4	4	13,365	22,651
ICI (United Kingdom)	5	5	13,240	20,817
Ciba-Geigy (Switzerland)	10	7	6,835	12,027
Rhône-Poulenc (France)	13	8	5,392	10,956
Montedison (Italy)	8	9	7,080	9,757
Norsk Hydro (Norway)	15	10	4,370	9,202
Akzo (Netherlands)	14	11	4,719	8,388
Solvay (Belgium)	18	16	3,590	6,894
DSM (Netherlands)	9	19	7,050	5,120
Atochem (France)	22	21	2,621	4,711
Orkem (France)	21	23	2,950	3,616
United States				
Du Pont[1]	1	1	30,560	32,917
Dow Chemical	6	6	10,679	16,682
Union Carbide	7	12	9,150	8,324
Monsanto	12	13	6,691	8,293
W.R. Grace	11	17	6,730	5,786
American Cyanamid	17	22	3,857	4,592
Japan				
Asahi Chemicals	16	14	4,230	7,672
Sumitomo Chemicals	19	15	3,492	7,002

[1] Includes oil

Source: Anizon, 1990

Here it should be noted that even in these areas where medium-sized and small firms are more prominent, each speciality product market tends to be dominated by very few firms. In the heavy bulk chemical industry there have for a long time been very substantial plant economies of scale for the major products. There are also important economies of scale to be gained in R & D, where the European chemical industry has great strength (Chapter 3).

II Bulk chemicals – organics and inorganics

Before World War I, the organic chemical industry developed on the basis of coal as the main feedstock, and the German industry, having invented the in-house R & D laboratory in the 1870s, led the world in R & D based on coal chemistry. From the 1880s onwards intensive in-house R & D was characteristic of both German and Swiss chemical enterprises and these firms dominated European markets even after World War I, although they were

Table 12.4 Acquisitions and mergers in world chemical industry, 1988

A Acquisitions and mergers by region

	No. of transactions	% of total
USA	723	56
W. Europe	404	31
Japan	24	2
Other	132	10
Total	1,283	100

B Acquisitions and mergers by location of acquirer

	No. of transactions	% of total
USA	613	48
W. Europe	520	41
Japan	56	4
Other	94	7
Total	1,283	100

C Acquisitions and mergers by type of business

	No. of transactions	% of total
Speciality chemicals	217	17
Pharmaceuticals and health care	211	16
Fabricated plastics and rubber	148	12
Tonnage chemicals	136	11
Consumer products	80	6
Other	491	38
Total	1,283	100

Source: CHEMTRAK, Kline and Co., Rue Froissart 89, 1040 Brussels

then joined by other European firms. Since World War II the organic sector of the industry has shifted from coal to oil as the primary feedstock; petrochemicals and the whole range of new products – plastics, man-made fibres, detergents etc – are based on this feedstock.

The major breakthroughs in hydrocarbon chemistry were made in the 1920s and 1930s. Based on the research on macromolecular structure, innovations came thick and fast – polystyrene, perspex, PVC, polyethylene, synthetic rubbers and nylon and all the man-made fibres. It is notable that all the major innovations were developed in the laboratories of large chemical companies and that most of the companies responsible for them still exist today. I G Farben (which was broken up after 1945 into Bayer, Hoechst and BASF), ICI and Du Pont were the most influential, but Union Carbide, Shell, Standard Oil and Dow all made major contributions to the development of petrochemical processes for the new intermediates and final products. The length of time (and hence the financial strength) required for pilot plant work to develop commercial production processes, and the skill and

Figure 12.1 Petrochemicals – the production pathway from basic feedstock to final product

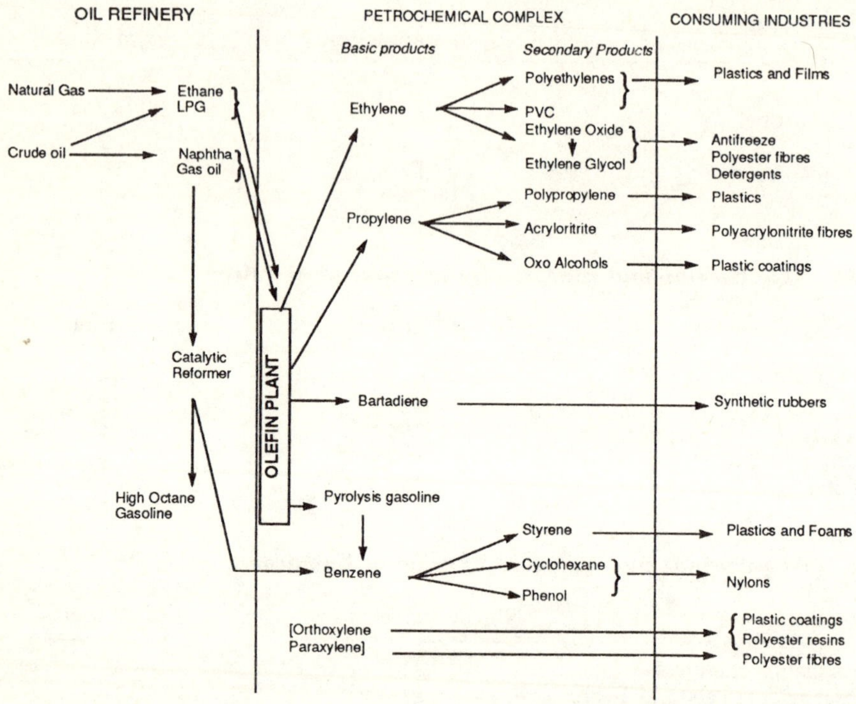

Source: OECD, 1979: xxvii

experience required to handle these difficult technologies, helps to explain the subsequent strong commercial position and the continuing dominance of large firms.

Figure 12.1 illustrates the long chain of processes from the basic petrochemical feedstocks – natural gas and crude oil – through intermediate to final products and shows how the three major intermediates – ethylene, propylene and benzene – are the basis of secondary products which are widely used and produced in bulk. This explains the close relationship between oil refining and the petrochemical sectors of the industry and why the major entrants between 1930 and 1960 were oil refiners. Since there is trade in intermediates, it also shows why the chemical industry is its own best customer and, in turn, why many firms are either wholly or partially vertically integrated. Downstream, where products multiply, firms rely for their raw materials on the large producers of the bulk secondary products – broadly speaking the major chemical and oil companies. Even where they are not vertically integrated, the supplier-customer links are strong.

Among the downstream products developed with petrochemicals were many new materials which in turn led to further innovations and the displacement of natural fibres, natural rubber and metals in an enormous number of products. This explains very rapid growth in petrochemicals in the 1950s and 1960s – annual growth rates of 12–15 per cent were not uncommon.

The continuing dominance of large firms in bulk petrochemicals is partly explained by developments in technology. Process development may have been originally based upon some significant innovation, but subsequently much is incremental resulting from learning and experience in the application of new technologies in plant design and construction. One result of this has been a very great increase in the capital intensity and the minimum efficient size of plants. In the early 1950s, a 30,000 tons/year ethylene plant was considered large; by the mid-1970s, 300,000 ton plants were being built. These vast, capital-intensive plants in turn had a major influence on the economics of the industry. The marginal cost of production was low, and there was therefore a strong incentive to maintain sales by selling well below average cost. This gave considerable potential for price instability, and the industry has in the last two decades experienced considerable excess capacity and 'soft' selling.

Given the high minimum efficient plant size, entry into the industry was possible only for large, strong, technologically-based firms such as the oil companies. It is notable that it was the chemical companies, not the oil companies, which made almost all the major product innovations. Indeed, some oil companies had difficulty in competing with the chemical companies when they tried to enter and sustained losses for some years, despite feedstock cost advantages, as a result of technical problems. Process plant came from the engineering companies, often custom built, but increasingly by the late 1950s turnkey plants became available for the simpler products. An increasing number of government-based firms, as well as the oil companies, also sought to enter the industry.

The bulk petrochemical industry today is thus highly integrated. Much of it consists of huge chemical complexes which are major growth points for associated activities. In 1980 there were in Europe about a dozen such large petrochemical complexes (Molle and Wever, 1984). The most famous example is Rotterdam. It has five oil refineries, three major petrochemical plants (Shell and Gulf naptha crackers and Esso's aromatics plant), two carbon black plants and sulphur production. The particular strength of Rotterdam is that it is both a chemical complex and a port complex. These complexes are so important that they are inevitably a major aspect of any government's industrial policy.

Mossmoran in the Shetland islands; Italian government aid for plants in the Mezzogiorno regions and new Greek proposals are all examples where governments have had much influence and given much aid. Such aid is subject to review by the European Commission under the state aids provisions of the Treaty of Rome, and the Commission has, throughout the last decade, struggled long and hard to prevent what is almost a zero sum game of subsidies developing between member states (CEC, 1988b).

The oil crises and recessions in the 1970s and 1980s brought a sharp reduction in growth rates. The petrochemicals income multiplier (the percentage growth for every 1 per cent GDP growth) dropped from 3 to 1.4, and, given the expansion already in the pipeline, led to excess capacity and substantial losses in many basic products. The first response was collusion among the industry giants. The idea of a temporary 'crisis' cartel for orderly capacity reduction and possible temporary price regulation under the supervision of the EC was mooted but ran into difficulties with the EC competition authorities. Nevertheless in polypropylene there were quite elaborate collusive arrangements with monthly meetings and concerted action to raise prices where possible. The EC Competition Directorate did not spare the firms involved. The leading polypropylene manufacturers were found guilty of price fixing by the European Court in 1986 and were fined heavily, the largest fines – over £5m – being imposed on Montedipe (part of Montedison – now Enimont), ICI, Shell and Hoechst.

The second response to overcapacity was 'restructuring' or 'rationalisation'. This involved the withdrawal of some, mainly large, firms from markets where they were not particularly strong. Much of the restructuring has been achieved by inter-firm agreements. In 1986, for example, ICI and Enichem formed a joint venture in PVC, where the Commission considered that the agreement restricted competition but elected to exempt it 'for a period limited to five years, primarily because, in the short term, it allows the activities of the two parent companies to be rationalised in the PVC sector, which is part of an industry suffering from serious structural over capacity in the whole of the EEC' (CEC, 1988b:69). The new group is the largest PVC producer in Europe with sales in 1988 of £639m.

By the mid-1980s, therefore, a combination of agreements and closures had brought about a significant reduction of capacity. In PVC, the sector which saw the most drastic rationalisation, the number of producers dropped from 28 to 18; the low-density polyethylenes dropped from 25 to 19 producers; high-density polyethylenes from 20 to 15 (Metz, 1984:874). The result has been a major and continuing concentration of the production in Europe. The recent formation of Enimont, a joint venture between Montedison and the state-owned energy conglomerate ENI, to create Italy's largest chemical company, is along similar lines.

With the return to profitability in these bulk chemicals towards the end of the 1980s, many began planning expansion. But it is unlikely that the large bulk chemical industry will see reasonable price stability without considerable further concentration. In late 1989 and 1990 there were signs that the same cycle was starting again, with recession pending and a fall in profitability of several of the major petrochemical companies, even before the rise in oil prices as a result of the Gulf crisis.

The 1980s have also seen the entry of three important new sets of players into the industry who add to competitive pressures and severely limit opportunities for expansion in Western Europe. The first of these are the Gulf states, whose investments in petrochemical capacity are now coming on-stream and are likely to build up through the 1990s. They have the advantage of cheap and readily available feedstock supplies, unlike the other new set of players, the newly industrialising countries of South-East Asia, whose fast growing markets have attracted investment from both multinational and domestic sources. In addition, the revolutions of 1989 will mean substantial investment to upgrade Eastern European capacity and exports. Together these developments augur a rapid expansion in world-wide capacity in bulk petrochemicals and increasing competition for the Western European industry.

Collusion and restructuring can thus offer only temporary relief to overcapacity and price instability. A third response has been to pursue technological opportunities through innovation and to increase the degree of specialisation by looking to develop niche markets and exploiting new

Table 12.5 World chemicals output, estimated product group breakdown, 1989 ($1,000m)

	Western Europe	North America	Japan	World total (including others)
Inorganics	31	19	12	91
Agrochemicals	24	18	10	83
Petrochemicals	119	105	62	414
Man-made fibres	14	15	10	75
Pharmaceuticals	60	44	38	192
Other chemicals*	92	74	58	275
Total	340	275	190	1,130

*includes specialities and dyestuffs, paints, detergents, cosmetics

Source: Chemical Industries Association quoted in Allen, 1990

materials. The main trend now is to develop and produce special differentiated products for special uses, often as customised products. The firms willing to rationalise have been getting out of commodity chemicals and expanding into the new, more diversified – and yet monopolistic! – specialised markets. Rhône Poulenc, Hoechst, ICI, Dow and Monsanto are examples of firms which have adopted a strategy of deliberate diversification into speciality chemicals while divesting themselves of some areas of bulk commodity chemical production.

The experience and development of the older inorganic chemicals is of great relevance to the changing pattern of the industry (Table 12.5). Inorganic chemicals are highly heterogenous – in product, process, value, scale of production, rate of change, degree of monopoly ... and so forth. They are not linked (as are organic) by a common raw material base, nor are they to the same extent by-products or joint products. As a result, firms can and do produce a limited range of products on a relatively small scale, many exploiting a specific technology related to a particular family of chemicals. The degree of vertical integration thus varies enormously.

The great firms of the years before 1939 – I G Farben (now Bayer, Hoechst or BASF), ICI, DuPont, Dow etc – are still the industry leaders in inorganic chemicals as well as organics. But in inorganics the businesses are more discrete, and there are more older, substantial, medium-sized firms. There are also many smaller firms, many of which work closely with their customers. Technical progress is rapid but less dramatic in that no novel inorganic chemicals have sold in the enormous quantities seen in the petrochemical industry. Nevertheless, there is continuing incremental change in process plant, and there are similar increases in scale and capital intensity for the older bulk products. There is also a great deal of development of more sophisticated and specialised products. These changes are leading to very substantial changes in the structure of the industry, which will be discussed in the next section.

III Speciality chemicals

'Speciality chemical' is an awkward but useful term used to describe an enormous range of products. Broadly, it covers products which are 'special' to the producer – that is, they are differentiated and the markets are imperfect. It includes many fine chemicals with small markets and products which are to some extent customised – that is, specially designed for individual (or small groups of) customers or differentiated in retail outlets through advertising (like cosmetics). In other words, it is a sector where each producer is attempting to make its products 'more special' and carve out monopoly niches for itself. Indeed, there are many similarities between pharmaceuticals and speciality chemicals – with markets highly differentiated and fragmented into specialist areas (see Chapter 13).

Speciality chemicals are produced both by the large chemical firms and by small and medium-sized firms. They are also produced by large firms whose chief activities are not usually seen as part of 'the chemical industry' such as the big mining and metal-processing firms and firms strong in refractories and ceramics. Increasingly highly specialised chemicals are also being produced by large user-firms in the electronic and other industries.

The application of sophisticated chemistry to some of these products has not only made them more R & D intensive but in several cases changed the structure of industry. Paints and packaging are good examples. The paint industry used to be highly fragmented, but is now largely in the hands of large firms such as ICI which has brought a new professionalism to meeting customer needs.

The rapid development of physical chemistry and its applications have also produced a flow of new speciality products, particularly new materials. New composite materials, epoxy glues and resins, special plastics and fibres all open up numerous new opportunities. Hoechst and Rhône-Poulenc provide two good examples. The German giant Hoechst became more concerned about low profits on commodity petrochemicals, and put increasing emphasis on specialities. In 1986 it acquired the Celanese Corporation (United States) for \$2.85bn – a huge step into a strong position in the American market – and the emphasis is now on new high-technology special products for an enormous range of industries and applications.

Similarly, a massive restructuring of the largely state-owned French chemical industry has been proceeding. Two changes are taking place; first the industry, encouraged by the government, is becoming more international and, associated with this, the lead firms – Rhône-Poulenc, Elf Aquitaine and Total CPF – are growing rapidly by acquisition. The important acquisitions are largely in speciality chemicals although there has also been some rationalisation of bulk chemicals. The policy is explicitly to develop companies large enough to compete in the Single Market of 1992. In January 1989 Rhône-Poulenc was less than half the size of the three German leaders in the industry and significantly smaller than ICI, Dow, Du Pont or Ciba Geigy. It spent some £1.5bn on major acquisitions in 1986–7. These included Union Carbide's agrochemical business (\$575m in 1986) and the basic chemicals operations of Stauffer (disposed of by ICI after the latter had bought Stauffer). In 1989, it went on to spend a further \$1.5bn on four organic chemical plants from Monsanto (in United States, Britain and Thailand)

Norddeutsche Saserwerke (a German synthetics company), Siliconas Hispanias, ICI's silicon operation and $800m for Connaught Biosciences (Canada). In early 1990 Rhône-Poulenc also acquired GAF–SSC (an American speciality chemicals company) for $480m, and RTZ's speciality chemicals business ($839m) and, more recently Rorer (American pharmaceuticals) in a complex deal valued at $3.2bn (*Financial Times,* 20 January and 11 July 1990).

This is by no means the end of the French restructuring. Orkem, which in the last few years had developed successfully in specialities and acquired Coates (British – inks and resins) and Bostik (American adhesives), in 1990 was split between Elf and Total, thus increasing the size and reducing the number of major French companies operating in this sector. Elf now concentrates on petrochemicals and fertilizers; Total on inks, adhesives, acrylics, glass and paints.

The merger and acquisition activity noted in Table 12.4 represents the other side of this reorganisation among larger chemical firms. They have taken over firms and disposed of subsidiaries 'not now strategically necessary', (which might mean that they had not achieved success in the new area or that the subsidiary was acquired in a package with a bigger merger and then sold off). Overall, this reshuffling has tended to increase concentration in existing activities in the hands of the larger multi-product firms, able to exploit innovations quickly and in many markets. When a new product area is identified as having substantial technical and market opportunities, a large firm with strong R & D and finance for product launch can move in at a speed and on a scale which no small (or, often, even medium-sized) company can contemplate.

This is not to deny that innovation is not, as always, an uncertain and risky business. A good example is carbon fibre. This is basically an acrylic fibre bound together with epoxy resins to make a light but very tough material. Carbon fibre was developed in the 1960s by the Royal Aircraft Establishment at Farnborough working in conjunction with Rolls Royce. Rolls Royce failed to use it for the RB 211 engine (but bankrupted itself in the process) and the new material was offered in 1970 to ICI for further development. ICI turned it down, but it was taken up by Courtaulds who supplied what was then a tiny market. Courtaulds licensed the technology to Hercules, an American chemical company with interests in defence equipment. By the time the licence expired (in 1979) Hercules had developed multiple uses for carbon fibres and turned to a Japanese producer, Sumitomo, for its supplies of acrylic fibre (the intermediate product). This contract gave the Japanese access to the technology and today the Japanese supply 80 per cent of the fast-growing world market in carbon fibre. Ironically, the Pentagon are now worried about over-dependence on Japanese supplies and are looking to an American subsidiary of ICI – Fiberite, acquired as part of the 1984 purchase of the Beatrice Group – to supply Hercules with raw materials (*Financial Times,* 15 August 1986).

Carbon fibre is a prime example of how Japanese firms, although originally much weaker in R & D than European or American, are seizing opportunities to enter the market for new materials, while also catching up in basic commodity chemicals. The example also illustrates the importance of hands-on experience in developing new materials. Because carbon fibre properties can be varied according to use, it requires the chemist and the engineer to work closely together in development. Both the upstream and downstream relationships are of considerable importance – upstream working with those developing acrylic fibres and epoxy resins; downstream with the aeronautical engineers and other engineers designing the products. In other words a major development such as carbon fibre is not a discrete project but a whole family of inter-linked projects – a complex system of developments in which user–producer relationships are critical. Even where overheads are shared, there is no hiding the substantial costs and the major uncertainties incurred in such developments and it is not surprising to find even large firms, such as ICI, being selective about getting involved.

Advanced materials of many types are being developed in all leading countries and are recognised as a growth point of great significance. Indeed advanced materials technology has figured with biotechnology and ICT as one of the three 'generic' technologies which have been the object of special government R & D programmes in almost every major industrial country. Some advanced materials development has been closely related to new developments in ICT, especially in computerisation of design and development and process-control technology. Japanese firms in electronics, information technology and the engineering industries (including automobiles and machine tools) are often to be found working in association with strong chemical firms. 'Chematronics' is not quite analogous to 'mechatronics', but there are many new developments in the electronics industry which bring chemical companies and electronic companies into close collaboration, especially in the ceramics area. For this reason if no other, the European chemical firms cannot afford to be complacent about Japanese competition in the 1990s. Some European firms have realised this and have located new R & D facilities and plant investments in Japan in the 1980s, in order both to facilitate access to this fast-growing market and to be closer to the new developments in advanced materials.

The customer–user collaboration found in new materials has become almost routine in some sectors of speciality chemicals. It is often a case of providing a service with products rather than the old style of providing a product with some supporting technical service. This type of innovative 'customised' marketing is growing fast: it is a knowledge-intensive activity and competitive advantage depends very largely upon scientific expertise and know-how. And, just as in major engineering contracting the sophisticated engineering companies can affect the competitiveness of their customers, so in chemicals, and in particular in advanced materials, international innovative and competitive success depends upon access to the most forward-looking and sophisticated customers.

What we see, therefore, in the speciality-chemicals sector is the strong impact of rapid technical change on both the structure of the industry and on the nature of competition. The intense technological activity, the new products, the new applications and the refinement of existing products is a continuous process. The industry as a whole is changing and growing. There is change in the scale of operation because so much more of the industry is subject to the dynamic economies of scale, in turn requiring large scale R & D and financial and technical resources to open up new opportunities. Continuous change also involves a high level of entrepreneurial activity –

activity by both large and small firms. But it is the large firms which are calling the tune. Firms with the corporate, technological and entrepreneurial capacity to innovate are growing, others are declining or being acquired. It is the smaller, old-fashioned firms that are being squeezed out; they provide the bread and butter of the mergers and acquisitions figures quoted in Table 12.4. But the wider application of more scientific chemistry is also opening up new niches and there are new small firms emerging to exploit these niches.

There is also the paradox of competition, which in many parts of the speciality chemicals business is intense, even though there are many strong monopoly/oligopoly positions. This paradox is a feature of dynamic industries. When the rate of change is fast there are often substantial dynamic economies of scale and scope – including the advantages of a multi-product portfolio to spread risks. But rapid change also offers opportunity to enter markets by innovation rather than by price. This applies to new applications, to new products, to rapidly expanding markets and to markets where some small differentiation will carve out a new share. The major firms in the industry are so large and so diversified that they are often seen as monopolists; however, with respect to the more fragmented speciality sectors, they should more realistically be seen as powerful *potential* competitors. This is important since it makes many of these markets contestable – despite patents, close customer–supplier relationships and successful market and product differentiation. Innovation and growth can, in fact, make many of the apparently strong and highly profitable monopoly positions merely temporary.

IV The impact of 1992

The European chemical industry is in better shape than most others to face world competition in the 1990s. Much that happened in the last 20 years has already anticipated 1992 and the Single European Market. It is already an international industry; much of the re-structuring and rationalisation that took place in the 1980s has strengthened its global position. However, the EC will face intense competition from the new petrochemical complexes in the oil-producing countries in the bulk-chemical area, and from the United States and Japan in the speciality chemicals, closely related as they are to new technological developments in other industries. Only sustained efforts in R & D and innovation can maintain European competitive strength.

The Single European Market is essentially a competitive concept. The aim of promoting competition involves not only enforcing competition policy by limiting collusion between firms and preventing anti-competitive mergers and monopolistic behaviour but also preventing governments from diminishing competition by nationalistic policies such as subsidy and undue favour (for example, by public purchasing policies) to local firms. It also involves the harmonisation of the regulatory system to eliminate barriers. Some regulatory differences are historical and reflect national attitudes; others are the result of government policies designed to protect or promote their national industries. Thus the Single European Market, while promoting competition within the EC, also prevents the pursuit by governments of national 'industrial policies', of a kind that may still be open to both the Japanese and American governments.

The trend in both bulk and speciality markets is towards concentration of production in the hands of a small number of firms. In the bulk markets most of the firms have long been multinationals and national governments have had little influence over their activities. This is now likely to extend to the speciality-chemicals area where, as the Commission's recent report on competition admitted, a number of product markets have become highly oligopolistic as 'acquisitions have resulted in a situation where a small number of highly diversified companies control sizeable market shares in several small markets ... The trend towards highly differentiated, highly segmented markets contributes to the reduction of competition and further increase in the already high profit margins in the speciality chemicals sector' (CEC, 1988b). Many of these oligopolistic markets remain, however, vulnerable to potential competition from new entrants, and this limits the degree to which individual firms can exercise monopoly power.

National European governments are (and have for some time been) powerless to prevent monopoly/oligopoly emerging among the major multinational chemical companies, although at the margin they may be able to impose conditions which modify monopoly power. Control, in so far as it can be exercised, has to come now from the European Commission and the fines imposed in the polypropylene case indicate that neither Commission nor Court will shirk responsibilities for preventing collusion. The new powers on mergers will give the Commission the right to investigate cross-border mergers above a certain size if it thinks it will lead to a diminution of competition. But competition authorities, however well organised, have always found it difficult to curb the cross-subsidisation and price discrimination which the many (often temporary) monopolies of giant firms make possible. Except in its reinforcement of the powers of the Commission to operate as the lead competition authority within the Community, 1992 is likely to do very little to change this situation.

The main impact of 1992 is therefore likely to be in terms of regulation of the industry rather than in competition. This has several aspects but among the most important are the testing and registration of chemicals, the handling of chemicals and protection of the environment.

The testing of chemicals is a major activity and neither the task of drawing up regulations and doing the testing, nor the problems facing those responsible for drawing up new Community regulations should be underestimated. There are between 50,000 and 70,000 chemicals already on the market, and some 500 more are designated every year, many of them new speciality chemicals. For each one, testing and evaluation can cost between $50,000 and $1 million (OECD, 1983). In addition, for regulators at the European Community level, there is the problem of agreeing standards when some countries are more sensitive than others, when attitudes are changing fast and when technology is itself moving so fast that the content and scale of the appropriate response are also changing.

The control and regulation of the handling of chemicals is, like testing, a much more complex business than it might seem. There are chemical regulations applied to manufacturers, regulations about transport and regulations about waste treatment and disposal (see Chapter 19). The difficulties as with testing, are increased by the flow of new substances and changed processes. Again, the problem of agreeing standards is difficult to divorce completely from the interests of particular countries – high standards give a competitive edge and low standards permit the survival of relatively inefficient plants. The two problems are deeply intertwined because there are no absolute criteria – there is no 'zero risk', and views differ on levels of acceptable risk.

The control of the use of chemicals because of their effects on the environment is no new subject. But the topic has come to the fore again with the greater scale of the industry, the greater recognition of the dangers and the development of Green politics. The chemical industry is well aware that standards are going to rise and that tougher controls are on the way. The interesting issue today is that this rise coincides with the growing importance of the EC. Already the Commission has sought to limit the use of nitrogenous fertilizers by imposing tougher standards on water quality (see Chapter 19). The phasing out of the use of CFCs in the interest of preserving the ozone layer indicates that a start has been made with the even more difficult problem of achieving global agreement on some of the greatest hazards. It is to be hoped that the EC will give a lead not only to European governments but to all industrial countries by setting exemplary standards in the areas of greatest perceived dangers. This does not diminish competition; indeed it may intensify it. But it defines new ground rules for competition which in the longer run can perhaps help harmonise environmental and growth objectives.

European firms are concerned that EC regulations should not be tougher than those met by their competitors. This can be a short-sighted view since tough safety and environmental standards need not weaken but can strengthen competitive power. Certainly, it is desirable to move to global standards whenever possible. European firms are also concerned by the extent to which, in response to Green politics, individual countries go ahead with *national* regulations within Europe before agreement has been reached on Community regulations. Differing national regulations 'are making it harder, not easier, to tackle environmental problems. At a time when authorities in the EC member states are meant to be moving towards Community level regulations, they are in fact creating chaos in the chemicals sector by burdening us with wildly differing national environmental rules' (Metz, 1989). This is an area where harmonisation and standardisation are sorely needed.

V Conclusions

In conclusion, we return to the problem of major innovations in a competitive market in which aid by individual governments is prevented because it distorts competition. Put differently, does 'the market' produce the best results in the chemical industry or is there a role for intervention? The firms that dominate this industry are large, diversified and international in outlook. During the course of the 1980s they have, on the whole, become larger and more international both in structure and (one assumes) in outlook. The key to their success in the past has lain in innovation, in their ability to generate new ideas, products and processes from internal resources, and to adapt, to assimilate and absorb those developments, whether generated in-house or by competitors. For the European industry the crucial issue is whether it can retain this flexibility at a time of heightened international competitive pressures on profit margins.

In general, it would appear that the European firms, alongside their global counterparts, are meeting the challenge. There are, however, some indications that they are not always as quick off the mark as their global counterparts. We end this chapter with two brief notes of warning.

The first relates to the science base. The ability of the chemical majors to go on generating new ideas in-house and absorbing and assimilating them from elsewhere will depend on the quality of the science base in Europe, the training of R & D personnel to the standards required, and the degree to which there are collaborative mechanisms for linking academic research in universities and research institutes with industry's needs. It is not clear that European governments have yet succeeded in creating the responsive and flexible higher education institutions to meet these requirements.

The second warning relates to the ability of firms to nurture in-house 'infant industries' which then provide the base for future growth. Traditionally this has been the way the chemical majors have grown, cross-subsidising the new from the old. It is worrying, however, to find the chairman of ICI voicing doubts as to how long that company should 'hang on' in an area as important as specialised thermoplastics (part of the new materials revolution) because of 'the uncertain profit prospects'. The Single Market of 1992 brings in the liberalisation of financial markets and the opening up of the (traditionally closed) banking sectors of countries such as Germany and France. This liberalisation, combined with the dominant Anglo-American business school model of decentralised operations and separate profit centres, could make obsolete the old-style, cross-subsidising chemical industry model. If it did, however, it could also destroy the European chemical industry.

13 Pharmaceuticals and biotechnology: perspectives for the European industry*

Margaret Sharp

I Introduction

Many early pharmaceutical products, such as Bayer's aspirin, were by-products of the chemical industry which was dominated by German and Swiss firms until World War II. The post-war period saw the rise of the modern research-based industry, characterised by extensive patenting and a drug discovery process akin to molecular roulette in which thousands of chemical compounds were screened in the search for new chemical entities with identifiable therapeutic effect. In 1935 when the first modern anti bacterial drug – Sulfonilamide – was introduced, May and Baker, its discoverers, had screened 693 compounds. By the 1960s, typically 8,000–10,000 compounds would be screened for every useful product launched.

Today, although there is still much uncertainty and risk attached to drug discovery, the hit and miss nature of those early days has passed. Increasingly drugs are being designed rather than discovered, they are being sought on the basis of a growing depth of understanding about the body's own biochemistry and immune system. Biotechnology, which has brought the cloning of substantial numbers of the body's own therapeutic proteins, is now recognised as an essential part of this revolution and, as we shall see, an increasing number of drugs are likely to derive from this route.

The purpose of this chapter is to consider the role of European firms within the context of developments in the world pharmaceutical industry. The next section looks at the contribution made by the different firms and different countries in Europe and their relative standing in this fast-moving industry. The following section considers developments in biotechnology and their comparative impact on the pharmaceutical industry. The final section considers some of the issues brought to the fore by the evolution of the Single European Market and the prospects for the European industry in the light of these developments and other trends in world markets.

*I am indebted to Dr Bryan Reuben, Simon Shackley and Professor Stuart Walker for comments on the detail of this chapter. Simon Shackley in particular provided much of the detail for Section IV.

II The European Pharmaceutical Industry

The pharmaceutical industry world-wide has a turnover of approximately $120 billion and numbers over 5,000 firms (*Scrip,* 1989). It is thus about one-tenth the size of the entire chemical industry but has a much higher rate of growth. It is characterized by a two-tier structure in which the largest 20 or so firms (by turnover) account for over 50 per cent of sales and 85 per cent of R & D expenditures. This two-tier structure is unusual in three respects. First, it is international – most of the top-20 firms have a presence in all important markets and there has been little attempt to keep national markets for indigenous firms in spite of a substantial public-sector dominance in purchasing. Second, in spite of the concentration of the industry in the hands of relatively few large firms, the largest of these, Merck Sharp and Dohme of the United States, has only 4.5 and a half per cent of world sales (Table 13.1). Third, until recently, there has been remarkable stability among the cohort of top firms – no more than one or two firms per decade have moved into the top 20. Firms tend to specialise in a limited number of therapeutic areas (such as anti-infectives, cardio-vascular, central nervous system etc). The structure of the industry is in fact highly oligopolistic and characterised by intense competition between the leading half dozen or so firms in each therapeutic area.

Table 13.1 lists the top-25 companies by 1989 sales and market share. The strong presence of European companies has been a long- term feature of the pharmaceutical industry – the German and Swiss companies who dominated the pre-1935 industry have kept their position ever since. The American industry, many of whose firms (like Merck) sprang from the expropriated assets of earlier German companies, emerged as an international presence in the early post-war period, building on their strength as research-based companies and establishing rapidly an international network of subsidiaries as a means of penetrating markets. The most prominent trend of recent years has been the relative slippage of the big American companies and the rise of the British industry – ICI, Glaxo, Wellcome, and Smith Kline Beecham – and most recently of Japanese firms. Twelve of the top-25 companies in Table 12.1 are European: only three are Japanese.

The leading drug markets are the United States (29 per cent of world total), Japan (20 per cent), West Germany (8 per cent), France (7 per cent) and Italy (7 per cent). As a whole the European Community (EC) is the largest market in the world (29 per cent), but is also a significant net exporter to the rest of the world with Germany, the United Kingdom and France all being sizeable net exporters (*Scrip,* 1989). There is surprising variation in per capita drug consumption, where Japan heads the league among major countries with consumption valued at $160 per head in 1989, compared to British consumption of only $50 per head. (But account has to be taken of price variations: in terms of quantity, France heads the league).

These variations in consumption reflect both national proclivities towards hypochrondria and institutional arrangements for the provision of prescription drugs. In Japan, for example, most doctors dispense drugs and have thus had built-in incentive to prescribe high-cost drugs, whereas the National Health Service in Britain now makes it advantageous for doctors to limit the cost and quantity of drugs prescribed. Indeed, in recent years many public

Table 13.1 The 25 largest pharmaceutical companies, 1989

Rank	Company	Country of origin	Pharmaceutical sales ($m)*	World market (%)
1	Merck	United States	5,485	4.5
2	Bristol Squibb	United States	4,230	3.5
3	Glaxo	United Kingdom	4,215	3.5
4	Hoechst	Germany	3,900	3.3
5	Smith Kline Beecham	United Kingdom	3,640	3.0
6	Bayer	Germany	3,405	2.8
7	Am. Home Products	United States	3,110	2.6
8	Ciba Geigy	Switzerland	2,960	2.5
9	Eli Lilly	United States	2,920	2.4
10	Sandoz	Switzerland	2,705	2.3
11	Pfizer	United States	2,640	2.2
12	Johnson & Johnson	United States	2,625	2.2
13	Takeda	Japan	2,580	2.1
14	ICI	United Kingdom	2,295	1.9
15	Eastman Kodak	United States	2,220	1.85
16	Roche**	Switzerland	2,163	1.8
17	M. Merrell Dow	United States	2,100	1.75
18	Upjohn	United States	2,050	1.70
19	B. Ingelheim	Germany	1,870	1.6
20	Schering AG	Germany	1,675	1.4
21	Wellcome	United Kingdom	1,640	1.4
22	Sankyo	Japan	1,615	1.35
23	Cyanamid	United States	1,575	1.3
24	Rhône Poulenc***	France	1,460	1.2
25	Fuji Sawa	Japan	1,435	1.2

* Ethical (prescription) drug sales only
** Does not include Genentech (taken over January 1990)
*** Does not include Rorer (taken over in 1990)

Source: Shearson Lehman Hutton, 1990

authorities, who ultimately pick up the bill for much medical expenditure, have been putting pressure on purchasing agencies to limit drugs bills, using a mixture of price controls and limited lists (which require use of low-cost generics or branded drugs). This trend is likely to grow with the increasing number of old people in the population and the accelerating costs of health care; many of the major new drugs of the 1960s and 1970s are now moving out of patent, so there will also be more generics to choose from. Indeed, many of the major companies, while voicing strong opposition to the use of generic drugs, are also expanding their own interests and production in this sector of the market. Generics used to be produced mainly by the smaller companies who could exploit a discovery once it came off patent, but 70–80 per cent of generics are now made by the major companies.

The outcome (at present) of the varying patterns of production, consumption and regulation in Europe is considerable variation in prices. Table 13.2 illustrates the range of prices for a number of top prescription drugs across Europe in 1988. Generally speaking the former West Germany and

Table 13.2 Pharmaceutical price differentials in Europe, 1988
Index: United Kingdom = 100

Drug	W Germany	France	Italy	Belgium
Ventolin (Glaxo)	209	n.a.	52	81
Zantac (Glaxo)	144	79	112	100
Tenormin (ICI)	139	39	85	52
Voltarol (CibaG)	50	46	n.a.	56
Adalat (Bayer)	117	52	81	78
Zovirax (Wellcome)	98	80	68	67

Source: Shearson Lehman Hutton: Quoted in *Financial Times* Supplement on Pharmaceuticals, 6 November 1989

Switzerland were high-price markets, the United Kingdom a medium-price market and France, Belgium and Italy low-price markets. One of the issues to be tackled under the Single Market is how these prices can best be aligned. It had been hoped that parallel importing from one Community country to another would solve the problem, and an important European Court of Justice decision in 1976 (No. 104-75) upheld the right to re-import. Although appealed against by the pharmaceutical industry, the ruling has been repeated and indeed become an important 'principle' underlying the current EC market. However, parallel importing has not grown as some predicted and has so far made less impact on price differentials than might have been expected. Packaging, language and varying brand names also help to keep markets separate.

The existence of separate price and regulatory regimes in the different European countries has encouraged another prominent trend in the post-war pharmaceutical industry – the establishment of subsidiaries. Most of the major American companies established a subsidiary in each of the main European countries in the post-war years, but especially in France, Italy and the United Kingdom where governments put pressure on them to do so. The United Kingdom was particularly successful in attracting R & D facilities from US companies. Indeed, the pharmaceutical industry is often quoted in the United Kingdom as an example of a successful inward investment strategy where foreign firms have matured and reinvested their profits in expanding and deepening their market presence (Brech and Sharp, 1984: Chapter 4). Others point to the industry as a prime example of Britain's failure to exploit its own comparative advantage. It is the quality of the science base and low cost of R & D that has attracted multinational companies, who exploit these assets to make (and remit) substantial profits to their home base.

The British industry, nevertheless, benefited from the 'role model' set by the American multinationals and the management cadre trained in their British subsidiaries helps to explain the growing importance of British firms within the European industry (Sharp, 1988). Both ICI and Glaxo have in the last ten years moved into the top-20 firms (by sales) in the industry, (Glaxo's best-selling ulcer drug, Zantac, carrying it to the number two spot behind Merck in 1989), while the 1989 merger of Beecham with the American company Smith-Kline, also brought that company into the top ten. These British companies are also among the world's most profitable. All three are ranked among the top-12 most profitable pharmaceutical companies with net profits as a percentage of sales of well over 25 per cent.

The justification put forward by the industry for high profit margins is high R & D expenditure, but independent observers have also frequently pointed to the even higher expenditures on marketing. As indicated earlier, the pharmaceutical industry is characterised by intense oligopolistic competition in a series of clearly defined market sectors. In static terms competition is about market share and typically focuses on product differentiation not price. In dynamic terms this translates itself into an intense battle to introduce new drugs of acknowledged therapeutic value and to interest the medical profession in dispensing them. A major new drug will be able to command a premium price as long as its patent holds and there are no close competitors in the market. A successful company is one which is able to command such premium markets for a reasonably long period of time over a range of products and to have new premium products replacing old ones as they lose market share. For their part competitors crowd in upon the profits derived from premium markets and aim to market different (in patent terms) but equivalent formulations as quickly as possible.

The keys to this competitive process are research and development and marketing; production and investment costs are relatively low. R & D is essential both for the discovery and development of new drugs and to underpin the firm's ability to move rapidly into competitors' markets. For most pharmaceutical companies R & D as a proportion of net output ranges from 12–20 per cent, averaging 16 per cent in 1988–89. Although already well above the average of other industries, the proportion devoted to R & D has been increasing steadily over the course of the last 30 years. This has reflected two main factors. First, the rising concern among the general public over the safety of drugs and medicines has led to increasing and costly measures for the testing of new drugs; second, the clear diminishing returns present in the traditional techniques for discovering and developing new chemical entitities – the molecular roulette that was referred to in the introduction – and the need for more research time and effort for each new promising discovery. Typically it now takes 12 years to bring a new drug to market, the time element in itself adding considerably to the costs of new drug development. While the average cost of developing any *single* successful new chemical entity (NCE) is about \$30m, it has been estimated that in the United States this rises to \$114m if allowance is made for the cost of failures and to approximately \$230m if allowance is made for the time element (using a 9 per cent discount rate) (Di Masi *et al.*, 1991). In the 1950s and 1960s – before the thalidomide scandals – it typically cost about \$10m in current prices to develop a new drug and the launch time was four or five years.

Companies are, of course, ever hopeful that the drug they are developing will prove a drug such as Zantac, Glaxo's very successful anti-ulcer drug which is currently attracting annual sales of over \$2bn and has spearheaded Glaxo's rise up the league tables. Companies, therefore, maintain a continuous flow of new products under development. Table 13.3 (A) ranks major European pharmaceutical companies by the number of new compounds under development. It will be noted that this ranking correlates with neither the sales nor the profitability rankings. Glaxo, high in the profitability stakes, is low in this table and ICI and Beecham are not ranked. Success, however,

Table 13.3A Europe's leading companies in terms of number of compounds under development, 1988

	Total	Compounds in R & D Own	Licensed
1 Rhône-Polenc	103	73	30
2 Roche	90	64	26
3 Ciba Geigy	88	59	29
4 Hoechst	83	67	16
5 Sandoz	71	51	20
6 Schering AG	58	50	8
7 Erbamont	58	48	10
8 Sanofi	56	45	11
9 Boehringer Ingelheim	56	42	14
10 Wellcome	52	41	11
11 Glaxo	51	33	18
12 Roussell Uclaf	49	28	21
13 Bayer	47	36	11

Source: *Scripp*, 1989

Table 13.3B Proportion of new chemical entities introduced by country

	1976–80[1]	1981–85[1]	No. in World Top-selling 50 drugs in 1990[2]
Former W Germany	16	16	10
France	11	7	0
Italy	14	7	0
United Kingdom	5	4	12
Switzerland	5	8	6
United States	25	25	18
Japan	11	22	3

Sources: [1]Reis-Arndt, 1987; [2]Barclays de Zoete Wedd, 1990

depends not on the number of new drugs under development but on their relative success when launched. Table 13.3 (B) shows that while the British companies may rank low in terms of compounds under development, they have a high ranking in terms of successful new drugs, with a disproportionate (in relation to the NCEs under development) number of top-selling drugs. Indeed, the British companies claim that one reason for their good profitability record is their success in the early screening phases of drug development and their strict policies of dropping products which seem unlikely to be successful. This contrasts with French and Japanese companies which have introduced many new drugs but mainly for local market needs; they were not sufficiently innovative to reach world markets. This is already changing, and the Japanese in particular are now developing innovative new drugs which, if successful, could have a substantial effect on the standing of the Japanese industry.

Europe's pharmaceutical industry, in common it should be said with their American and Japanese counterparts, is thus confronted in the 1990s with four major problems:

(i) the secular rise in cost of R & D as public authorities require longer and more complicated testing of drugs;
(ii) the depreciating value of patents as patent life (normally 20 years) is eroded by testing procedures; typically a patent will only have seven years to run when a new drug is launched;
(iii) pressure from public authorities to cut health care costs, with the cost of drugs being often a prime target;
(iv) the rise of the generics which benefit from the (effectively) short patent life of many new drugs and the public authorities' search for cheaper alternatives to expensive branded drugs.

It was against this background in the 1980s that biotechnology made its debut and was seen as a vitally important new technology for both the pharmaceutical industry and, more generally, medicine and health care. There were hopes that it would provide the answers to these problems – a new quick route to drug discovery where the products, being replicas of naturally occurring proteins in the body, would rapidly gain approval and market access. Like many new technologies, biotechnology has not lived up to early hopes; indeed, it has proved tantalisingly slow in commercialisation. These 'teething troubles' should not, however, detract from the major impact it will have in the long run.

III The impact of biotechnology

Biotechnology by its broadest definition is 'the application of biological organisms, systems and processes to manufacturing and service industries' (ACARD, 1980). In this sense biotechnology has been around since the New Stone Age when humankind first learnt the art of cross-breeding plants and domesticating animals and first began to appreciate the qualities of yeast. For many centuries broad empiricism sufficed as technology, but by the beginning of the 20th century this was replaced by a more systematic attempt to sift, screen and categorise the role and variety of micro-organisms existing in the natural environment and to exploit those that had useful application – penicillin being a prime example. The aspirations of microbiologists and enzyme chemists were, however, overshadowed by developments in synthetic chemistry and never realised their potential. There remains, even today, much to be exploited from this 'second-generation' biotechnology.

The new or 'third-generation' biotechnology dates back to the early 1970s when two breakthroughs in molecular biology – the discovery that restriction enzymes could be used to cut and splice genes (recombinant DNA), and the use of cell fusion to develop hybrid cells of desired characteristics which would multiply themselves rapidly (hybridomas) – heralded the coming of genetic engineering. Using the methods of recombinant DNA the genes of micro-organisms such as yeast could be 'reprogrammed' to produce useful proteins. As the micro-organism multiplied itself so it became, in effect, a mini-factory for producing (cloning) the protein.

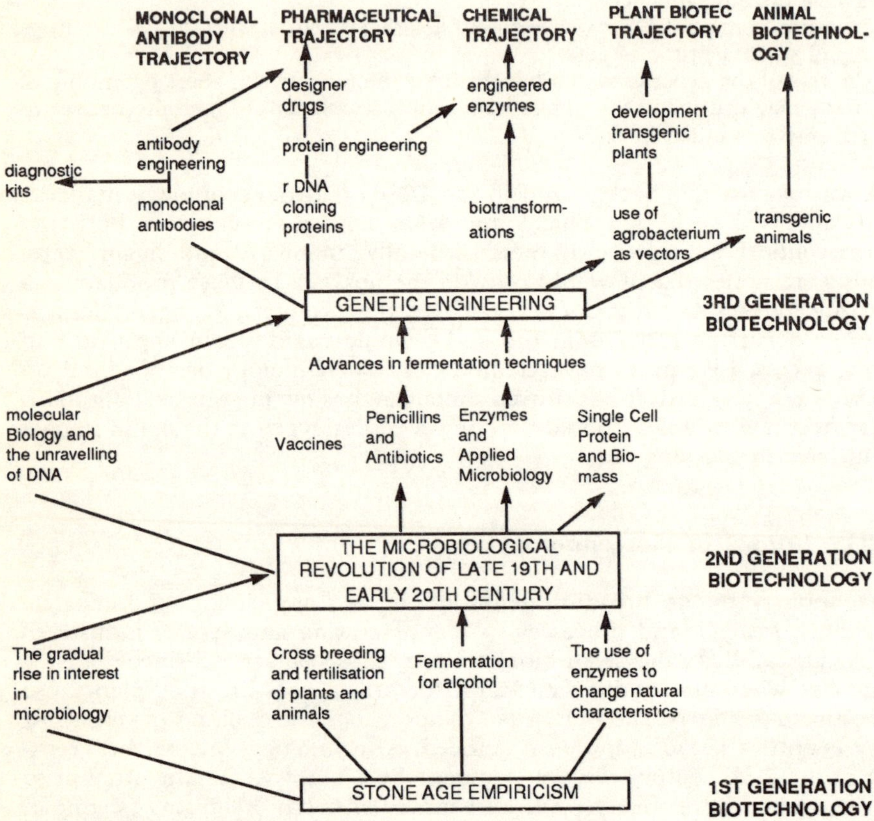

Figure 13.1 The three generations of biotechnology and the main areas of development of the new biotechnology
Source: Sharp (1991)

The applications of these radical new techniques were rapidly appreciated. By the early 1980s genes had been cloned for producing proteins of recognised value such as insulin, human growth hormone and Factor VIII (for treatment of haemophilia) and scale production was underway. These developments had also opened the way for the emergence of a whole new range of protein drugs based on naturally occurring proteins in the body's immune system – proteins such as the interferons, the interleukins (with potential for cancer treatment), tissue plasminogen activator (TPA – used to dissolve blood clots in arteries) and erythropoietin (EPO – used to treat kidney failure). Hybridomas led to the development of monoclonal antibodies (MABS), which, having the antibody characteristics of specificity to a particular antigen, had wide application in separation processes, diagnostics and therapeutics.

Applications outside pharmaceuticals also rapidly became apparent. It was not long before the techniques of recombinant DNA were being applied to plant genes, enabling the rapid development of new hybrid plant species incorporating desired characteristics such as resistance to drought or frost. Genetically altering the characteristics of fungi is leading to the development of new fungal bio-pesticides. And just as penicillin and other antibiotic strains can be altered and strengthened, so it is possible to develop new strains of micro-organism to help destroy toxic wastes. Similarly, a mixture of genetic engineering and bio-transformations is opening important new routes to old (and new) chemicals.

Figure 13.1 illustrates developments in biotechnology and identifies the five main 'trajectories' of development from genetic engineering – monoclonal antibodies, pharmaceuticals, plant biotechnology, animal biotechnology and chemicals. The dominant trajectories have been the pharmaceutical and agricultural developments, with the pharmaceutical innovations in the early period of development constituting the 'leading edge' technology. The different areas of developments in biotechnology have now moved firmly in different directions with their own specialised technology and applications. In the long run the potential impact on agriculture is huge, offering new opportunity for another 'Green Revolution'. In this chapter, however, we concentrate primarily on developments in health care and pharmaceuticals; developments in its application to agriculture, the food industry, chemicals, mining and pollution control are not discussed (for a wider discussion see OTA, 1984 and 1988; and *Biofutur,* 1990).

Close linkage to the science base and uncertainty about technical and market viability are two characteristics of the early development of biotechnology which help to explain another prominent feature – the swarming of many small new biotechnology firms on the east and west coasts of the United States. The very large sums of money poured into academic research in the life sciences in the United States help to explain the early and continuing dominance of the United States in this new technology (Irvine *et al.*, 1990), and the small new biotechnology firms are *par excellence* an American institution – in many respects an extension of the academic research laboratory where American scientists, brought up in an entrepreneurial culture, have gambled on their skills and knowledge to make themselves millionaires. They were in part also a copy-cat phenomenon – after the success of the semiconductor firms in Silicon Valley, the venture capitalists were only too anxious to branch out into new pastures.

Table 13.4 Biotechnology: pharmaceutical products in R & D

	1987	1989
Pre-clinical tests	484	592
Phase I clinical	47	75
Phase II clinical	52	72
Phase III clinical	26	38
Under approval	16	20
Approved	3	7
Total	628	803

Source: Marcel, 1990: Table 4

These new small biotechnology firms had considerable advantages over the established pharmaceutical companies. Because they grew from the academic science base in biology they had immediate links with that base, whereas the pharmaceutical companies' traditional links were with chemistry and pharmacology departments. As a result, the small firms were readily able to mobilise the cross-disciplinary teams – the geneticists, micro-biologists, protein chemists and biochemical engineers – necessary to take a project from workbench to pilot plant production; and they were able to try out ideas and take risks at which the larger firms would baulk. Until they were clear that this new technology was 'going places', it was much easier for the large firms to buy themselves into the technology by placing contracts with, or buying an equity stake in, one of these small firms. On all sides it was a marriage of convenience – for the pharmaceutical and chemical companies it was a quick and reasonably risk-free way of gaining access to the technology; for the venture capitalists, a new and potentially fertile field of development; for the academics, an easy source of funding with which to pursue leading-edge research. Moreover, it was a phenomenon which fed upon itself. Once launched, all parties needed to 'talk big' about potential success to keep equity values high and encourage further investment. Hence the hype, the over-optimism and the disappointments.

Neither Europe nor Japan experienced the same swarming of new firms in the development of biotechnology. The reason for this lies partly in culture – academics in both areas frequently still spurn a personal involvement in commercialisation; and partly in institutions – European science in particular tends to be institute- rather than university-based, with employees paid on a full-time basis. Nor are the venture capital markets as large or mature as those of the United States. But a good part of the explanation lies in the later response – European firms did not really wake up to what was happening in biotechnology until the early 1980s by which time it was clear that although there were only limited profits to be made from the first generation of protein drugs (the interferons, interleukins etc), biotechnology had radically altered the route to drug discovery and was therefore a necessary competence to develop in-house. Initially, large firms built up enough competence merely to equip themselves with a 'window' on the technology; investment in earnest began only in 1984–5.

Table 13.4 gives some indication of how few new products have so far emerged from this new technology. By the end of 1989, including insulin, only seven new products based on recombinant DNA technology had been approved for pharmaceutical use. Insulin, human growth hormone, interferon, OKT3 (a monoclonal antibody product used to offset rejection in kidney transplants), TPA, EPO, and most recently interleukin-2 (which has gained approval in the EC but has been turned down in the United States). Contrast this, however, with the numbers of biotechnology-based drugs in the pipeline – nearly 800 – and it is apparent how deeply biotechnology is now affecting the process of drug development. Current sales of these products amount in total to some $1.2bn, $900m of which is in the United States, but markets are expected to grow strongly over the next few years, with sales valued at a minimum of $10bn by the end of the decade (Marcel, 1990:52). (This compares with current world sales of all pharmaceuticals of $120 bn.)

Beside drugs, the use of monoclonal antibodies in diagnostics is now widespread with sales currently valued at approximately $2bn and growing by 6–7 per cent a year (Policard, 1990). Since they are products that can be used *in vitro*, diagnostics are not subject to the lengthy test procedures of drugs or vaccines, and many small companies have entered this segment of the market, whereas the cost and complexity of nursing a therapeutic product through its ten years of testing deters most small companies. Even when the small company pursues the development of such products, they are left dangerously exposed on the one hand to the whims of their financial backers, and on the other to finding partners among the major pharmaceutical companies prepared to market their products (Dodgson, 1990). Indeed, as biotechnology has matured many of the more creative of the new biotechnology companies have been bought up by larger firms. The most prominent of these takeovers was that of Genentech, widely regarded as the strongest and most go-ahead of the new biotechnology firms, by Hoffmann LaRoche in January 1990. But there had been many earlier marriages. Among European firms, Glaxo had taken over Biogen's European laboratory in 1987, Bayer acquired two small companies, Molecular Therapeutics and Molecular Diagnostics, which later formed the basis of its West Haven laboratories, and in 1989 Rhône-Poulenc took over Canada's leading new biotechnology firm Connaught Biosciences.

The development that today drives the pharmaceutical industry is the technique of protein engineering, namely using genetic engineering techniques not just to clone known proteins, but, in the process, to change their characteristics in such a way as, for example, to eliminate unwanted side effects – in effect creating a slim-line peptide targeted to a specific functional task. Although techniques of protein engineering are still experimental, there is now confidence that it is a feasible route to what is (confusingly) called the 'second generation' of biotechnology drugs and that by the early part of the next century rational drug design based on knowledge gained through biotechnology will be accepted as the way forward.

The emergence of protein engineering, which clarified both the technological route and the likely timeframe within which new products would emerge in the pharmaceutical sector, proved the trigger to substantial investments by the private sector. Interest, therefore, focuses on the large chemical/pharmaceutical companies. Table 13.5 lists these companies and ranks them by sales and R & D expenditures. It is very difficult to gauge what part of these expenditures is being devoted to broadly defined biotechnology, par-

Table 13.5 Europe's major pharmaceutical companies

Firms ranked by sales	Sales and R & D 1988–89			
	Sales ($m)	R & D ($m)	R & D as % of sales	R & D personnel numbers
1 Glaxo (UK)	4,215	530	12.6	4,800
2 Hoechst (FRG)	3,990	610	15.3	4,400
3 Smith Kline Beecham (UK)	3,640	625	17.2	4,500
4 Bayer (FRG)	3,405	465	13.7	2,000
5 Ciba Geigy (Switz.)	2,960	490	16.6	4,300
6 Sandoz (Switz.)	2,705	430	15.9	3,000
7 ICI (UK)	2,295	295	12.9	3,200
8 Hoffmann LaRoche (Switz.)	2,163	540	25.0	3,000
9 B. Ingelheim (FRG)	1,870	375	20.1	3,500
10 Schering AG (FRG)	1,675	240	14.4	2,150
11 Wellcome (UK)	1,640	298	18.2	3,500
12 Rhône-Poulenc (Fr.)	1,460	280	19.2	3,000
13 Sanofi (Fr.)	1,045	195	18.7	1,400
14 Astra (Sweden)	945	205	21.7	1,600
15 Erbamont (Italy)	810	145	17.9	2,000
16 E Merck (FRG)	690	90	13.0	1,300
17 Knoll (FRG)	645	95	14.7	n.a.
18 Boots (UK)	630	75	11.9	750
19 Akzo (Netherlands)	610	120	19.7	1,500
20 Solvay (Belgium)	605	115	19.0	1,100
21 Kabi (Sweden)	590	81	13.7	n.a.
22 Fisons (UK)	540	90	16.7	1,500
23 B Mannheim (FRG)	530	100	18.9	n.a.
24 Novo-Nordisk (Den.)	510	80	15.7	1,000
25 Ares Serono (Switz.)	420	70	16.7	600

Source: Shearman Lehman Hutton, 1990

ticularly given the increasingly pervasive use of biotechnological methods in drug discovery. A recent statement by the research director of ICI indicated that approximately 25 per cent of that company's pharmaceutical research was devoted to biotechnology. On that basis at least eight European companies are investing more than $100m a year in biotechnology, and many more sums between $50m and $100m a year.

In building up their capabilities in biotechnology, the European majors are looking to the United States as much as to Europe. Hoffmann LaRoche's purchase of Genentech in January 1990 is perhaps the most dramatic example of this trend, building as it did upon Roche's already extensive research facilities in the United States. Ciba-Geigy, another Swiss firm, has research laboratories in the United States and the United Kingdom, as well as its extensive Swiss research facilities; in addition, in biotechnology it has two joint ventures, one with Chiron, one of the larger American new biotechnology firms, and one with Corning Glass on diagnostics; and it has research contracts or linkages with six other new biotechnology firms in the

United States and three in Europe. Bayer, the German chemicals/pharmaceuticals group has similar links. Besides its own internal research institute at Mannheim, it has a special relationship with the Max Planck Institute for Plant Breeding in Cologne, and owns three separate research centres in the United States – the Miles Cutter laboratories at Elkhart, NJ; the Mobay Research Center (agricultural/veterinary) in Kansas, and the West Haven Research Center, Connecticut, established in 1985 to concentrate on molecular biology and immunology. Bayer also has close links with the universities of Rochester and Yale and had a research link-up with Genentech to develop Factor VIII. Rhône-Poulenc's purchase of Connaught Biosciences (Canada) and Rorer (United States) is also interesting and reflects that company's desire to build itself (largely via acquisition) into a major multinational chemical/pharmaceutical company (see Chapter 12).

These examples illustrate well the way European biotechnology is now developing – the big chemical/pharmaceutical companies, after several years of sitting on the fence, are proceeding with major investments, and in so doing, spinning a complex web of relationships both within Europe and across the Atlantic. One of the main purposes of these links is to access academic research. There is no denying that in most areas of biotechnology American academic science is ahead of its European counterpart, and large European firms are exploiting to the full the open access to this research gained through the medium of the new biotechnology firm or through research contracts placed directly with university departments.

It is noteworthy that the collaborations are almost always with small firms and not with other large firms. Research collaborations with competitors are not favoured in the pharmaceutical industry – intellectual property rights (patents etc) are too important. Many small firms for their part have aspirations to become major pharmaceutical firms but know that they cannot afford on their own to develop and market new drugs and have therefore to find partners among big firms. The marriage between large and small firms has therefore been to mutual convenience.

It is also worth noting that, to date, collaboration with Japan has not been an important element of European or American firm strategy. Japan, however, is an increasingly important player in the pharmaceutical industry and its companies have been more active than European companies in forging links with American new biotechnology companies and accessing the American science base by this route (and by direct research contracts with American academic groups). We have already noted Japan's increasing contributions to drug discovery. Japan is also important as a market, and a number of European companies, like their American counterparts, are establishing R & D departments in Japan, partly to ease the path of market entry and partly to keep a watching brief on Japanese developments. Some have also set up marketing collaborations with major Japanese companies, and these downstream collaborations between major companies, which often amount to market sharing arrangements on a global basis, are likely to become an increasing feature of the industry.

IV Pharmaceuticals, biotechnology and 1992

For the European pharmaceutical industry the pressures identified at the end of Section II remain real. Biotechnology has to date done nothing to relieve cost pressures; indeed rather the reverse – it has added to the costs of R & D and vastly complicated the whole patenting issue. These complications are exacerbated by the European Commission's commitment to complete the internal market and the clear intention to make sure that the pharmaceutical sector does not, as in the past, manage to sabotage attempts at harmonisation. To date, as we saw, the sector has thrived on the fragmentation of markets and the discriminating pricing regimes which separate regulatory authorities have helped maintain. This will no longer be the case, although it seems likely that it will be in 1999 rather than 1992 that the single market is achieved in this sector. Debate centres around three issues – the regulatory framework and moves towards the creation of a European Medicines Evaluation Agency as a single common regulatory body; patents and the issues of intellectual property rights; and rules covering the price regime and the harmonisation of the various European systems. Those are considered by turn.

Regulatory issues and drug licensing

The criteria for evaluating the safety, quality and efficacy of new drugs have been laid down in the Community since 1965 and developed through a series of guidelines issued by the Commission. These relate to such issues as pre-clinical testing procedures, manufacturing protocols etc. In theory, therefore, the regulatory procedures for testing drugs in one country should be acceptable in all member states. In practice this seldom happens. There are in fact three separate sets of procedures:

(a) *the biotechnology/high technology procedures* instituted in 1987 whereby a submission is made to only one member state. This is copied to the EC's Committee on Proprietary Medicinal Products (CPMP), which, in consultation with member states, then issues an opinion on approval which may or may not be accepted by member states. In practice, on the relatively few occasions when this procedure has been used, member states have accepted the CPMP's decision. To date (late 1990), only seven biotechnology drugs have been approved by these procedures.

(b) *The multi-state system* (used for all new non-biotechnology drugs) first came into operation in 1975. It requires an application for approval to be submitted in the first instance to just one country. If it is approved then the papers may be passed on to two other countries (before 1983 it was five other countries) and if they both approve the drug is authorised throughout the EC. If one country approves the drug but others object, the CPMP is asked for an opinion. The CPMP's opinion is not binding on member states, and in practice this multi-state procedure has frequently run into objections.

(c) *The review of old drugs* A major problem for the EC regulatory authorities has been to review the many thousands of old drugs which were licensed before the member states and the EC established the basic regulatory requirements. The Commission requested agencies to complete a review of old drugs by 1990. This has been extremely difficult in those countries which have a large number of products on the market, such as Germany (with ten times as many licensed drugs as Denmark and five times as many as the United Kingdom or the Netherlands). While some authorities (for example, the United Kingdom) used the opportunity of reviewing drugs to clear the market of insufficiently safe or effective drugs, others (for example, France, Belgium and the Netherlands) saw the review process as merely administrative and have in effect nodded through all existing drugs.

The major firms are in fact keen to see the high-tech procedures ((a) above) work, for they constitute a precursor to a common European licensing procedure. From industry's point of view a single approval procedure obviously has major cost and timing advantages. In practice, as the relative failure of the multi-state route indicates, member states cling jealously to the right of approval and it is likely to be some time before a common licensing system is accepted.

A further complication for biotechnology-based products is the influence of the Green Party's views on genetic engineering, for they view it as an unwanted and dangerous technological development. In Germany, for example, for some time they prevented Hoechst from using a new insulin plant it had built. (This has recently been given the go-ahead under Germany's new Gene Law.) In the European Parliament they inspired a proposal for a five-year moratorium on the release into the environment of genetically manipulated organisms which was only narrowly defeated. Fear of what such proposals would mean for their activities has led a number of German companies to shift the focus of their biotechnology work from Europe to the United States (Fishlock, 1990). For their part, the major companies are lobbying hard against mooted restrictive legislation which they think would inhibit the development of biotechnology. The EC have already restricted the veterinary use of bovine growth hormones and delayed a decision on Bovine Somatotropin (BST), a hormone which increases cows' milk yields. There is also discussion of adding a 'proven market need' clause to approvals for all biotechnology products.

Intellectual property rights

Patents, as indicated in Section II, are a key issue for the pharmaceutical industry. Once disclosed in a patent, the composition of a new chemical entity can be copied and replicated by all comers; therefore patent protection is vital if companies are to cover the R & D costs of bringing new drugs to market. The increasing cost and complexity of the regulatory requirements have not only added to R & D costs, but lengthened development time and eroded effective patent life. The United States have responded to this by permitting drug companies to apply for an extension to patents of up to five years. In Europe, patenting is still primarily the function of national governments, although mutual recognition under the European Patent Convention of 1977 exists between most (but not all) EC countries. An increasing number of firms are also filing patents centrally with the European Patent Office which then automatically holds for countries who are signatories to the Convention. In most European countries (and with the European Patent Office) patents last for 20 years. The Commission has proposed to extend the period of patent protection for new drugs to make up for the loss of patent term as a consequence of lengthy regulatory evaluation (typically a loss of 10–12 years). They have suggested using a 'supplementary protection certificate' to add to the effective patent life (EPL). The maximum EPL would be 16 years and no certificate would last for more than 10 years. It has been suggested by some commentators that the changes in patent law should have been associated with policies to assist generic drug manufacturers, as in the United States (for example, reducing the amount of data required to be submitted prior to marketing).

As with licensing regulations, biotechnology complicates the scene considerably. The patenting position in Europe differs from that in the United States and makes it harder to get patent protection for biotechnology products in Europe. First, in the United States a 'period of grace' operates which effectively gives inventors/discoverers a year in which to decide whether or not to pursue a patent and does not preclude publication of details during this period; in Europe, any discovery that has been published is not patentable. This has inhibited the patenting of a number of major scientific discoveries, notably monoclonal antibodies. Second, in Europe the interpretation given to the 'non-obvious' criterion of a patent has been to exclude the patenting of naturally ocurring entities, whether cloned or uncloned. Emphasis has therefore been on process rather than product patents, but these, too, are less satisfactory, since process patents can more easily be copied and adapted. The EC Commission has put forward a draft directive which, while leaving unaltered the European Patent Convention, seeks to strengthen the protection offered to biotechnology, in particular by proposing that living entities should not, *ipso facto*, be excluded from patentability. If accepted, this would bring the European position somewhat closer to that of the United States. Neither system could be said to be ideal – except for the lawyers who are exploiting the opportunities for objection and litigation.

Towards harmonising drug pricing

As we have seen in Section II (Table 13.2) prices of drugs vary considerably in the EC, with high prices in Germany, Denmark and the Netherlands and low prices in Spain, Portugal, Italy, France and Greece. As a first step towards the long-term aim of harmonising the prices of drugs in the EC, a directive on the transparency of pricing arrangements has been proposed. This suggests that the authorities of member states must adopt criteria which are open to public scrutiny and fair. The directive does not go so far as establishing uniform criteria for setting prices, and it is still very much under discussion. In no sense could it be said that there was agreement on this issue, and it is likely to become highly contentious.

V Conclusions

The implications of these developments in regulation and control for national and corporate strategies towards pharmaceuticals are far reaching. For national governments, in this, as in other areas, the focus of policy-making is shifting to Brussels. Safeguard clauses in both the EC Treaty and the Single European Act, reserving the rights of national governments to act when public health or safety is at risk, will provide ground for rear-guard actions on issues of sovereignty, and for this reason, if no other, 1992 is, for pharmaceuticals, only the beginning of the end of the fragmented regimes, not the end itself. Nevertheless, by the turn of the century it seems likely that we will see in the European Community a single European licensing and patenting authority and a market for drugs in which differentiation results from the multiplicity of health care systems rather than from the distinctions of national boundaries and price control systems.

For the pharmaceutical companies there are both gains and losses from this scenario. The prime loss will be the ability to act as a discriminating oligopolist between markets and the benefits of profit and public clout that go with this. The gains will come in terms of rationalising the unduly fragmented European market structure. Those who gain will be those who seize the opportunities for cost rationalisation in both geographical and production terms; those who lose will be those who for long have benefited from inertial demand in home markets and fail to recognise the train of events. The industry will remain, as now, international. The gainers are as likely to be American (or Japanese) as European multinationals. The moves towards concentration and mergers, of which we have had a foretaste with Smith Kline-Beecham and Rhône-Poulenc-Rorer, will continue unabated.

The impact of biotechnology will be to accelerate rather than slow these changes. Given the size of its pharmaceutical industry and its strength in basic science, Europe has played a surprisingly small part in the emergence of biotechnology, whose evolution has been spearheaded by the academic groups and the small biotechnology firms in the United States. With the large pharmaceutical companies now moving to exploit these developments, Europe has the chance not only to capture a greater share of the value added but to encourage and promote the further development of the technology. There are two key factors which will influence outcomes: first, the quality of the science base; second, the regulatory environment.

The first factor, the quality of the science base, remains the responsibility of national governments not the Commission. It cannot be stressed too strongly that biotechnology is a new technology closely tied to its scientific underpinnings. Its development needs both creative thinking from that underpinning and a steady stream of well-trained scientists across the full range of skills in the life sciences (Irvine *et al.*, 1990). If the higher education institutions of the Community cannot meet these needs, the pharmaceutical industry will take its R & D and production facilities – and their value added – to the United States or Japan.

The second factor, the regulatory environment, is now primarily an issue for the Community rather than national governments, although national governments retain the power to influence, delay and, for that matter, to destroy EC initiatives. The industrial lobby is pressing for patent changes and regulations which are recognised Community-wide and are encouraging not discouraging in their requirements. This does not mean lax regulations. If biotechnology is to take root and flourish in the Community, what is needed, more than anything else, are clear and rapid decisions on this issue. As the tough gene laws in Denmark (and now Germany) demonstrate, industry can cope with tough regulations: what would kill all prospects for biotechnology within the Community would be a protracted period of uncertainty and delay.

This chapter has demonstrated that the futures of the pharmaceutical industry and biotechnology go hand in hand. The pharmaceutical industry is a high value-added industry and the strength of its firms has brought great benefit to Europe. Biotechnology, as a new technology, is about maintaining and adding value to the output of this industry. In a world in which capital, labour, knowledge and skills are increasingly mobile, there can be no foregone conclusion that, because Europe's pharmaceutical industry is strong, so it will necessarily assimilate new ideas and grow in strength. Europe has to demonstrate that it can provide the skills and the environment necessary to encourage the flowering in its midst of this new technology. Otherwise the skills, the knowledge and the value added of this important industry may flourish instead in other places.

14 Technological change and the internationalisation of construction in Europe

David Gann

I Introduction

Shifts in the nature of demand for buildings, together with technological and organisational innovations in the construction process are leading to radical changes in European construction industries. The transition from craft-based processes towards engineered solutions including the use of assembly techniques is increasingly evident. The international 'leading edge' now looks very different from traditional, provincial construction industries and the gap between the methods and technologies used in the production of high-technology buildings and conventional offices and housing appears to be widening.

National construction industries are changing, and an international construction system is emerging. In Europe, firms originating in one country are increasingly building under contract in other countries, seeking to expand operations and increase size through joint ventures, acquisitions and mergers. In Britain and more recently the rest of Europe, American project management techniques are being adopted. At the same time, Japanese developer/constructors have begun to operate in Europe, increasing the potential for introducing more industrialised methods of construction. The construction sector is distinct from service sectors because it entails the transformation of materials and components on site. It may also be distinguished from manufacturing because the final product is built at a fixed location where it is to be used. For these reasons, the labour, materials and plant required in the construction process are highly mobile. The installation of equipment in buildings now accounts for a growing proportion of the value of construction work. Firms from electronics, telecommunications and engineering industries are entering the construction arena, introducing new technologies and methods of working.

This chapter focuses on recent developments in commercial building (offices, retail developments, leisure complexes and light industrial business parks) which have been at the forefront of these changes. Demand for such buildings has grown fast over the past 10 years with the transition from an economy based upon mass production and consumption to the information economy, typified by new types of buildings. The need for new office facilities, emanating first from the rapidly growing financial services sector, has led to the development of so called 'intelligent buildings'. Technical and organisational changes developed in the construction of such projects are beginning to diffuse into other types of building work. For this reason, an examination of office developments offers an insight into possible future trends across the entire construction sector.

The chapter begins by describing changing European construction markets. This is followed by an account of differences in the construction industry among countries, assessing relative strengths and weaknesses of various national systems of construction. We shall assess the emergence of an international construction system in Europe and discuss the main characteristics, particularly competition, the innovation process and changing industrial structure. Finally we examine the role of the European Community and the implications for industrial policy in the European construction sector.

II European construction markets

In 1987 the EC construction market was valued at about £250bn, amounting to some 32 per cent of total construction work done in the developed market economies. Table 14.1 illustrates how this was divided between the 12 member states; the largest contributors were Germany, France, Italy, and Britain. For comparison, the American and Japanese construction markets were both rather smaller at approximately £160bn each.

Construction markets are highly cyclical, and national trade cycles do not always coincide. In 1990 the British industry entered a severe recession and in several other EC countries there was a slow-down in construction; but in Germany activity continued at a high level. It is therefore difficult to generalise about Europe as a whole. Industrial, private commercial and office-building activity grew particularly rapidly between 1981 and 1988, by an average 7 per cent per year. In most countries, house building recovered in the mid-1980s after several years of decline. Renovation and modernisation grew rapidly in the period 1980–88, rising from 31 per cent of building output to 39 per cent. Public-sector construction work in the EC member states grew more slowly. Repair and maintenance is the largest area of activity in the larger EC economies and the Netherlands (Figure 14.1). They already have well-developed infrastructures and building stocks. Non-EC countries also have sizeable construction markets. The total value of construction work in the EFTA countries was £54.5bn in 1987 (Table 14.2) but Eastern Europe is perhaps the largest potential construction market.

In many East European countries construction accounts for a larger proportion of GDP than in Western Europe (Figure 14.2). Only Japan has a similar high proportion. But East European industries and infrastructures are in need of modernisation and in many countries housing is in poor condition. The rapidly changing political and economic climate in Eastern Europe makes it difficult to predict how construction demand will be affected in the future. If trade with the West increases and East European countries open their economies assisted by Western aid and investment, it is possible that construction demand will increase substantially. This could provide opportunities for Western European construction firms to expand their operations, transferring their more advanced technologies to the East. But it would also raise questions about how less-developed economies can best develop their own construction technologies and building materials industries.

Table 14.1 Value of EC construction work, 1987

Country	Value (£bn)	(%)
Former West Germany	79	32.5
France	42	17
Italy	39	16
United Kingdom	35	14.5
Netherlands	12	5
Spain	12	5
Denmark	8	3
Belgium	7	3
Greece	4	1.5
Ireland	2	1
Portugal	2	1
Luxemburg	1	0.5
Total	243	100.0

Note: The data presented in this chapter are based on contract values, not on net output. They should be used with caution. It is impossible to obtain directly comparable contract data for construction because industries organise their accounting procedures differently in each country. For example, in some cases double counting may occur owing to contractors and subcontractors each registering the same work, and the scale of activities may be skewed by one or two very large special contracts.

Source: Euro Construct – Proceedings of Conference on European Construction, Vienna, 1988

Table 14.2 Value of EFTA construction work, 1987

Country	Value (£bn)
Switzerland	13.5
Sweden	13
Austria	10
Norway	9
Finland	8*
Iceland	1*
Total	54.5

*Own estimates

Source: Euro Construct – Proceedings of Conference on European Construction, Vienna 1988

The development of the Single Market after 1992 also seems likely to lead to a new wave of investment in buildings and infrastructures in the EC that will boost construction demand, including demand for new types of 'intelligent buildings'. There are also likely to be substantial EC structural and regional development funds (of up to £57bn) available to Spain, Italy, Portugal and Greece, to be used for infrastructure investment, while France, Germany and the Netherlands are committed to major new investment in railways.

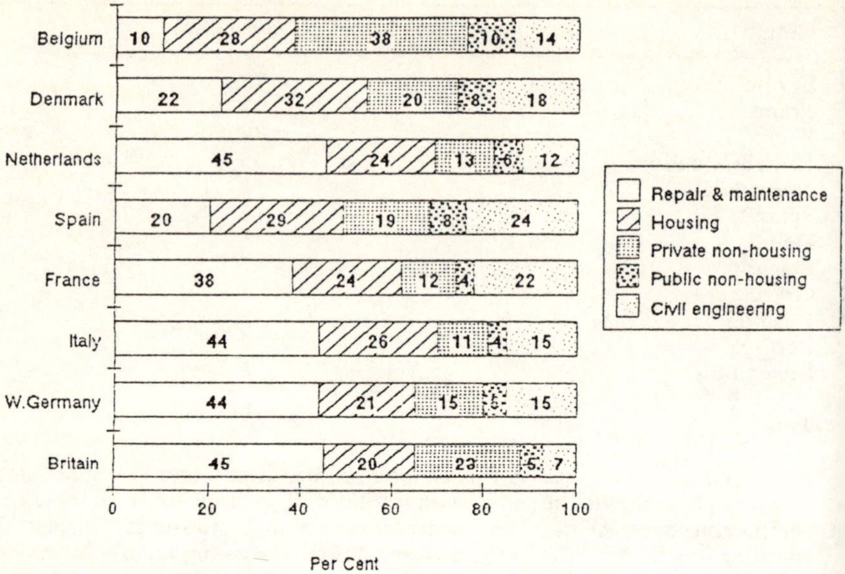

Figure 14.1 Percentage of construction output by marker sector, 1987
Source: BSRIA/Euro Construct – Proceedings of Conference on European Construction, Vienna, 1988

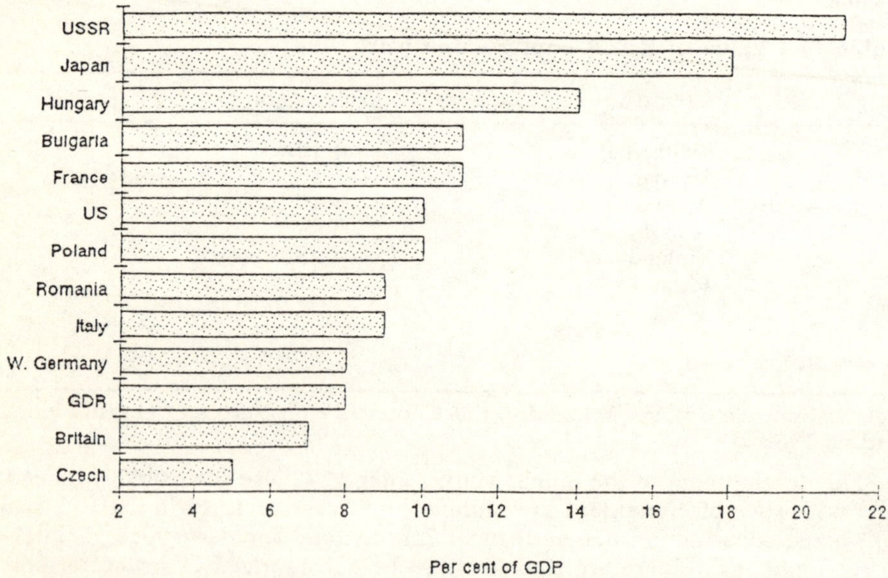

Figure 14.2 Construction activity as a percentage of GDP, 1989
Note: data for non-East European countries is 1987
Source: Camargue Communications Market Research Unit and Euro Construct – Proceedings of Conference on European Construction, Vienna, 1988.

Finally, global environmental issues are likely to make a huge impact on the construction industry. For new building work, more stringent energy regulations together with legislation to limit pollutants in the building process may lead to the use of new building materials and the adoption of new construction techniques. Repair, maintenance and refurbishment workloads may also be boosted by the need to bring the existing building stock up to modern standards for energy conservation.

To sum up: the prospects are for strong demand on the construction industry in the 1990s all over Europe, but with the caveat that the industry can be heavily affected by cyclical fluctuations. To meet peaks of demand will create great pressures for organisational and technical change as well as high mobility of resources.

III European national construction industries: labour and training

The construction industries of each European country have developed historically along different paths based upon craft skills, locally available supplies of building materials and the demands imposed by climatic and geographical conditions. There have therefore been marked regional differences in the style of buildings and their methods of construction. Regional fluctuations in demand have led to a high level of mobility of construction labour within countries. But the differences between construction methods in each country, institutionalised in legal frameworks of regulations, procedures and practices, have tended to reduce labour mobility across international frontiers. Problems associated with transporting bulky materials have also generally lessened the drive towards a more internationalised industry. Paradoxically the large building-materials suppliers have been among the first to become more international through a strategy of ownership in other countries rather than by exporting materials; the British firm RMC is a classic example.

Construction is labour-intensive and the performance and capabilities of different national construction industries are heavily dependent upon the skills of their workforces. The innovation and learning processes are interrelated. In particular, informal methods of learning by doing on-site – often in gangs – contribute to incremental changes and adaptations that facilitate the successful completion of projects. Workers often have high levels of tacit knowledge concerning particular materials which can prove invaluable in solving problems of 'unbuildable' design details, but which are rarely fed back to designers or made available more generally. However, methods of acquiring skills are in a state of flux: first because of structural changes within the industries – for example changing specialisations, the growth of self-employment, technical change and changes in paths of communication – and second, because of changes outside the industry in the provision of training, such as the shift in British government policy to employer-led training.

The learning process retains a semblance of traditional craft practices with parts of the industry clinging to the master–apprentice relationship and time-served experience as the only viable method of training. Apprenticeships and formal training with certification are more firmly adhered to in the regulated labour markets of Scandinavia, Germany and France. Trade unions work alongside employers and government to maintain up-to-date training programmes and ensure construction workers are properly qualified. In Germany, a skilled workforce has been maintained by adapting traditional training structures to a modular approach. Thus these countries have managed to sustain a higher level of skill among their workers than Britain over the last decade. In particular, Sweden is very strong in management expertise, and the German and French industries are strong in construction engineering skills. However, even in these countries the question remains whether the training systems can be adequately adapted to provide the skills needed by an industry facing rapid technological change. In Britain the quality and quantity of training has been eroded by the casualisation of employment. The number of traditional craft apprenticeships has dwindled and been replaced by youth training schemes and short specialist courses to train for semi-skilled work. No clearly identifiable and workable approach to training is evident despite the erosion of apprenticeship schemes. This situation may seriously jeopardise the future competitiveness of the British construction sector.

At the craft and operative levels, the construction industry throughout the world (with the possible exception of Scandinavia) suffers from the problem of only being able to attract school leavers with the lowest academic achievements. The physical harshness of work, its poor record on health and safety and its poor image contribute to recruitment problems. Firms, denuded of indigenous workers during the 1960s, had little choice but to import labour: Turkish and Italian workers were employed in Germany; Spanish and Portuguese, Algerian and North African workers in France; Italian labour in Switzerland and Irish workers in Britain. Internal migration is also important – the 1980s' construction boom in south-east England drew an influx of construction labour from across Britain, particularly the North and Scotland. Shortages of skilled operatives were experienced in many countries despite the use of migrant labour. They occur either because not enough skilled labour is being trained and encouraged to stay in the industry or because skill needs have changed and those in the industry are no longer equipped to carry out the tasks required. In Japan, persistent skill shortage is one of the main forces driving the trend towards the industrialisation and automation of construction, including the development of construction robotics.

Labour markets are changing in two important respects: there has been a shift to self-employment; and increased use of specialist subcontractors. These have come about partly because firms have tried to reduce costs by shedding directly employed labour and partly because of the demand for increasing technical sophistication. The changes have been most pronounced in Britain where well over half the work-force is now employed casually. Even in more regulated labour markets such as Scandinavia, France and Germany there has been a trend towards self-employment. An oligopolistic industry has emerged in north European countries where a relatively small number of large national contractors are at the top of a pyramid of medium-sized regional contractors and a base of many small local builders, specialists and the self-employed.

Despite attempts to industrialise construction, the building process has remained labour-intensive relative to many other branches of industry. The site-specific nature of production and the bespoke product are often cited as the reasons for low levels of mechanisation. But construction's general inability to innovate using industrial methods probably has more to do with structural problems within the industry itself (Ball, 1988). Construction in southern European countries such as Portugal, Greece, Spain and parts of Italy is very localised and craft-based. In northern Europe higher levels of industrialisation based on prefabrication were achieved in the reconstruction programmes after the two World Wars and in Britain with the slum clearance programmes of the 1960s.

In Eastern Europe under the centrally planned economies, the state pursued more vigorous attempts to industrialise the construction process. Systems were imported from Western Europe, but often in such a way as to give little choice over design and as in Britain, poor workmanship led to these industrialised techniques becoming discredited. The social unacceptability of poorly designed high-rise flats was an additional factor. However, other European countries such as Denmark, Sweden and Finland persisted with the use of industrialised components with better designs and superior skills. In part, the problems of building in and for harsh climatic conditions spurred the use of off-site fabrication methods. But, as we shall see below, a new stage in the use of industrialised, componentised construction techniques characterises the emerging international construction system right across Europe.

IV The emergence of an international construction system

International construction activity dates back to the age of empire and the large energy and transportation projects (often with military interests) of the 19th century. During the 1950s and 1960s international construction orders expanded in Africa and Asia with development aid programmes geared towards infrastructural, mining and industrial projects. Growth in overseas construction moved into a new phase in the 1970s with the rise in demand from the Middle East, associated with newly found oil wealth. Many petrochemical complexes were constructed, and the size of the market opened possibilities for firms from newly industrialising countries (NICs), such as the Republic of Korea, to enter the international system. By the end of the 1970s the Middle East was the fourth largest construction market in the world behind the EC, the United States and Japan. Construction in the Middle East peaked in 1982 with 42 per cent of all international contracts awarded in the region. By 1983 the boom was over and the locus of international construction activity shifted to the Asian markets (Strassmann and Wells, 1988).

Despite the entry of firms from NICs into the international arena, firms from the industrialised countries maintained their domination. Between 1979 to 1985, American firms captured a 36 per cent share of total overseas work, Western European firms 39 per cent, and the Japanese nearly 15 per cent. By the mid-1980s a group of powerful multinational construction firms from these countries had emerged. Since the 1960s the European Commission has made efforts to liberalise cross-border trade in contracting, but with remarkably little success until the downturn in Middle Eastern and developing country workloads meant that the big international firms had to search for new markets. Firms such as Holzmann and Hochtief of West Germany began to facilitate projects through the provision of finance and involvement in real estate and speculative development. The location of work shifted away from developing countries to construction projects in the industrialised countries:

in 1985 almost one-third of total overseas work won by the top 250 international contractors was in the developed countries.

In contrast to the experiences of international contractors with their provision of design and management expertise to developing countries, international trade in building materials has always been characterised by intra-developed country activity. In 1986, international trade in building materials amounted to $120bn with exports from industrialised countries to other developed countries accounting for 60 per cent. The former West Germany, Japan and the United States produced 30 per cent of global exports and bought 25 per cent of all imports (Drewer, 1990). In Britain, the construction boom between 1983 and 1988 sucked in a large volume of imports, ranging from bulk materials such as cement and bricks, to high value-added components such as mechanical and electrical equipment. Britain imported twice the value of construction goods and services exported; the deficit in the balance of trade in construction amounted to nearly 18 per cent of the total trade deficit in 1988.

The development of long-term strategies is not easy in this industry. In addition to the cyclical problems, firms operate within an environment dominated by trade restrictions, building and planning regulations, different legal frameworks and constraints over access to finance, information, skilled labour and other resources. Firms are also constantly balancing workloads between portfolios of projects based on fluctuating domestic markets and those based on fluctuating overseas markets. It is therefore difficult for them to commit resources to building an international presence on the basis of long-term plans.

American firms have generally been able to maintain their competitiveness where complex organisational skills and technological expertise are required. They have particular strengths in management and design, but are weaker in utilising technologies on-site in the construction process. This is despite the strong research orientation of a few groups, such as the Bechtel group, which has an R & D department to rival that of the large Japanese contractors.

Japan modernised its construction sector by importing American and European technologies and through developing a strong research base. It quickly gained a reputation for excellence, particularly in design/build projects. Unlike their European and North American counterparts, Japanese contractors play an important and leading role in innovation. The big six contractors each have large research laboratories which carry out long-term research projects, often in collaboration with manufacturers. Extensive research is carried out into electronic systems, construction robotics, prefabrication, the use of new materials and grand projects such as 'intelligent cities' and ocean, underground and space cities. The ministry of construction and MITI encourage these efforts and were involved, for example, in negotiations with the Australian government to extend their 'Intelligent City' programmes to Australia. Japanese contractors are prepared to invest in developing their engineering expertise to gain international technological leadership, in the hope that they will be regarded as engineering developer/constructors capable of taking on any project, anywhere in the world. Japan therefore spends 3 per cent of turnover on R & D, compared with 0.5 per cent spent by the United States and Britain.

We have already discussed some of the strengths and weaknesses of European construction industries. French firms are often at an advantage because they receive government backing from export credit guarantees, aid packages and soft loans. German contractors tend to focus on highly complex projects such as hospitals and airports, but they also specialise in maintenance and aftersales services on completed projects. German contractors had an 80 per cent success rate in winning international contracts where they had been involved in early design and consultancy. This compares with 63 per cent for the Japanese, 50 per cent for French and Italian firms, 43 per cent for the United States, and only 13 per cent for British firms (Murphy, 1983). One reason for the lower success rates of British and American firms is the independent role played by consultant engineers. German consultants tend to be more closely linked to specific construction firms.

Swedish, Danish and Finnish contractors have a great deal of expertise in prefabricated systems building. The Swedish construction industry is advanced in the use of IT, particularly CAD/CAM, when compared with the rest of Europe and Japan; the highly organised system of codification for materials and components has helped promote this advantage. The British construction industry had much experience of building in the ex-colonial countries, but now has closer links with the American construction industry. North American concepts of project management have been imported into Britain and spread rapidly since 1982. Thus, a number of large British contractors have considerable project management expertise, but British firms generally lag in technical know-how, lack investment in R & D and training and generally suffer from a short-term approach to financing.

V Changing patterns of demand and supply

The change in emphasis from manufacturing to the service sector in developed countries meant a shift from attempts at raising productivity of blue-collar factory workers to increasing productivity of white-collar office workers. This, together with the need to compete using information transferred in real time, has driven the demand for new types of intelligent office building. Large multinationals such as Ford and IBM increasingly require integrated communication systems, connecting offices sited in different locations across Europe and the rest of the world. Ford, for example, need networked design offices linking CAD/CAM systems in different countries (Davies, 1989). Such technical requirements give rise to the development of new 'Euro-offices' – offices that function in a similar manner in whatever country they are located. The Single European Market appears to be as much a response to these changes as the driving force of change itself, but 1992 is likely to act as a catalyst to further change. From the supply side, developers producing speculative offices are also providing standardised office facilities at different locations. Mitsui Real Estate Corporation, for example, uses this as a method of tying their tenants into renting their particular style of offices across Japan.

A new wave of fixed capital investment began in Europe with the rapid growth in development of commercial office space between 1983 and 1989. This expansion was precipitated by the needs of the international financial services sector and the growth of new telematic services (see Chapters 8 and

9). In 1981, Britain was the leading country in Western Europe in the number of installed network terminating points (Sauvant, 1986; Morgan and Sayer, 1988). Deregulation of the telecommunications infrastructure in 1984, and of the City in 1987 – 'The Big Bang' – added to demand for building and office automation systems. London's office stock was technologically obsolete in terms of meeting users' growing demands: new office building amounted to a replacement of existing stock of only 1 or 2 per cent per year. A great deal of re-equipment created rapid growth in the 'retrofit' markets which needed skills and technologies distinct from the new building or maintenance markets.

During the 1980s, the pattern of British investment shifted, spurred by a change in planning restrictions in 1987 when a new class (B1) of general purpose commercial/industrial buildings was created. This in part reflected the need for high-technology industries to have more office-based activities such as software programming, corporate IT centres, research laboratories and other light industries. But it also facilitated a new wave of deregulated speculative development by large multinational property investors, investing in out-of-town office campuses. The trend towards decentralisation of office space has been driven by difficulties in finding sites in cramped inner city areas and commuting problems experienced by office workers. The growth in IT services is facilitating the trend to decentralisation. But the main location for intelligent buildings is limited to the nodes in the new information networks, usually in the financial centres.

By the end of the 1980s the boom in speculative office development in Britain was spiralling rapidly downwards with the collapse of several major developers. Rising interest rates and a glut of office space were the main catalysts. At the same time office construction work began to grow rapidly in the financial centres of continental Europe. Business and governmental realignments in anticipation of the Single European Market, the dissolution of East–West boundaries and growth in global capital markets stimulated demand. The emphasis in construction work had shifted from London to Frankfurt, Paris, Milan, Madrid and, more recently, Berlin.

Technical change in construction has typically been a piecemeal process in which European contractors have only played a minor part. Innovations developed by materials and component manufacturers or design teams have generally been the main source of change. But recently the demand for high-technology buildings has played a leading role in driving the rate of change. This has resulted in technical innovation in both products and processes. Buildings have become complex in terms of the sophisticated mechanical, electrical and electronic equipment installed in them. They have to be viewed as 'active systems' – in many ways similar to ships – rather than passive, monolithic artefacts (Gann, 1989, 1990). The production of such complex systems has placed new demands on the construction sector resulting in a shift from traditional craft practices to an engineering approach of component assembly. Design and construction teams are increasingly using CAD systems to enable such engineered solutions to coordinate their work.

In order to manage such complex projects, some general contractors have adapted their role to become construction managers, solely providing management (and sometimes design) expertise. They employ no labour directly, but coordinate chains of specialist subcontractors. In terms of this

Table 14.3 Top-20 European contractors and their exports as a percentage of turnover, 1989

Rank/Company	Country	Turnover £bn (rounded)	Exports as % of turnover	No of employees ('000s)
1 Bouygues	France	4.5	28	69.6
2 BICC	United Kingdom	3.8	49	46.0
3 Tarmac	United Kingdom	3.5	17	32.1
4 SEE	France	3.3	37	63.2
5 Trafalgar House	United Kingdom	3.2	45	29.1
6 Skonska	Sweden	2.5	20	29.5
7 Philipp Holzmann	Germany	2.5	49	28.7
8 Dumez	France	2.5	73	35.9
9 SAE	France	2.4	41	25.7
10 Spie Batignolles	France	2.2	34	38.8
11 Italstat	Italy	2.0	n.a.	n.a.
12 Wimpey	United Kingdom	2.0	26	17.4
13 AMEC	United Kingdom	2.0	18	30.1
14 Beazer	United Kingdom	2.0	51	20.5
15 GTM Entrepose	France	1.8	22	30.7
16 Hochtief	Germany	1.7	36	26.2
17 NCC	Sweden	1.7	17	19.3
18 Costain	United Kingdom	1.4	32	13.7
19 John Laing	United Kingdom	1.4	n.a.	12.8
20 Taylor Woodrow	United Kingdom	1.3	25	9.2

Source: Building, 1990

type of organisational change contractors have been extremely innovative: many different forms of construction management have emerged ranging from fee management to design/build. Such ideas were first used by American contractors who also pioneered the fast-track technique of construction, aimed at speeding up the process, and overlapping design and construction to allow building work to commence early. There are however numerous problems associated with the use of this technique (Gann, 1991).

Technical and organisational changes stimulated by the demands for high-technology buildings have led to the entry of engineering, electronic and telecommunications firms into the construction sector. IBM and Siebe recently founded a joint-venture company – RIBS – to move into the intelligent building market. In Germany the electronics firm Thoma has successfully penetrated the lift industry. Competition from highly skilled specialist engineering firms has intensified with the growth of supply and fix subcontracting. Firms such as Josef Gartners of Germany, specialising in curtain-walling systems have begun to compete successfully in several European countries and in the Far East. They have been particularly successful in Britain, despite a large cost disadvantage, owing to their high quality product, engineered by a well-trained and skilful workforce. Similar examples can be found in many other value-added construction components such as micro-electronic building management systems, air-conditioning and heating equipment, lifts and escalators and structural components.

In response to this new technological competition firms have regrouped, diversified or merged: for example, BICC in Britain has brought together a number of specialist construction, engineering and electronics firms. Bouygues, the largest European contractor, has shares in a private television channel and grain milling. Mergers are taking place across industries as firms prepare to tackle large infrastructural projects such as the upgrading of water supply to meet more stringent environmental standards. For example, Dumez, the French civil engineering group and Lyonnaise des Eaux, the major French water and waste company, merged in July 1990 to create a group with a turnover of £5bn (*The Economist,* 1990). Specialist firms, concerned that they are too small and lacking in local knowledge of regional markets, have attempted to increase their size and expand international operations through mergers and acquisitions. European contractors are increasingly buying into companies in other countries, setting up joint ventures, and entering into *ad-hoc* agreements to bid for individual contracts. Table 14.3 illustrates the top-20 European contractors and their exports as a percentage of turnover.

British and French companies appear to have been most active in the merger stakes: Bovis has recently acquired a French building firm John Mowlem has acquired a French scaffolding group and secured majority holding in a Spanish firm in the same business (Cooper, 1989a; Taylor, 1989). Laing Homes began discussions with possible partners in the Netherlands and is considering entering the German, French, Spanish and Italian markets through joint venture or takeover. Wimpey has operated in France for several years, and several other British housebuilders are already there or thinking of setting up there.

Several companies are seeking to expand in Spain and Portugal (Cooper, 1989b) and British construction interests in Spain are already substantial (*Building,* 1988). German companies such as Hochtief are expanding their British operations. Firms are also entering joint ventures with counterparts from Japan and the United States (for example, the Anglo–American management company Bovis-Schal).

VI The role of the European Commission

The emergence of a European construction system raises many questions for European policy-makers. For example, will a unified market give rise to increased competition from American and Japanese construction organisations? If so, will this depress profits and therefore reduce expenditure on R & D by European firms, thus further weakening their competitiveness? Will the divisions between leading-edge and traditional construction work diverge to the extent that some regions will benefit from high-technology buildings because they have particular expertise, while other regions will lose out?

The European Commission is influencing the direction of change on several fronts, for example through regional aid funds, through developing transport and infrastructural strategies and through regulations for energy efficiency, safety at work, and in codes, standards and practices.

In the area of competitive tendering for public works, the aim is to 'blow apart' the national protectionism which surrounds the vast majority of public-sector contracts within the EC. These regulations came into force in 1990 and stipulate that all contracts valued at more than £100,000 must go out to competitive tender and be advertised throughout member states (Cooper, 1988). The Commission is also working on a standard form of contract to unify conditions across the member states.

A major initiative is being taken by the Commission to provide the framework for thousands of new standards and codes of practice concerning construction products. Its aim is to provide for the free movement, sale and use of construction products which are 'fit for their intended use and have such characteristics that structures in which they are incorporated meet certain essential requirements'. These essential requirements are similar in style to the functional requirements of the Building Regulations in force in England and Wales but rather wider in scope. They relate to mechanical resistance (stability), safety in case of fire, hygiene and health, environment, safety in use, protection against noise and energy economy (*National Council of Building Material Producers and Building Magazine,* 1988:15). This directive is supposed to be implemented by member states by 1992, but in reality it will be many years – perhaps decades – before the directives are agreed. The broad scope for definitions of what is 'fit for intended use' would only seem to add to confusion.

The directive covers standards and performance requirements for design, building materials and components and certification of innovative products: all of which will have to conform to new CEN (Comitè Europèen de Normalisation) standards or prove that they comply with other standards that meet EC requirements. Inevitably political and commercial pressures affect the standardisation process as well as purely technical issues (Chapter 11). The German Standards Institute (DIN) is pressing for rationalisation of the one hundred cement standards which exist in Europe now (Fisher, 1989). DIN has been responsible for introducing more standards than any other country in the EC, and other states suspect that Germany's influence in European standard-setting bodies may be too strong. DIN standards are highly respected through the world, but the director of DIN was reported as agreeing that DIN standards have been used successfully by German exporters as a 'competitive tool'.

The European Commission is also concerned with directives to promote construction labour mobility. Such directives together with those on skill certification are proving very difficult to agree. Well-qualified, well-paid German craftsmen such as electricians and plumbers have been reported as expressing anxiety that mutual recognition of craftsmen could lead to their jobs being lost to craftsmen from other countries willing to accept lower pay (Goodhart, 1988). But progress is likely to be slow. It took 17 years of laborious negotiations to agree community directives on the qualifications and training necessary for architects to practice throughout the Community.

VII Conclusions

The three most striking features of international construction activity in Europe since the mid-1980s are: rapid technological innovation; organisational changes in the construction process; and industrial restructuring in consequence of the changing nature of competition brought about by technical and organisational change. There are two inter-related ways in which construction firms have adapted their competitive strategies to respond to these conditions. First, they have sought to extend their markets by offering different services – of particular interest here is the shift into construction management. Second, they have attempted to reduce their costs by adjusting labour inputs through organisational changes; by adopting new materials and components; by shifting production off-site through the use of prefabrication; and by utilising more capital equipment.

Industrial policy-makers are left to grapple with several important questions. First, in future, governments in most European countries are likely to have less influence on the size and type of construction demand than in the past. Second, at a European level, some regions are likely to be more economically attractive (Benelux, southeast England, France and southern Germany) and are likely to benefit more from technology transfer and expertise within the international construction system than less-attractive regions. In consequence, uneven development is likely to be reinforced, and disadvantaged regions will remain even more dependent upon national governments and Brussels for aid.

Third, the changes discussed in this chapter are associated with the general growth in international trade and the demands made by multinational companies. While the removal of trade barriers may reinforce these trends, the divisions that exist within Europe are only to a limited extent the result of trade barriers and are unlikely to be greatly affected. Demand for construction output varies across Europe, not so much because of barriers, but because of differences in preferences, habits, climate and culture. Trade liberalisation is likely to have a primary impact on the major firms who are already operating internationally; it may well do little for the majority of small and medium-sized building firms operating on a local scale. Differences between the 'leading-edge' firms in construction and the rest could thus be exacerbated, adding to the problems of uneven regional development.

Finally, growing concern about environmental issues is likely to have a number of consequences for the construction industry and its choice of technologies. For example, concern over the resource waste may counter the quickening obsolescence in buildings. Technologies associated with energy conservation (see Chapter 17) are likely to expand to deal with the consequences of global climate changes, with new and specialist skills required among engineers, managers and designers. To date, the response to environmental pressures by the European construction industries has been slow and cumbersome, even though this is a sector where it could take an imaginative lead.

It is by no means certain that the European construction industries will be capable of developing the new technologies and skilled workforces needed to meet the challenge of international competition in a sector which has become rejuvenated by the influx of firms from other industries. Nor is it certain that the traditional construction industries will be in a position to respond to demands of producing more environmentally friendly and safe buildings and structures. The European Commission could play an important role in supporting innovation and raising the levels of skills and environmental standards in order to meet these challenges.

15 Mechatronics and the machinery industry

John Bessant and Bill Haywood

I Introduction

The engineering industries are involved in the production of a range of machinery for use in other sectors of the economy and in this role play a key part in determining the productivity level in manufacturing. They are simultaneously the main producers and consumers of advanced manufacturing technology and thus their experience of adoption and implementation of technological change is to a considerable extent indicative of the level of development of industry as a whole.

This chapter deals mainly with ISIC 382 – Machinery (NACE 32). Characteristic of this industry is a relatively low average batch size – mass production accounts for less than 20 per cent of all output and the vast majority of products are made in batches of less than 50 items. This has important implications for the nature of technology and production organisation which has grown up within the industry and which is geared to respond to the pressures of high variety and customer-specific small-batch production while also trying to take advantage of scale economies whenever possible.

Information technology has offered new solutions to many problems of the industry, especially in batch production. The diffusion of computer-related equipment ('mechatronics') has been the main feature of technical change in the last 20 years. 'Mechatronics' is a broad term (originally coined in Japan), which describes a range of manufacturing innovations involving combinations of mechanical and information technologies. This group includes computer-numerically controlled machine tools, robotics and Flexible Manufacturing Systems (FMS), as well as a variety of equipment for measurement and test, storage/handling and transport. These types of equipment are also diffusing rapidly in the automobile, aircraft and electronics industries. The automobile industry in particular has been one of the biggest users of robotics and other machinery for automation. This chapter therefore should be read in conjunction with Chapter 16 on automobiles. The key enabling technologies have been:

- low cost microelectronic components (microprocessors, memories, interface management, chips, etc.) which enable 'intelligent' control of equipment;
- advanced computer network technology (local area networks, fibre optic communications etc.), which enable interconnected systems to use distributed rather than simply hierarchical control;
- rapid accurate response actuators (such as direct drives for robot arms);

- advanced sensory systems;
- Computer-Aided Design (CAD), which enables a closer integration of design and manufacturing and greatly assists engineers by the display of graphics and calculations.

The emergence of these elements has had the effect of accelerating a long-term trend towards integration within manufacturing systems which offers radical improvements not just in local *efficiency* of particular operations but in overall *effectiveness* across the entire manufacturing function. In other words they enable a shift from 'substitution innovation' – doing what was always done a little better – to doing things in totally new ways. We will discuss this theme in detail in Section II of this chapter.

As has been shown in Chapter 2, several European countries, particularly Germany and Italy, have performed relatively well in production and export of machinery in the 1980s (Table 15.1). In fact the EC is the biggest producer and exporter of these products (Table 15.2).

But whether the EC continues to keep pace with the world competition in the 1990s will depend increasingly on the adoption and effective use of 'mechatronic' technologies. Although mechatronic systems have radically increased the range of activities which can be linked together in an automated system, their contribution to improved effectiveness is dependent upon the organisational context in which they are used. Flexible systems which are constrained by inflexible and rigid organisational arrangements have repeatedly been demonstrated to be less effective than those which are well matched to the organisational context. Jaikumar's (1986) comparison of the use of FMS technology in the United States and Japan, for example, showed that Japanese firms are able to make essentially similar physical systems perform in a way which is at once both more productive and more flexible than their American counterparts. The rate of growth of Japanese production and exports in the 1980s was much higher than that of either the EC or United States (Table 15.2).

The diffusion of mechatronics should be seen within the wider framework of the 'mis-match' between old institutions and the new information technology (Perez, 1983). The work of the Tavistock Institute already showed in the 1960s the importance of relating social and technical change with the concept 'socio-technical systems'; McLoughlin and Clark (1988) have developed this in relation to work satisfaction.

In this chapter, we concentrate on the diffusion of FMS (Section III) and robotics ((Section IV) in the EC, Japan and the United States as a major indicator of relative competitive strength in new technologies. Finally, in Sections V and VI we show differences in patterns of work organisation and policy implications.

II Flexibility in manufacturing: the evolution of machinery systems

Since the machinery industries typically manufacture a mixture of products in batches of varying size, flexibility is extremely important. However, as Gerwin (1986) suggests, there are many types of flexibility, some of which have short-term effects and some of which have longer-term implications. A

Table 15.1 Production of machinery by country (Current prices, ECUbn)

	1980	1989
EC	115.5	209.8
Belgium	3.2	5.2
Denmark	2.0	3.8
Former West Germany	47.7	93.0
Greece	0.1	0.2
Spain	4.3	6.5
France	15.4	23.9
Ireland	0.2	0.5
Italy	15.6	35.3
Luxemburg	0.2	0.2
Netherlands	3.8	5.9
Portugal	0.3	0.5
United Kingdom	23.5	34.9

Source: CEC, 1990b,

Table 15.2 Production and trade in machinery (Current prices, ECUbn)

	1980	1989
EC production	116	210
USA production	114	172
Japan production	53	148*
EC exports extra-EC	35	57
EC imports extra-EC	11	26
Net exports extra-EC	23	30

*Japan 1988

Source: CEC, 1990b

variety of studies confirm this view. Krafcik (1987) suggests that we need to differentiate between complexity and flexibility, where he defines complexity as the product/part variation at a given moment in time, whereas flexibility is the ability to shift product/part mixes over time in dynamic fashion.

In practice, short-term *response* flexibility refers to ways of directly improving the flexibility (service, variety etc.) associated with a particular product or market. But in the longer term *range* flexibility is important and has implications for investments in equipment, in labour and in production organisation, since making the firm capable of a wider range of product flexibility involves changing or adding to its manufacturing resource base (Slack, 1989).

One important way of reducing the problem of flexibility is to integrate more functions into a single machine or system. In other words, by increasing the complexity of operations which one machine can perform we extend its range flexibility. Instead of moving between several different machines, each capable of a single operation, the process is integrated into a single machine or process stage. We can see this trend towards increasing flexibility through integration in the case of the development of metal-working machine tools.

During the first half of the 20th century engineering manufacture essentially elaborated on two themes, with machines becoming more complex and much faster. Among other consequences, this led to a gradual substitution of machines for people when and where it became possible to mechanise standard operations in the manufacturing sequence. However, there were limits to this substitution until the development in the 1950s of *numerical control.*

Numerical Control (NC) technology for machine tools evolved out of a programme for the United States Air Force in the late 1940s and essentially provided a way of describing the entire complex process of producing a part in terms of a mathematical expression. This code, when read by a suitable device, would translate into a variety of control actions in the machine tool. The form this generally took was a paper tape with holes punched in it, but later versions substituted magnetic tape for paper tape and later still solid-state memory chips which controlled the machinery operations.

In NC any change in the parts to be produced no longer required extensive physical changes to the way in which the machine was set up but only a change in the control programme. Thus, for the first time, the flexibility of general-purpose machinery could be combined with the precision and accuracy of special-purpose equipment. In the process it became possible (though not necessarily desirable) to replace the human component as an interface translating the design information into machine activities, with some form of information-processing device (Noble, 1982).

While the principles of NC were established in the 1950s, the technology did not diffuse widely until the 1970s when low-cost programmable control based on the microprocessor became available. (Jacobsson and Edquist (1986) suggest that the development of the microprocessor reduced the costs of the controller by about 50 per cent). This opened up the development of the concept of Computer Numerical Control (CNC), and once the basic machine operations could become incorporated as part of a control programme, it was only a matter of time before other functions such as tool changing or part manipulation also became automated. Technical development at this time was characterised by a rapidly increasing integration of functions associated with the machine tool, leading to highly sophisticated, multipurpose machining centres complete with a range of support functions such as tool change, head change, transport and manipulation, often using robotics or automated guided vehicles (AGVs), all under computer control.

This progressed to the idea of Direct (or distributed) Numerical Control (DNC), in which more than one tool (plus associated functions) could be grouped into a manufacturing cell under the overall control of a host computer, which would be responsible for work scheduling, routing of products and parts, monitoring status, feeding in new programmes and so on. Essentially this was bringing many different machines into a configuration in which they behaved as if they were a single, complex and highly integrated machine.

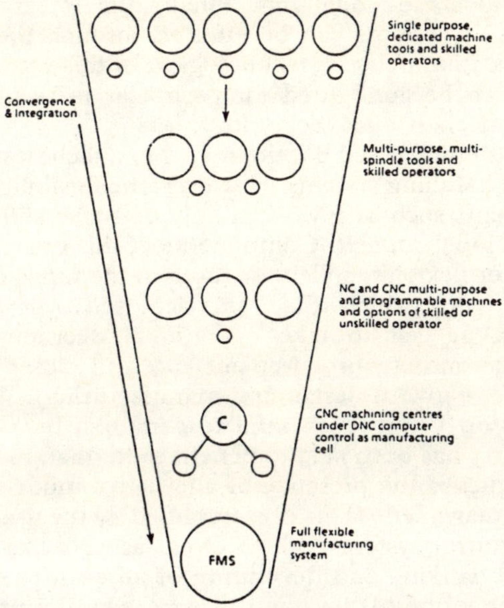

Figure 15.1 Convergence and integration in metalworking production

From this the step to a Flexible Manufacturing System (FMS) was a very short one. In general, a working definition of FMS would be some combination of machine tools, handling systems, transport systems and ancillary equipment whose overall control and management is under some form of hierarchical computer control. Figure 15.1 highlights this pattern over time, gradually integrating more and more functions within and between machines. Such integration of technical elements is given greater potential when used in conjunction with changes in production methods and in the pattern of work organisation.

Integration on its own is, of course, not enough to offer greater flexibility. What is also needed is reprogrammability – the possibility for changing the ways in which that integrated configuration behaves without extensive intervention to reset it. This can be accomplished rapidly – by exchanging magnetic tapes or memory chips or directly on-line from a control computer – and it can also be prepared in advance, further reducing the set-up time. A key feature of mechatronics is that, since all elements in the system operate under the same basic type of control, it becomes possible to link them together and configure them into a networked control system. This raises the problem of open interface standards as in other computer-controlled systems; these are discussed at the end of this chapter and in Chapter 11.

We have dwelt in some detail on machine tools and their development into flexible manufacturing systems. But the same pattern can be seen in other areas – for example, transport and handling. Here the emergence of industrial robotics or 'intelligent' guided vehicles exhibits similar convergent patterns.

In each case there is a gradual integration of mechanical functions followed by a blending of control and operation (usually based on microelectronics) to create a powerful stand-alone unit. Finally the integration into larger manufacturing cells or systems can be effected through the exploitation of electronic communication network technologies. In this way systems of varying sophistication can be configured for particular tasks using the various building blocks of mechatronics technology.

The foregoing has emphasised developments in a technical area associated with machinery and handling systems. However, the flexibility offered by new technical developments such as FMS can only be successfully exploited in a supportive organisational context. Components of this organisational change would include greater flexibility of labour (both in the range of skills available and the flexibility with which they are deployed), flexibility within a manufacturing area or cell to take operational decisions (for example, changing schedules, routings, procurement etc) and closer functional integration (for example, between design and manufacturing). We will return to these issues in Section V but it is useful to comment here that one reason why Japanese industry has been able to benefit more than most from systems based on mechatronics is the presence of alternative models for production organisation and management. This is exemplified by the use of 'just-in-time' and total quality control systems, which stress aspects like cellular organisation, multi-skilled working and devolution of authority to the operational level. This is probably one of the main reasons why the productivity gains from the diffusion of robotics, described in Sections III and IV below, appear to have been greater in Japan than either in the United States or the EC.

III Flexible Manufacturing Systems

The first FMS was developed in the United Kingdom by Theo Williamson in 1962 for the Molins Company, a small-batch manufacturer of machinery for the tobacco industry. His System 24 (on which the company still holds the patent for the name and design of a 'Flexible manufacturing system') was able to machine blocks of aluminium into a variety of parts with short set-up and changeover times. However, as with NC technology, it was not until the development of cheaper and more powerful control options that the technology of flexible manufacturing really came into its own in the 1970s.

One of the biggest problems in studying flexible manufacturing technology is the range of definitions which people use. Like Humpty Dumpty in 'Alice in Wonderland', the label is often used to mean whatever people want it to, with the consequence that the label 'FMS' has been equally applied to a single, relatively simple computer-controlled machining centre and a 30-machine tool factory! For the purposes of this chapter we will use the following broad framework of definitions (see also the Glossary).

Flexible Manufacturing System (FMS) – a combination of two or more CNC machine tools and an automated transfer system under supervisory computer control via some form of DNC linkage;
Flexible Manufacturing Cell (FMC) – a combination of two or more CNC machine tools and an automated transfer system but not under DNC control;

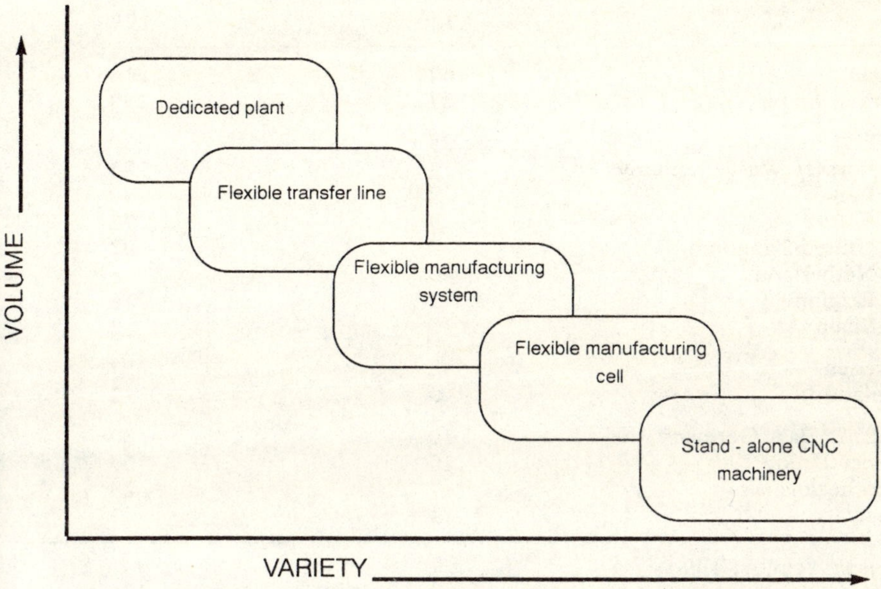

Figure 15.2 Volume of production versus variety of output

Flexible Manufacturing Unit (FMU) – a single, multifunction CNC machine tool, but with automated load/unload facility;
Flexible Transfer Line (FTL) – a multi-machine layout including several CNC machine tools and other specialist pieces of equipment all under supervisory computer control, used in high-volume industries like vehicle manufacture as an alternative to a dedicated transfer line (see Figure 15.2).

Diffusion of FMS accelerated during the 1980s: whereas there were perhaps 300 systems world-wide in 1980, this number had more than trebled by the end of the decade (UN/ECE, 1989). According to the International Institute for Applied Systems Analysis (IIASA) (Tchijov, 1989) even allowing for saturation effects within the machinery sector, there is likely to be a population of up to 3,500 systems by the year 2000 and this makes no allowance for non-metal-working applications.

As might be expected, the advanced industrial nations (Japan, the United States and Western Europe) account for the great majority of systems but an increasing number are to be found in developing nations and in Eastern Europe. Table 15.3 presents recent data on world-wide diffusion. Although these are the best statistics presently available, it must be stressed that there are so far no standardised international official statistics in this field and these estimates are less reliable than those for robotics in Section IV. In particular, they probably *understate* the level of diffusion achieved in the EC and the United States by 1980. However, the figures are probably credible in showing a growing lead of the EC and Japan in the 1980s and relatively slow diffusion

Table 15.3 FMS by country

Country or area	Number of FMS 1980	1989
Japan	69	213
United States	27	139
EC	25	310
Former West Germany	15	85
Italy	2	40
France	3	72
United Kingdom	3	97
Netherlands	0	8
Belgium	2	6
Spain	0	2
Sweden	5	37
Finland	0	12
Former East Germany	10	30
Soviet Union	10	56
Czechoslovakia	7	23
Hungary	1	7

Source: Tchijov, 1990

in the United States. With the proliferation of system type mentioned earlier has come a range of opportunities for smaller-sized users to enter into FMS technology. Whereas early FMS applications were developed for the high-value, high-complexity business typified by the aerospace sector, simpler and smaller versions (including flexible manufacturing cells as well as FMS) are finding widespread application in engineering. Table 15.4 highlights the key differences between the more complex and the simpler systems.

> ... the higher number of CNC machines combined with a large part family usually results in such a complexity of systems co-ordination (e.g. routing, scheduling, tool management) that high software and planning costs cannot be avoided. The only way to get the relative costs down is a modular systems structure and standardised software modules. The benefits of standard systems are already clearly visible in compact systems [Ranta and Tchijov, 1988].

In the early period of FMS diffusion it quickly became clear that even large and experienced firms were unable to supply a turnkey FMS package – reflecting two features of the market. The first is the highly specific nature of customer requirements which means that each system has to be customised for a particular user – a feature which has subsequently encouraged the development of simpler, compact systems which can be sold as a (relatively) standard product and do not require the extensive consultancy, user-education and long-term project involvement of large and complex FMS installations. FMCs accounted for over half the total American market by 1987.

The second is that few suppliers have the required breadth of experience and resources to be able to offer the complete package required for a large complex system. Even large manufacturers of computers, systems software,

Table 15.4 Illustrative characteristics of different system types*

Compact (FMC)	Complex (FMS)
2–4 machine tools	>15 machine tools
Conveyer transport	Automated Guided Vehicles (AGVs)
Possibly automated store/retrieval	Automated storage/retrieval
Computer control by PLC linked via LAN to local PLCs	Computer control by Mainframes (VAX or equivalent) back-up computer as well
$3m, of which	>$10m, of which
Machine tools = 50–5%	35–40%
Handling/transport = 15–20%	15%
Control and communications = 20–5%	25–30%
Planning and training = 10%	15–20%
Relatively limited expansion	Open expansion potential

*The actual configuration of systems can of course vary considerably, depending especially on skill availability (Bessant and Haywood, 1988)

Source: Ranta and Tchijov, 1988.

machine tools or other equipment have been forced to adopt alternative strategies to supplying the emerging market for FMS (Bessant, 1988). These include one or more of the following options:

- strategic alliances, joint ventures and mergers which bring together the relevant capabilities of different firms;
- specialisation into supply of a particular component of systems;
- specialisation in supplying smaller, general-purpose systems (for example, FM cells);
- moving into the new market of systems integration and project management.

Data from the Federal Republic of Germany (Lay *et al.*, 1990) and France (Margirier *et al.*, 1986) confirm the strong presence of this trend within European engineering. Essentially we are seeing a broadening out across the manufacturing spectrum of options for flexible manufacturing so that relatively high-volume users can exploit the technology of 'Flexible Transfer Lines' (FTL) while small-batch and prototype producers can also make effective use of flexible manufacturing cells and units. This reflects again the point that different users require different kinds of flexibility.

This proliferation is beginning to affect sectors beyond metalworking, such as woodworking, plastics and clothing. But the main concentration remains in metalworking. Around 80 per cent of systems were used in metalworking and of these 80 per cent were in cutting, 15 per cent were in sheet-metal work and only 5 per cent in flexible assembly systems (Bessant and Haywood, 1988). More recent data reveal that this pattern changed very little in the late 1980s (Tchijov, 1990).

Table 15.5 Robot population in selected countries

Country or area	Robot* numbers 1981	1988
Japan	21,000	176,000
United States	6,000	32,600
EC	4,600	43,000
Former West Germany	2,300	17,700
United Kingdom	713	5,034
Italy	450	8,300
France	790	8,026
Netherlands	n.a.	845
Denmark	51	349
Belgium	242	1,231
Spain	n.a.	1,382
Sweden	1,125	3,042
Finland	35	545
Soviet Union	n.a.	59,218
Hungary	n.a.	89
Czechoslovakia	n.a.	5,961

n.a. = not available
* Reprogrammable robots only

Source: ECE, 1990

IV Diffusion of robotics

Like FMS, robotics also began to diffuse widely in the early 1980s in parallel with the diffusion of CNC and FMS (UN/ECE, 1989) (Table 15.5). Some of these robots were for use in FMS, but most were in the automobile industry for welding and painting. Although diffusion continues at a fast rate (over 20 per cent per annum in many countries), there has been some slackening of growth. Much of this can be attributed to saturation effects, as obvious applications for robotics (such as in the car industry) have been filled. In the early 1980s the American auto industry accounted for as much as 40 per cent of all robot investment. Many experts still consider that present levels of application are very low, being held back by lack of suitably sophisticated and intelligent technology. This is particularly true for the assembly area, where lack of suitable sensors – for vision, touch etc. – mean that robots have so far only found limited application.

In general there is a trend away from volume of applications to variety, with much of the new growth coming in new application areas, including some in non-manufacturing areas such as construction (see Chapter 14). There has also, as with Flexible Manufacturing Systems, been a discernible trend towards use by smaller firms, especially for complex applications which exploit the flexibility rather than the labour-saving potential of robotics.

Thus the trends in the use of robotics technology are towards increasing diffusion across sectors, firm sizes and applications, with assembly automation set to become a much larger user in the medium-term future. Indeed, since the development of the SCARA (Selective Compliance Assembly Robot Arm) robots in Japan for general-purpose assembly tasks this area has experienced the most rapid growth, both in Japan and in other countries. Future diffusion to other sectors and to smaller firms will depend critically on improvement in the flexibility of robots – since this will increasingly enable them to be cost-justified on the basis of factors other than labour cost. This progress is primarily limited by technological factors; robots remain, at present, relatively unintelligent and insensitive tools which are not yet capable of the necessary judgement and adaptation for carrying out most tasks in manufacturing.

Among key technological developments currently being explored are:

- direct-drive robots, in which motors are directly connected to arm joins rather than via gears and transmissions;
- advanced sensors, especially in image processing for vision systems;
- improved, more sensitive grippers;
- artificial-intelligence-based programming so that robots can 'learn' from their own mistakes.

R & D is going on in all of these areas, and several expensive and dedicated applications already exist. The real challenge is to create a 'second generation' of general purpose robots with these characteristics; if and when this emerges, the diffusion of robotics is likely to broaden considerably.

V International differences in diffusion of flexible automation and patterns of work organisation

Making every allowance for definitional differences, Japan is well ahead of both the EC and the United States in the diffusion of robots and in their manufacture. Some earlier studies suffered from a broader definition of robots used in Japan but Table 15.5 includes reprogrammable robots only and Figure 15.3 shows that the Japanese lead is not just in numbers of robots but in numbers per employee. On the same basis Japan is also well ahead in FMS, although the gap here is narrower and the rate of adoption in the EC and the United States is apparently higher.

There are notable differences in manufacturing strategies employed by American, European and Japanese firms which help to explain these gaps. They reflect not only different environments and markets but also different approaches to development and investment in internal factory capability. For example, Japanese manufacturers are placing great emphasis on short product life cycles and reduced time to market for new products – as distinct from Western producers who are worrying more about delivery responsiveness and short-term costs.

Within Europe there are also notable differences, although often less marked than can be observed in East–West comparisons. Clearly such differences in overall manufacturing strategy will exert an influence over which

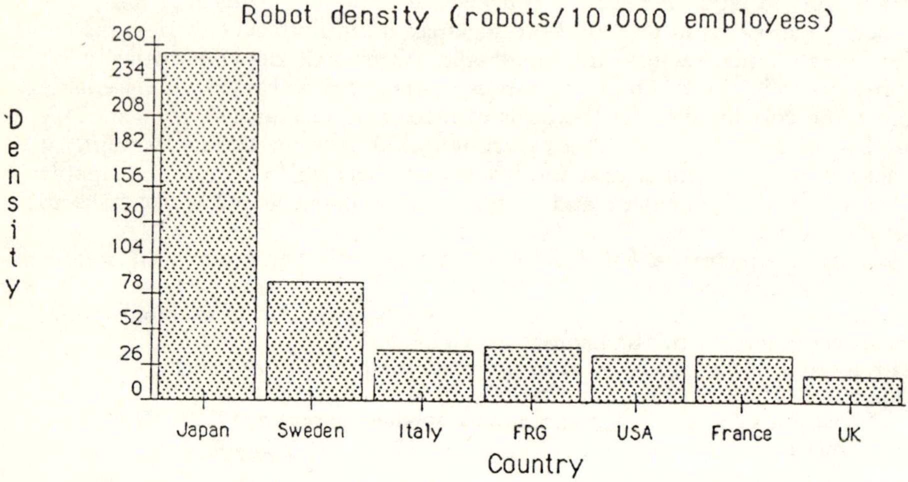

Figure 15.3 Intensity of robot use, 1987
Source: Tidd, 1990

technologies are adopted and the motives for which they are deployed. A second important difference emerges when we consider the question of linkage between producers and users in the engineering industries. With a high level of production and consumption of advanced automation, it is clear that close links between customers and suppliers play an important part in product innovation – as observed in other sectors (Shaw, 1987; von Hippel, 1988; Rothwell and Gardiner, 1983).

Of considerable importance here is the Japanese industrial structure which facilitates learning-by-doing in the design and use of capital equipment. As Scibberras and Payne (1986) point out, the Japanese machine-tool industry is one of the largest users of its own products and this applies also to its FMS and robotic activities. The lessons learned in using new technology on such a large scale can thus be absorbed into subsequent product technology. Jaikumar (1986) argues that a major difference between Japanese and American FMS users is that Japanese systems are constantly subjected to revision and development using the production floor as a kind of development laboratory for further improvement. Clearly this kind of learning is facilitated in firms which are both users and suppliers of such technology.

Another differentiating feature in East–West comparisons is the different approach taken to production organisation and management in Japan. Over a period of 40 years, patterns of human resource utilisation have shifted dramatically and now stress multi-skilling, teamworking and flexible working practices to a much greater extent than in the West. In general, Japanese firms have sought to develop flexibility first and foremost through human resource development (for example, within 'just-in-time' programmes) and only later have shifted to the use of automation. By contrast Western users, especially in the United States and the United Kingdom, have attempted to automate uncertainty out of the system through increasingly complex control software (and more recently, with the application of artificial intelligence). These differences in manufacturing philosophy and strategy are reflected in the relatively rapid adoption of FMS in the EC and the United States and the much more rapid adoption of robotics in Japan.

There are marked differences in the degree of control system sophistication sought which also show up among European nations – for example, our study of Swedish and British FMS applications suggested a much higher expenditure on control systems in the United Kingdom (Bessant and Haywood, 1988) than in Sweden. To some extent this can be explained by the availability of skills: flexible human-centred systems depend on the availability of a skilled pool of labour and where this is absent (as in the United Kingdom compared to Germany or Sweden) the options are much reduced and automation may be the only viable alternative.

However, it is important to stress that considerable research has indicated the importance of managerial choice in determining pattern of work organisation associated with mechatronic technologies (Sorge *et al.,* 1982; Boddy and Buchanan, 1984; Bessant, 1984). The model which originated with Ford and Taylor may represent one option but others have now been developed and demonstrated to have greater economic potential as alternatives. What determines the selection has less to do with technology than with a political process of negotiation and choice (Wilkinson B. 1983). As Jones and Scott (1987) point out in their analysis of FMS in the United Kingdom and the United States, it is possible to find the same basic technology associated with several different work organisation arrangements in different firms.

Analysis of the American position (Dertouzos *et al.,* 1989) suggests serious weaknesses at the level of the firm in terms of management attitudes towards production strategy and organisation (see Chapter l). The inability to take a longer-term view, the preoccupation with traditional approaches based on hierarchical control and division of labour and an adversarial culture characterising intra- and inter-firm relationships have all had a deleterious effect on performance, especially in contrast to Japanese competitors. These problems have been compounded by a tendency to over-rely on technology (narrowly defined to cover machinery and computers rather than also to include skills and organisation patterns) to deal with an increasing competitive threat.

With the rise in *glasnost* has come much more information about the Soviet and Eastern European experience with mechatronics, and this throws some of the above points into sharp relief. Measured by diffusion statistics alone things appear to be moving, with significant levels of application of these key technologies (Tables 15.3 and 15.5). But closer inspection (UN, 1990) reveals a serious problem of under-utilisation. Baranson (1990) attributes this to a system which is inflexible, engineering-driven and dominated by a 'command economy' in which there is little customer orientation or feedback. This creates an environment in which the dominant manufacturing strategy increasingly emphasises price rather than non-price competition.

Recent evidence from the Soviet Union concerning flexible automation reports that 'no less than a third of the 50,000 industrial robots produced between 1981 and 1985 had not performed even one hour's work. A sample inspection made by the People's Control Committee of the Soviet Union in 1985 showed that the annual return on introducing 600 robots, at a cost of more than 10 million roubles, was a mere 18,000 roubles' (Glaziev, 1990).

There are also numerous examples of very bad experiences with FMS and robotics in the West. So it is clear that unlocking the full potential gains offered by expensive automation investment involves much more than simply making the decision (and finding the capital) to adopt the technology. Examination of cases in a variety of studies confirms the view that successful firms recognise that they need to make significant and far-reaching changes to the way production is organised and managed. Using radical technologies like mechatronics requires a similarly radical degree of organisational change along a number of dimensions including the skills profile, the functional and hierarchical structure, the philosophy of management and control and the underlying culture of the organisation. Across this broad spectrum of flexibility in organisation most EC countries are in the middle but some are closer to Japan or offer alternative types of flexibility.

Some EC countries are clearly in a strong position to exploit flexible automation technologies – for example, the former West Germany now has a relatively high rate of use of FMS, robotics etc. and can build upon its position of strength as a specialist machine builder. Its ability to develop alternative working practices and obtain higher levels of responsiveness and flexibility will depend less on the question of skills availability – with around 70 per cent of the workforce possessing some vocational qualification the former West Germany is well-endowed and the union with East Germany will swell still further the labour supply in skilled personnel – than on managerial attitudes towards changed working practices.

Scandinavia has also demonstrated leadership in moving towards alternative models – ever since the Volvo experiments at Kalmar in the 1960s. Recent developments (Persson, 1988) suggest that on a broad front a new model for manufacturing organisation and management is emerging which marries highly skilled people, looser and more flexible work arrangements with high levels of Advanced Manufacturing Technology (AMT).

In the United Kingdom, on the other hand, as in the United States, there are many barriers to change. In addition to the serious limitations imposed by the shortage of skills and inadequate investment in training, there is a strong traditional management culture which may not be capable of reacting fast enough to the challenge for greater flexibility and responsiveness. The dominance of the Ford–Taylor model in the United Kingdom and the United States has been remarked upon elsewhere and it is significant to contrast this approach – which stresses a high level of functional differentiation and division of labour – with other European approaches which embody more integrated forms. The concept of 'Technik' as a unified view of manufacturing as a total system is one which is present in most continental European languages and cultures but lacking in the United Kingdom and the United States.

VI Conclusions and policy implications

Two key points emerge from the analysis. The first is that mechatronics is clearly not simply a collection of machines and computers but a much broader system including a key organisational component. Successful users are, above all, those firms which recognise this and manage such innovations as total systems. Contextual factors – like the layout of production, the choice of products to be made, the linkages between different support functions, the skills and work organisation of those operating the system – all have a bearing on how well it will perform. Indeed, the benefits of FMS investment often come more from the organisational changes which it provokes than from the narrow set of physical equipment which is installed (Jaikumar, 1986; Tidd, 1990).

Second, is the point that mechatronics is a *configurational* technology (Fleck, 1988), evolving and developing as a result of interaction between users and producers and during the long-term process of implementation. The centrality of organisational learning in this implementation process needs to be recognised. In this respect Europe stands at something of a crossroads with regard to AMT. Investment in the physical elements of AMT has already been extensive but the promise offered by such innovations is often not fully realised. At the level of the firm there needs to be a shift to a longer-term strategic perspective on manufacturing, within which choices about the appropriate use of mechatronics technology and the design of matching organisational forms can take place. In particular, firms need to recognise that, in the emerging manufacturing environment, competitive advantage will depend increasingly on knowledge and skill and on intensive technical co-operation and networking, both with users and with suppliers.

There are several ways in which the EC can help to develop a suitable climate for such strategies: one is to reduce barriers to inter-connection by promoting open standards. However, having a specification for OSI is not sufficient since that was established primarily for general communications between items of IT. The requirement is for a specific manufacturing version, and the front runner for this is the General Motors Manufacturing Automation Protocol (MAP). MAP is not a GM product but a standard specification which suppliers have largely begun to adopt. Most major automation manufacturers, including Fanuc of Japan, have announced that they will support it and semiconductor firms like Intel have also begun producing chips to enable them to implement it. Its likelihood of becoming a *de facto* standard appears high as a result of this and also because of backing from other agencies.

But many critics of MAP argue that its adoption for all but the largest systems is like using a sledgehammer to crack a nut – it is too sophisticated and powerful. For most cells some form of proprietary local area network (such as Ethernet) may be sufficient. The emerging strategy for many firms is one of 'islands of automation' based on technology cells – but with care being taken to build 'hooks' into the cells so that they can be joined together at a later stage.

A second role for the EC and national governments is in the promotion of new applications of mechatronics technology, especially in smaller-sized enterprises where its potential for improving flexibility may well represent an important aid to their achieving competitiveness in an environment which holds new opportunities for small-scale production. Since effective exploitation of mechatronics depends on taking a total systems approach, the provision of advice and consultancy support to underpin such investments is a valuable element of any promotion policy. Although the climate in Europe in recent years has swung away from direct financial support for investments in AMT, the scale of investment required and the costs of reconfiguring the organisation to take advantage of the technology are such as would benefit from government support. In a recent study conducted in the United Kingdom evaluating such a support scheme, it was clear that the major benefits had been in moving firms on to the learning curve earlier and further up than they would otherwise have gone. In an environment which is likely to stress such learning as a source of competitive advantage in future, this is an important outcome for policy and should be encouraged (Rush *et al.*, 1990).

A final policy issue concerns education and training. This chapter has repeatedly stressed the role of human resources in providing key flexible linkages in mechatronics systems. The extent to which this can be developed will depend critically on the level of skills available in the labour market and places a strong emphasis on training. Occupational structures are changing considerably as a result of the introduction of FMS – with a general cutback in the numbers of unskilled and semi-skilled functions such as loading, unloading, transportation and progress chasing. A consequence of the declining numbers of direct operators and the need to maintain high levels of utilisation and flexibility is that skill requirements change. Not only is there a need for higher levels of skill but also for greater breadth and flexibility to move between and across skill boundaries (Senker, 1985). Multiple skilling at all levels is increasing in importance and the portfolio of relevant skills is also changing, with increasing emphasis on preventive action rather than direct intervention, on diagnosis and problem-solving and on planning and programming.

The implementation of systems change requires good industrial relations and involvement of the whole workforce. The major industrial relations issues surrounding FMS have relatively little to do with redundancy. Rather, they are concerned with changes in working practices as workers become more flexible. As Ebel (1985) points out, sophisticated control over processes may well be incompatible with the informal processes which workers use to achieve high average performance but retain considerable control over time and stress at work.

Certainly the changes required by alternative working arrangements to promote flexibility have not always been received without opposition. Attempts to secure functional flexibility, for example, imply major reconfiguration of traditional representation and bargaining patterns – such as the shift to single-union worker representation. This has led, on occasions, to resistance and protracted negotiation with trade unions, especially in countries like the United Kingdom with a tradition of multi-union representation. Consequently, the success of advanced manufacturing technology depends upon a combination of training and advanced social policies as well as on the narrower issues of technological and financial inducements to invest.

16 Globalisation of the automobile industry: the challenge for Europe

Andrew Graves

I Introduction – restructuring of the automobile industry

The automobile industry, now a hundred years old, is often regarded as the main engine of industrial growth in the 20th century. Its production techniques, and in particular the assembly line, have also had a profound influence on the organisation and technology of other industries and services. The industry's requirement for materials and components spread far and wide throughout the mining, petrochemical, engineering and electronics sectors. Today's automobile, for example, contains over 12,000 separate parts provided by a diverse range of suppliers. The automobile industry remains an important and dynamic sector, even though it has now been displaced by the electronics industry as the largest and fastest-growing major industrial sector. After a period of structural adjustment in the 1970s and early 1980s, its rapid growth resumed in the late 1980s. But the environmental problems associated with pollution and traffic congestion place a question mark over its long-term future.

Europe is the world's largest producer of automobiles with 12.2 million units produced in 1989 compared with 8 million for North America and 8.2 million for Japan. Automobiles also account for over 12 per cent of the value added by the Community's manufacturing industry. The European automobile industry (including components) employed over 1.8 million people in 1988, approximately 6.2 per cent of manufacturing employment; and more than 3.5 million in related supplier and service industries. Europe is also the largest market with 11.8 million new car registrations in 1989 compared with 10.9 million in North America and 3.9 million in Japan.

The outstanding change in the world supply of automobiles after the post-war recovery of European producers was the enormous surge in Japanese production in the 1970s and 1980s. This is shown in Table 16.1, which indicates that in the 1980s Japanese production overtook American production despite the far smaller domestic market. It also shows Japanese producers now account for about a quarter of current world car production. If Japanese car production from overseas transplants are taken into account, Japan's share of automobile supply would be even higher. By 1988, five Japanese manufacturers were among the top-15 world producers of automobiles (Table 16.2). Japanese car production in the United States by 1989 was 2.2 million units and in the EC was 240,000. The latter is expected to

Table 16.1 World automobile production, 1950–1994 ('000s)*

	1950	1960	1970	1978	1982	1986	1989	1992	1994
North America	6,950	7,001	7,491	10,315	5,860	8,890	7,835	7,510	7,935
Western Europe	1,110	5,120	10,379	11,321	10,270	11,805	13,749	13,670	14,454
Japan	2	165	3,179	5,748	6,887	7,810	9,052	8,967	9,009
United Kingdom	523	1,353	1,641	1,223	888	1091	1,299	1,399	1,583
World	8,168	12,985	22,755	31,226	27,027	32,848	35,692	36,190	38,371

*Both Japan and United States have a much higher production of light vans and trucks than the EC. These are not included in this table. In 1988 Japan produced 4.4 million LCVs and trucks, United States produced 5 million and the EC 1.6 million.

Sources: Data for 1950–86: Society of Motor Manufacturers and Traders' year books, 1990: Data for 1989–94: DRI World Automotive Forecast, cited in *Financial Times,* 18 September 1990.

Table 16.2 Leading world automobile producers, 1988 (million cars)

	1988
General Motors	5.7
Ford	4.2
Toyota	3.1
Volkswagen	2.7
Peugeot	2.2
Nissan	2.0
Fiat	1.8
Renault	1.8
Honda	1.5
Chrysler	1.2
Mazda	1.0
Vaz(Lada)	0.7
Mitsubishi	0.6
Hyundai	0.6
Daimler-Benz	0.6

Source: United States Motor Vehicle Manufacturers Association; cited in *Financial Times* 18 September 1990.

Table 16.3 Value of Japanese trade in automobiles with rest of world ($m)

Year	Exports	Imports	Auto balance	Trade balance
1981	18,445	326	18,119	8,740
1982	18,017	347	17,669	6,900
1983	19,535	400	19,134	20,534
1984	21,899	461	21,438	33,611
1985	25,402	538	24,863	46,099
1986	32,945	1,069	31,875	72,743
1987	35,693	2,031	33,662	79,706

Source: White Papers of Japan, 1987

rise to half a million units by 1992 and to over 1 million by the end of the 1990s.

The recent performance of Japanese automobile exports is even more extraordinary. Table 16.3 shows the value of Japanese foreign trade in automobiles with the rest of the world. Japanese penetration was greatest in the American domestic market, while the lower share of the European market was due in large part to the non-tariff barriers and trade agreements limiting imports into Europe. This is especially true in the cases of France and Italy, where Japanese penetration has been restricted to under 3 per cent of the market and to 2,500 cars respectively. Nevertheless, Japanese market share in Western Europe is now almost 11 per cent, on a par with Ford and GM whose individual shares are of this size (Table 16.4).

Tables 16.5 and 16.6 indicate the value of exports and imports of cars between Japan and the United States and Europe as well as the value of respective automobile trade balances. While restrictions have afforded the European manufacturers a temporary respite from direct Japanese competition, it is now apparent that substantial adjustments in manufacturing performance will need to take place to maintain a competitive industry. The importance of this restructuring becomes even more acute as the leading Japanese manufacturers are now locating facilities within the EC.

Even more than in the case of machinery (discussed in Chapter 15), growing Japanese competitiveness in automobiles has been largely due to a new manufacturing philosophy. In this case, the development of a new 'best practice' in manufacturing has enabled leading Japanese producers not only to concentrate on high volumes but also develop a high degree of flexibility in model runs, together with high productivity and high-quality products. As Western producers close the productivity and quality gap with the Japanese and begin to implement some Japanese manufacturing techniques, Japanese producers are driving the industry forward through technological and organisational innovations.

Parallel to the organisational changes which are now sweeping the car industry in its attempts to meet world 'best practice' in productivity is a technological transformation which is occuring through a combination of factors on both the supply and demand fronts. On the supply side a range of new technologies is being offered by traditional suppliers and new entrant companies especially in the areas of electronics and new materials. On the demand side several factors are combining to stimulate innovation, including changing customer preferences regarding performance, style and fuel economy, and government policies on safety and environmental emissions. In addition aggressive innovative marketing strategies on the part of some Japanese corporations are forcing Western automobile companies to rethink their product development strategies.

In all automobile-producing countries increasing resources are being invested in new products and new production technologies by both manufacturers and suppliers. The early fruits of the substantial increase in both R & D and patenting by the Japanese are now reaching the market-place and will increasingly pose a threat to the one area where the European industry has been dominant. Strengths in styling, design and engineering were the cornerstone of the European challenge to the American domination of world autom-obile markets in the 1950s. European manufacturers provided specialised

Table 16.4 Market shares in Western Europe, 1989

	Volume (000s)	Share (%)
Volkswagen (Audi, Seat)	2,021	15.0
Fiat (Lancia, Alfa Romeo)	1,991	14.8
Peugeot (Citroen)	1,704	12.7
Ford	1,562	11.6
GM (Opel, Vauxhall)	1,488	11.0
Renault	1,392	10.4
Mercedes-Benz	434	3.2
Rover	412	3.1
BMW	377	2.8
Volvo	266	2.0
Japanese	1,457	10.9
Other	374	2.8
Total	13,478	100.0

Source: Financial Times, 22 January 1990

Table 16.5 Japanese trade in automobiles with the United States ($m)

Year	Exports	Imports	Auto balance	Trade balance
1981	9,313	50,3	9,262	13,312
1982	9,588	43,6	9,544	11,464
1983	10,564	35,5	10,528	18,182
1984	12,634	33,0	12,601	33,075
1985	15,597	20,3	15,576	39,485
1986	20,862	45,8	20,816	51,402
1987	20,953	88,3	20,864	52,090

Source: White Papers of Japan, 1987

Table 16.6 Japanese trade in automobiles with Europe ($m)

Year	Exports	Imports	Auto balance	Trade balance
1981	3,356	273	3,083	12,122
1982	2,885	302	2,558	11,464
1983	3,596	360	3,236	12,048
1984	3,564	419	3,145	10,811
1985	3,718	515	3,203	12,625
1986	6,436	996	5,440	18,973
1987	8,630	1,907	6,723	22,590

Source: White Papers of Japan, 1987

products offering different technical solutions to an expanding range of design requirements related to strongly diversified consumer demand. Their market share doubled during the period at the expense of American producers. This dominance was built not just on in-house capabilities but upon strategic alliances and cleverly managed sub-contracting to suppliers, design firms and styling houses. The Japanese producers have recently challenged this dominance, building upon their existing advantages in more efficient product development with shorter model cycles.

This chapter explores the sources of Japanese competitive strength in the automobile industry and its implications for European producers. In the following sections we examine first the importance of organisational change with regard to Japanese manufacturing techniques and the emergence of the 'lean production system'; second, new features of the Japanese challenge in the 1990s, including Japanese investments in R & D, reductions of lead times and new changes in product design and styling; and third, the implications of these changes for the European producers and their responses, including the development of pan-European efforts in applying new information technology to automobiles and traffic systems. We conclude with a discussion of some of the policy issues facing European government and industry.

II The sources of Japanese competitive strength

It was to understand the changing structure of the automobile industry and to predict the implications for its future organisation and development that the four-year International Motor Vehicle Programme (IMVP) was set up in 1986 by MIT (Womack *et al.*, 1990). The MIT study found that the world automobile industry was undergoing a major revolution as the era of mass production gave way to a new 'best practice' in manufacturing and product development which it called the 'lean production system'. There was a two to one performance gap between 'mass' and 'lean' production systems, this gap existing in all areas from new model design and development, through supply to manufacturing. Leading lean producers invest not only half the effort in manufacturing and design but also offer twice the number of products, at lower volumes per product, using half the inventories and producing half the defects involved in typical mass production. Although it originated in Japan, the lean production system has also been identified in other regions of the world.

The lean production system

The Toyota production system lies at the heart of lean manufacturing. It was pioneered by Kiichiro Toyoda, founder of Toyota, and perfected by Taiichi Ohno, a Toyota engineer, over a 20-year period to improve efficiency through a process of incremental improvement based upon the elimination of waste and the harmonisation of production flows and communication. An essential element of this philosophy is the elimination of inventories, buffer stocks and components in the plant and the return of some responsibility from management to supervisors and line-workers. A traditional mass-production

Fordist plant is characterised by a hierarchical management structure where discipline is imposed through the speed of the production line and production targets set by management. In the Toyota production system, production workers and engineers are organised in multi-skilled teams and are directly responsible for quality, having the capacity to stop the line if necessary. Work is organised around the two guiding principles of 'just-in-time' and 'total quality control', and workers and engineers are encouraged to initiate their own improvements to production efficiency. New cooperative assembler–supplier relations are also crucial to the functioning of this system.

The origins of the just-in-time concept are unclear. Cosumano (1985) has argued that Ohno credited Kiichiro Toyoda with developing just-in-time production and parts delivery when he set up his Toyota factory in 1937 to receive materials for one day's needs at a time. The assembly plant built only the number of vehicles corresponding to the output of the machine shop. Gradually Ohno and Kiichiro worked together to eliminate bottle-necks and cut inventories through reversing the process-information and conveyance flows and by limiting production to small lots, so that work on one station would not have to wait for the completion of work on others. This also involved educating suppliers to coordinate daily deliveries, thus eliminating warehousing and intermediate inspection. Similar techniques have been described in other Japanese industries (Schonberger, 1982). It is also argued that just-in-time has cultural origins related to Japan's lack of resources and industry's need to optimise space and minimise waste. Understanding the origins of and influences on just-in-time practices is important for the debate about their transferability to other world regions and cultures. Mass production on the other hand was developed in the cultural context of the United States where natural resources were abundant, land was not an impediment to plant design and construction and an unskilled immigrant labour force of great cultural diversity was available.

One major difference between mass production and lean production relates to the management of time and workers' skills. In mass production one worker uses one machine and always performs the same task. This invariably leads to idle time and irregularities in the flow of work. In the Toyota system, production is rationalised by revising cycle times in direct relationship to production needs, running machines at full capacity and giving workers a range of tasks to perform. This allows unnecessary buffer stocks and idle time to be eliminated. In Japan workers are also trained to maintain and repair their machines, thus minimising down-time. This is in contrast to traditional job demarcation in the multi-union Western context of labour relations.

The lean production system highlights inefficiencies and exposes problems as and where they occur and concentrates efforts on the immediate rectification of faults. This enables the maintenance of smooth-flowing and high-quality production. The traditional mass production 'buffered' system tended to conceal inefficiencies and reduced the input of ideas from management and workers. Built into the Japanese system of lean production is therefore a dynamic for continual improvement and change. A detailed audit of auto assembly plants around the world was undertaken by Krafcik with the aim of comparing productivity levels and quality performance (Krafcik, 1989). More than 90 plants representing 24 assemblers in 14 countries were assessed, representing about half of world assembly capacity. A methodology was

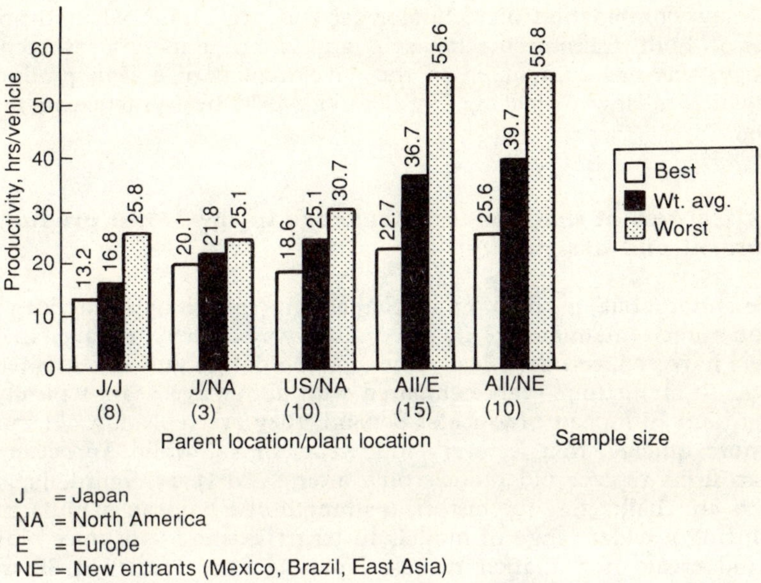

Figure 16.1 Regional productivity performance
Source: Krafcik and MacDuffie, 1989

developed to enable reliable comparisons to be made taking account of varying plant scales and product complexity.

Productivity performance data by region is shown in Figure 16.1. It indicates that European and North American plants required, on average, 118 per cent and 49 per cent respectively more effort to undertake the same manufacturing activities as an average Japanese plant. However, new entrant manufacturers (Mexico, Brazil and East Asia) were almost on a par with average European productivity levels and some plants outperformed their European competitors. Quality performance data indicate, as with the productivity data, that Japanese manufacturers again outperform their competitors and that large performance differentials occur within as well as between the regions. This indicates that there are relatively poor-performing Japanese manufacturers in Japan and relatively high-performing manufacturers in the United States, namely Japanese transplants and some traditional North American producers. The success of the Japanese transplants in North America is used by Krafcik to argue that Japanese manufacturing techniques are transferable outside the cultural and social context of their homeland.

Krafcik's findings challenge the traditional notion that high degrees of quality in automobile manufacture can only be achieved through extensive quality inspection and detailed re-work to the detriment of productivity, and that high-productivity manufacturers will always be forced to pay a 'quality price' for their productivity performance levels. It should be noted that while four North American and one new entrant appear in the high-productivity/high-quality sector, only Japanese plants (including one American/Japanese transplant) appear as 'world class performers'.

The basis of Japanese competitive strength in the automobile industry has therefore not been size, low wages or higher investment as often argued but rather a new combination of technological and organisational innovations. With its in-built training mechanisms and continuous improvement of technology, this has culminated in the development of a lean production system which is now displacing Fordism as world best-practice in manufacturing.

III New features of the Japanese challenge in the 1990s: product development and design

Japanese automobile production is challenging traditional American and European vehicle manufacture in three other ways. First, leading Japanese producers have reduced their lead times for the design and development of a new car to about four years compared with about five years typically for American and European producers. Second, they are replacing old models much more quickly, that is, every four years or so, while American and European firms replace old models only after 8–10 years. Third, Japanese producers are challenging the historical strength of European manufacturers by producing a wider range of models to target existing consumer requirements and create new market niches. Japanese firms launched 80 major model changes between 1981 and 1987 compared with 40 by European companies. Here, two features of this new Japanese challenge are analysed: research, development and engineering; and styling and design. What this analysis shows is that as American and European producers attempt to close the gap in productivity and quality, Japanese producers are investing increasing resources in product development in order to maintain comparative advantage.

Japan's automotive R & D effort overtook that of the former West Germany in the early 1970s and the whole of Western Europe by the mid-1980s, as summarised in Figure 16.2. If current trends are maintained Japan will overtake the United States by the early 1990s. The leading European producers, Volkswagen and Daimler-Benz, are now being outspent by Toyota and Nissan.

What is also striking is that if the output of the R & D system is measured in terms of patenting in the United States, Japanese firms have also overtaken GM and Ford (See Chapter 4). By 1985 Toyota, Nissan and Honda were all taking out more automotive patents in the United States than Ford or GM. Each of the Japanese automobile companies was taking out approximately the same number of patents as Bosch, Daimler-Benz and Volkswagen combined (Graves, 1987).

The roots of this Japanese technological challenge are not to be found simply in the quantitative difference between R & D expenditure or the areas of technologies being targeted, important as these factors may be. Rather they stem from qualitative differences between the way R & D activities are selected, how those activities are organised within the company and how R & D professionals are trained and managed.

Within the Western system of automotive R & D, different groups and company divisions are usually responsible for individual projects as they move

Figure 16.2 R & D in the motor industry

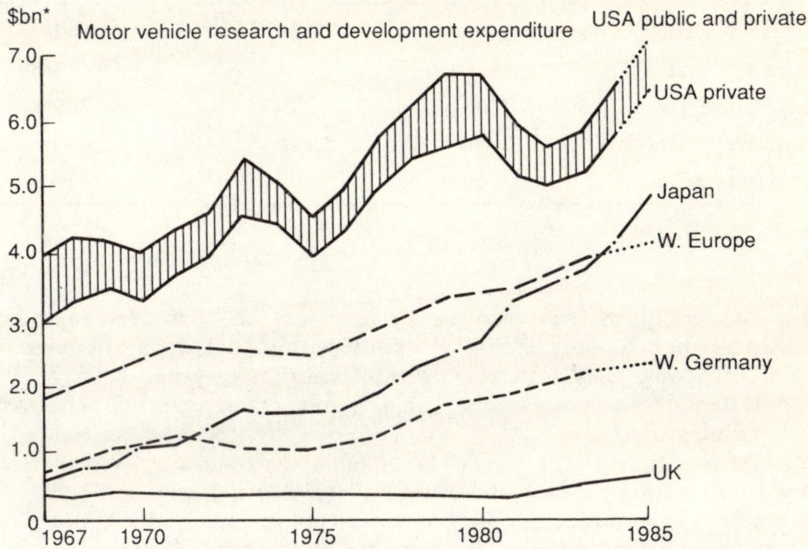

through the system from R & D towards application. As a consequence the Western system is top heavy, with a 'polarised' organisation which employs highly skilled researchers and engineers in R & D and an increasingly de-skilled workforce on the shop floor.

Little 'organic' integration exists between the two groups, resulting in huge problems of coordinating the myriad of new technologies being developed for industrial production. In the Japanese system production technicians and workers are closely involved from an early stage in the fine-tuning and improvement of new technologies. Projects are often initiated by ideas from individual employees and are constantly evaluated, involving production engineers and marketing strategists from the outset. Furthermore, this means that when an innovation is introduced into production there are few rejections or disruptions.

Crucial to the efficiency of Japanese R & D is the training and management of engineers. Strong leadership of a development project is considered very important. In Toyota, for example, the teamleader (called a *Shusha*) manages a team whose priority is the success of the project from start to completion. This saves time spent on inter-departmental coordination, and the resulting simultaneous engineering and close collaboration cuts the costs and lead times for new product developments. Western project development teamleaders are expected to be technical experts in contrast to their Japanese counterparts who are expected to be managers of people. In addition, the Western teamleader is often in a weak position relative to senior management with regard to the final outcome of a project. Their personal advancement

Table 16.7 Average model age and volumes, 1987

	Japan	North America	Europe
Number of models	73	36	47
Average age (years)	2.1	4.6	4.6
Average annual vol/model*	120	230	200
Cumulative vol/model*	504	2,116	1,840

*Thousands of units

Source: Sheriff, 1988

within the company will also be related less to their performance as teamleaders than to their technical ability within that department and the number of people they supervise. The promotion of Japanese R & D engineers is dependent upon factors such as a good track record of participation in successful projects.

Further evidence of the efficiency of Japan's dynamic system of R & D comes from work by Clark and Fujimoto (1988) between 1983–87. They have studied 29 product development projects in Japan, the United States and Europe and conclude that Japanese producers take 47 months to design a new vehicle compared with 60 months in the United States and Europe. In addition, the Japanese producers used 1.7 million engineering man-hours while American producers took 3 million and European producers employed 3.2 million to develop a similar vehicle. Furthermore, they claim that American and European firms employed at least 900 engineers during the development process, as opposed to fewer than 500 by their Japanese counterparts. The efficient organisation of R & D not only requires fewer engineers over a shorter amount of time to complete the same task but also results in a much lower staff turnover.

The importance of combining the Japanese R & D system, with its large difference in engineering effort and 30 per cent reduction in development time, with the 'lean production' system means that Japanese producers are able to make strategic choices regarding economies of scale. For example, Table 16.7 shows that average volumes per model are almost half that of the American and European producers. As the average age of each Japanese model is approximately two years, half that of their Western competitors, then only 500,000 vehicles are produced, compared with almost four times that number in the United States and Europe. This flexibility of model numbers and volumes enables the Japanese producers not only to exploit the increasingly fragmented markets of the world but also to generate their own niches in the market-place. In an era characterised by rapidly changing technology, it is important for these innovations not only to be incorporated into the latest vehicles but also to be offered to customers in the shortest possible time.

Leading Japanese producers are now not only challenging the Western volume producers but are investing heavily in new products and production lines for luxury and sports cars. The introduction of the Honda 'Legend', the

Nissan 'Infinity' and Toyota 'Lexus' in the luxury car segment, has heralded a significant competitive threat to leading European luxury car producers.

Strength in styling, design and engineering was the cornerstone of the European challenge to American domination of world automobile markets in the 1950s and 1960s (Altshuler *et al.,* 1985). Design was an integral part of company portfolios – it was as important to the volume producers like Volkswagen, Fiat and Austin-Morris as it was to specialised companies like Jaguar and Mercedes-Benz. The battle for market domination through design originality was also driven by an explosive growth in R & D activities by both producers and component suppliers – who were playing an increasing role in the design of new products (Graves, 1988).

The function of design is to transform technology into substance. Styling adds flair and shapes the exterior form. Design requires an overall understanding of the technology incorporated in a vehicle – new materials, structural requirements, manufacturing techniques and motoring standards. These are the constraints which determine the designers' freedom to link technology to customer needs. Design has an increasingly important role to play, not just as the means to engineer technological advance into a product but also as a way in which to penetrate regional markets through incorporating characteristics adapted to local requirements. It is then that styling has a role to play – in adapting these products to specific consumer needs, adding the colours and shapes. The current importance of styling and design is reinforced by a reduction in the duration of competitive advantage based on technological innovation. For example, radical innovations such as turbo-chargers, anti-locking brakes and four-wheel drive systems brought relatively short-lived monopoly rents. Even where patents are taken out, competitors are increasingly quick to adopt new technological advances. This means that styling and design capabilities are major factors stimulating product differentiation and maintaining competitive advantage. An interesting point here is that design and styling cannot easily be patented.

Japanese producers have made determined efforts to build up competence in styling and design and to integrate an efficient and original styling and design strategy into their production organisation. It is now often the turn of Europe and the United States to learn from the Japanese in styling and design.

The application of new electronics-based technologies and materials has demanded a radical change in the way that styling and design activities are integrated in overall production. Designers are working hard to incorporate plastics and composite materials to replace remaining metal parts and mechanical components such as fuel tanks, bearing cages, engine and carburettor parts. The use of plastics has increased from an average 40 kilograms to more than 100 kilograms per vehicle in only five years. The potential savings to be reaped through weight and cost reduction and therefore fuel consumption are considerable, as are the benefits accruing from corrosion-free and durability characteristics. The Fiat 'Tipo' represents a significant stride in this direction through offering a rear door manufactured entirely of composite. This lead is stimulating other stylists and designers to innovate in the area of new shapes based on the flexibility characteristics associated with new materials. However, it must be of growing concern to European producers that the Japanese have already established themselves at the forefront of new materials technologies and are poised to incorporate these advances in their own designs.

Different design strategies are constantly evolving among car companies. A few companies such as Daimler-Benz undertake styling and design themselves and have organised in-house design and engineering teams for that purpose. This 'single source strategy' maintains market niches through strong model identification and avoids products being identified as coming from famous design houses. A 'dual-sourcing strategy', such as that employed by Fiat, Volkswagen and Peugeot, involves use of a combination of in-house and sub-contracted design, and a close relationship will be built up between the design house and car company. This strategy generates competition between in-house teams and sub-contractors to enhance originality. The Japanese companies employ a 'triple-source strategy', using in-house design facilities together with off-shore company-controlled design subsidiaries as well as independent design houses. This aims to strengthen in-house competence and to adapt products in order to capture regional markets.

So far, no single strategy has become dominant. Out of all the producers, the Japanese have so far shown themselves to be most skilful at adapting rapidly to existing markets and, more importantly, to creating new market niches. The reason for the success of the Japanese in this respect stems partly from the inherent conservatism of the European design process. This has been linked to their very slow model replacement cycles and a desire (based on bitter experience) not to shock consumers with too radical design changes. With much shorter product replacement cycles, the Japanese are able to introduce model changes more gradually and to experiment by developing incrementally what often turn out to be quite radical concepts.

The implications arising from this are twofold. First, the Europeans will be increasingly pushed to reduce their own product cycles to keep technological advance in step with consumer acceptance. Second, European producers may well end up having to rely on the Japanese to bring consumers 'up to speed' on new styles and designs. By following trends set by the Japanese, they may well be more able to make greater steps in design change – but at the expense of losing their lead. Therefore, the European automobile manufacturers may well become imitators rather than trend-setters in the realm of styling and design – as is occurring in the sphere of technology.

IV The European response: is lean production transferable?

There are conflicting arguments posed in the literature about the transferability of Japanese lean production techniques to Western societies. As we have seen, Krafcik (1989) has argued that the existence of highly competitive leading producers in the United States – both Japanese transplants and new American manufacturing facilities – attest to the transferability of Japanese methods of 'lean' production outside of the cultural context in which they originated. It is becoming clear that some American multinationals, particularly Ford, are learning lean production techniques from their Japanese partners, in this case Mazda, and that GM has learnt some important manufacturing lessons from its partner – Toyota – at the NUMMI plant in Fremont, California (Womack, 1988).

However, Dankbaar (1990) argues that the situation is more complex and that Krafcik's isolated examples of success in the United Kingdom and the United States can be explained by the fact that they are greenfield sites, with hand-picked personnel and a unique management team aware of new procedures and how to implement them. He goes on to say that efforts to introduce the Japanese model into other countries will at best meet only limited success as long as management fail to take account of the social context in which the plants are operating. Dankbaar cites evidence of a range of factors in the Japanese case which cannot be reproduced easily in the European environment. These include: in-firm vocational training, reduced labour mobility and short-cycled, machine-paced labour on a continuously moving assembly line. European abilities to assimilate the Japanese model, he argues, will be inhibited by entirely different industrial policies, labour relations and national systems of education and training. He goes on to make the case for a new European model of production as illustrated by the 'German–Scandinavian model', which is characterised by 'the influence of socio-technical thinking, which is completely absent in the Japanese model ... and efforts to uncouple labour from the motions of machinery and to introduce stationary work places ... with longer work cycles' (Ibid.). Dankbaar argues that 'the most extreme example of this move away from the Fordist assembly line is the new assembly plant of Volvo in Uddevala, where a small group of workers is assembling a complete car on one spot' (Ibid.). However, he is inconclusive about the outcome of this model and points out that Japanese producers in Europe may be quicker to learn from it than their Italian, French, Spanish or British counterparts.

Generally, the leading Japanese producers have located on greenfield sites away from traditional automobile centres. In North America, Nissan located its plant at Smyrna in Tennessee; Honda at Marysville, Ohio; and Toyota at Georgetown, Kentucky (Table 16.8). In the United Kingdom, Nissan has located at Washington, Tyne and Wear; Honda at Swindon; and Toyota at Derby (Table 16.9). These locations have two major advantages for the lean production system. First, they avoid the company having to deal with strong trade union presence, being able to choose a single union deal or none at all, and being able to introduce to a workforce new methods without them being encumbered by past practices associated with mass production. The last point reflects Japanese practice in its own social context where management and in-house labour unions develop a corporate loyalty and group solidarity. Second, they bring advantages of being able to introduce new lay-out, equipment and work stations tailored to the optimisation of space and new production practices.

This investment trend in production has been paralleled by the establishment of Japanese R & D and technology development centres in Europe (Table 16.10). Although their scale is still tiny compared with facilities in Japan, they do imply an effort by Japanese companies to placate European fears about establishing simply 'screwdriver' plants for assembly. More importantly, these development centres indicate an attempt not only to attune existing Japanese models to the European environment but perhaps later to develop entirely new models to be built at their European production facilities. In addition, these R & D centres could play an important role in coordinating technology acquisition from suppliers to meet local content requirements.

Table 16.8 Japanese transplant production facilities in North America

Firm	Date	Location	Employment	1989 Production	Announced capacity
Honda	1982	Marysville, OH	8,800	351,670	360,000
	1989	East Liberty, OH			150,000
		Alliston, ON		86,447	100,000
NUMMI[1]	1984	Fremont, CA		192,235	340,000
Toyota		Georgetown, KY	7,000	151,150	240,000
		Cambridge, ON		20,859	50,000
Nissan[2]	1983	Smyrna, TN	3,250	238,640	480,000
Mazda	1987	Flat Rock, MI	3,500	216,200	240,000
Diamond Star[3]	1988	Bloomington, IL	2,900	91,839	240,000
CAMI[4]	1989	Ingersoll, ON	2,500		200,000
SIA[5]	1989	Lafayette, IN	1,700		120,000
Total				1,349,000	2,520,000
Engine Plants					
Honda		Anna, OH			
Nissan		Smyrna, TN			
Toyota		Georgetown, KY			

Notes: 1 GM/Toyota Joint Venture; truck line being added
2 Second assembly line being added
3 Chrysler/Mitsubishi joint venture
4 GM/Suzuki joint venture
5 Subaru/Isuzu joint venture

Source: Womack *et al.*, 1990

Table 16.9 Japanese transplant production facilities in Europe

Firm	Date	Location	1988 Production	Announced mid-1990s capacity	Potential additional capacity
Nissan	1980	Greece		10,000	
	1986	Washington, United Kingdom	57,000	200,000	200,000
	1988	Barcelona, Spain	76,000	150,000	
Honda	1981	Swindon, United Kingdom		140,000	260,000
		Longbridge, United Kingdom	4,000	40,000	
Toyota	1992	Burnston, United Kingdom		200,000	200,000
Toyota/VW	1989	Hannover, Germany		15,000	
		Lisbon, Portugal	14,000	15,000	
Isuzu/GM[1]	1986	Luton, United Kingdom	35,000	80,000	
Suzuki[1]		Linares, Spain	22,000	50,000	
		Esztergom, Hungary		50,000	
Mazda[2]		?			100,000
Mitsubishi[2]		?			100,000
Total assembly			208,000	950,000	860,000
Engine Plants					
Nissan		Washington, United Kingdom		200,000	200,000
Honda		Swindon, United Kingdom		70,000	330,000
Toyota		Shotton, United Kingdom		200,000	200,000
Total engines				470,000	730,000

Notes: 1 Excluding the assembly of Land Rovers
2 New plants under discussion

Sources: Womack *et al.*, 1990

Table 16.10 Japanese development centres in Europe

	Development Centre	Established	Place	Engineers Japan.	Engineers Euro.
Toyota	European Toyota Technology and Design Centre	Sept 1989	Brussels	47	15
Nissan	European Technology Centre (UK)	July 1987	Sunderland	48	15
	European Technology Centre (Belgium)	July 1989	Brussels	33	10
Honda	R & D Centre Europe	June 1985	Frankfurt	47	15
MMC	European R & D Centre	April 1989	Frankfurt	11	4
Mazda	European R & D Representative Office	November 1989	Frankfurt	12	10
Subaru	Liaison Office in Europe	August 1980	Frankfurt	5	4
Suzuki	Liaison Office		Brussels	4	0
Daihatsu	Liaison Office		Brussels	4	0
Isuzu	Liaison Office		Brussels	2	1

Source: Author's interviews

In the new transplants and joint ventures, Japanese managers have generally kept key positions at senior levels and have maintained overall control. The evidence suggests that this is not because Japanese companies have been unwilling to integrate Western senior management into their corporate structures but that cultural, educational and management philosophy makes it difficult to do so. Other cases of success such as GM and Toyota at NUMMI, and Ford and Mazda in Hermosillo, Mexico can be explained with respect to the crisis situation which left previous management with no alternative but to make radical changes.

The Japanese have therefore created with their new investments their own special social contexts which, where successful, are located in areas of high unemployment with specific economic and political incentives. Some argue that the Japanese system is more humane because 'the Japanese have shown that there is an important difference between working hard and working efficiently, and between work that is a challenge and work that is a bore' (Jones, 1990). But it should be remembered that with just-in-time production discipline, the line imposes a pace of production on workers, in spite of their new responsibility, which can lead to growing levels of stress and antagonism. Since lean production implies relatively more investment in skills and in-firm training, workers' future negotiating capabilities might be expected to change. A lean production system may well be more vulnerable to the type of labour discord that has characterised industrialisation in Western societies.

Furthermore, there are other factors which may impede the diffusion of lean-production techniques, as shown for example in the cases of Motor-

Iberica and Nissan in Spain and Rover and Honda in the United Kingdom, where learning from Japanese partners has been slow. The former case shows that even location in areas of high unemployment does not imply that a workforce will accept radical change. Furthermore the overall company identity is unclear, with Spanish and Japanese managers responsible for different sections and some reluctant acceptance by Spanish workers of their new Japanese managers. Finally, the lack of a greenfield site has led to fragmented facilities and the prevalence of Spanish former mass-production work practices. The case of Rover Group and Honda re-inforce these arguments. Their relationship has been at the vanguard of attempts to introduce Japanese management and organisational techniques into the car industry in Europe, through learning from the experience of a leading Japanese partner. While Rover has had considerable success in improving its productivity and quality, albeit beginning from a low base, it appears to have done this by exploiting a different product development and production philosophy than that which is practised by Honda. Oliver and Wilkinson (1988) argue in their book the *Japanisation of British Industry*, that Rover has been incapable of creating a reliable supplier network operating with a full just-in-time production system and go on to argue that even 'more problematic for Rover is the creation of a new company culture ... and a failure on the part of the company to bond workers to the company above all else and to give them aspirations identical to those of the company' (Oliver and Wilkinson, op. cit.).

Japanese companies tend to see workers as the solution to their problems and not as 'the problem' as so often perceived by Western companies. If workers are seen as the problem, then a production philosophy is developed to eliminate labour throughout the system or treat it as a semi-skilled commodity as in the Fordist model. The Japanese take great care to ensure that the political and economic environment of the host country and location are conducive for transplanting the lean-production system. Since social context is so important to the diffusion of Japanese production techniques, the appropriate environment may be easier to create in North America and the United Kingdom at the moment and possibly more difficult in countries such as Sweden and Germany.

The Scandanavian–German model, mentioned above, is based upon an anti-hierarchical and highly-skilled team concept, which offers an alternative to the Fordist moving assembly line (MacDuffie and Krafcik, 1988). However, it does not yet produce the level of productivity *and* quality that the lean-production system produces. Two factors may explain this. First, while line discipline has largely been dispensed with, the Scandanavian–German model does not appear to have introduced an alternative dynamic system of 'best practice', as the lean-production system has with the just-in-time discipline. Second, in contrast to the lean-production system in Japan, where intense efforts are made to minimise absenteeism among production workers, for a number of socio-political reasons it can run as high as 30 per cent in the case of some plants in Sweden. These levels may prove to be a severe impediment to the operation of the German–Swedish model, as it requires a high level of capability-building in the teams over a long period of time.

V The PROMETHEUS project

European manufacturers therefore face severe competition from Japanese firms that have higher productivity and have reached, and in some instances even surpassed, European standards of product design. These firms are also investing heavily in R & D so that they are threatening to open up a technology gap as well as a productivity gap.

The PROMETHEUS project, which is part of the EUREKA programme, is Europe's attempt to respond collectively to this technological challenge. It involves shifting the focus of innovation from the automobile *per se* to the transport system of which it forms a part. Its aim is to develop automobile and communications technologies so that the automobile becomes part of a more efficient and safe integrated traffic network. PROMETHEUS research concentrates on five areas: artificial intelligence, driver navigation assistance, vehicle-to-vehicle communication, vehicle-to-environment communication and intelligent on-board computer hardware. Thus it also amounts to an effort to give Europe a competitive advantage in automobile electronics, which will constitute as much as one-quarter of the value of an automobile by the end of the century.

PROMETHEUS is predicted to cost $800m over several years. The research phase from 1989 to 1992 is costing $430m, of which 40 per cent is being contributed by (and spent in) Germany, 25 per cent by France, 20 per cent by Italy, 10 per cent by Sweden and 5 per cent by the United Kingdom. In addition to the vehicle manufacturers, more than 70 research institutes and universities, and over 100 electronics and supplier companies are involved. An important characteristic of PROMETHEUS has been the close interaction between the automotive industry, component suppliers and the scientific community. At some stage, decisions will have to be made on how to involve American and Japanese companies to ensure the compatibility and integration of PROMETHEUS technologies if and when they are commercialised.

It remains to be seen whether the PROMETHEUS project will have any impact on Europe's industrial competitiveness. However, it has other agendas. It is also a response to the growing concern about the increase in European road traffic and its harmful effects in terms of accidents, congestion and environmental pollution. In 1984 in the former West Germany alone there were 333,800 accidents incurring injury and 8,100 deaths on the roads, and in Western Europe as a whole there were 1.37 million serious accidents involving 1.81 million injuries. The broad intention of PROMETHEUS is to halve road deaths in Europe, currently around 50,000 per annum, by the year 2000. In terms of environmental damage, the Federal German Environment Ministry has estimated that road traffic accounts for 58 per cent of CO_2 emissions, 54 per cent of NO, 43 per cent of hydrocarbons, 10 per cent of dust and diesel particulates, together with an unquantified level of noise pollution.

The PROMETHEUS programme thus represents an interesting example of a technological initiative that seeks to reconcile economic, health and environmental objectives. It will need to be watched carefully to ensure that the automobile firms' interests in increasing the numbers of automobiles on the roads, and in gaining protection from Japanese competition, do not gain precedence over the broader social goals.

VI Conclusions: Japanese competition

Through the development of the lean-production system, the Japanese car industry can now be viewed as the most efficient competitor in the world. Both European manufacturers and suppliers need to adopt significant structural changes to meet this new 'best practice', since they can no longer hide behind trade barriers owing to the global nature of the Japanese production system. The global strategy of Japanese car companies not only helps them to avoid the instability of exchange rates, but also enables them to leap-frog foreign protection. As the trade balance between Europe and the United States is at the present time in favour of the European producers, any attempt to restrict Japanese exports from their transplants in North America could have serious economic repercussions.

This means that Europe has to develop clear strategies to bridge the gaps that we have identified in productivity, quality, product development and model range. This strategy needs to be based on existing European strengths, for merely to imitate the Japanese could leave European producers lagging behind.

The European Single Market is challenging the car producers to consider how they move from national producers largely dependent on sales in their home market to becoming truly European firms. The move towards a pan-European market will require new forms of collaboration and alliances in production, component supply and technology development. At the present time only Ford and General Motors can be viewed as truly European companies, with facilities spread throughout the whole of Europe. PSA and Renault have production facilities in four European countries, Volkswagen in three and Fiat in one. Jones (1990) argues that in the early 1990s Europe will have its first trade deficit in motor vehicles for over 30 years. This will become a concern because motor vehicles are such a large part of inter-regional trade. Even if Japanese exports were to stabilise in unit terms, the move upmarket by the Japanese and the expected import of components to start up their new plants will increase the value of Japanese imports into Europe, irrespective of overall quotas and tariff barriers.

This shows the complexity of the trade-balance issue. Europe's future competitiveness will have to depend less on protectionist policies than on developing its own technological strength and its own manufacturing improvements to increase productivity and close the gaps in product quality, product development lead times and technical and environmental performance. It has become crucial to overcome the structural fragmentation of the European technological base and achieve a greater concentration of resources in R & D on a pan-European basis. Europe's historical strengths in technology are still not sufficiently integrated into the automotive sector as they are in Japan and increasingly in the United States.

A clear EC strategy will be needed to cope with the Japanese competitive challenge. The components of such a strategy would also need to include at least the following: a requirement that Japanese transplants undertake high levels of local R & D and ensure high percentage value-added local content; Japanese company participation in European collaborative R & D such as the PROMETHEUS programme; maximising the learning opportunities and demonstration effects from Japanese investment through strategic joint-venture agreements and business alliances; and finally, wider opening of the Japanese market in automobiles and other sectors to European producers.

Notwithstanding the difficult question of Japanese entry into a post-1992 Europe, there are two other major issues which will need to be addressed within a common policy framework. The first relates to environmental pollution and related concerns and regulations regarding vehicle emissions. (Boehmer-Christiansen, 1990). The second issue concerns the need for improved traffic management systems, and in this area, based on 'PROMETHEUS', Europe is already making a determined attempt to spearhead advances in the development of 'smart cars' and 'intelligent highway systems'. They could provide an opportunity for Europe to take a lead in new directions for technical change which might harmonise environmental with competitive industrial objectives.

Part IV
Energy, Environment and Regulation

Preface to Part IV

So far in Parts II and III we have concentrated on the issue of the technological competitiveness of European firms in a variety of industries and technologies, but the last part of Chapter 16 has broached some more fundamental topics. As we argued in Chapter 1, there are some environmental hazards which transcend the struggle for competitive advantage, and in every industry there are problems of quality standards, which go beyond those of technical connectivity discussed in Chapter 11. In Part IV we look in more detail at some of these complex problems, which are as much political and social as technical and economic in nature.

The production, distribution and consumption of energy are at the heart of many of the most difficult environmental problems – the greenhouse effect, radioactive waste disposal, ocean pollution, fossil fuel depletion and acid rain among the most serious. These hazards cannot be dealt with purely at the level of the nation–state or at the level of the EC. We therefore start Part IV by placing the EC energy industries in a global context. As John Surrey's chapter makes clear, long-term strategies are indispensable for the EC both for environmental reasons and for security of supply.

Although we may hope for a gradual transition to the widespread use of cleaner and renewable technologies in the supply of energy, there will inevitably be continuing problems of waste disposal. Frans Berkhout shows both the ubiquity of these problems and the extraordinary difficulties of establishing effective and acceptable regulatory regimes. As in the case of law observance in general, it is informed and alert public opinion at all levels which is the only reliable means of resolving these problems, rather than Draconian penalties or intrusive policing.

This conclusion is relevant for all issues of environmental policy and it is therefore instructive to examine the political process leading to various types of legislation, regulation and standard setting. Chapters 19 and 20 provide striking and contrasting examples in the case of acid rain and food quality. The contrast between the experience of two member countries – the United Kingdom and the German Federal Republic – serves both to reinforce the point about well-informed public opinion and to illustrate the great difficulties of harmonisation of policies and standards at the EC level. There are also contrasts between industries. Whereas Chapter 13 suggested the probability of a rather positive outcome of the EC harmonisation process in the cases of pharmaceuticals, the position in the food industry gives far less cause for optimism.

These varying experiences lead to more general conclusions about the 'democratic deficit' within the EC institutions to which we return in Part V.

17 The EC in the global energy context

John Surrey

In global terms the EC is a large energy market, heavily dependent on imported fuels and thus a major producer of acid rain and greenhouse gases. Energy is therefore a crucial policy area for the EC and the EC is important for international energy questions, chiefly with regard to oil supply security and global warming. Under the Single European Act the Commission is trying to create a unified energy market stretching from the Baltic to the Adriatic. This chapter examines the policy issues as the Commission takes over more responsibilities in the energy sector.

There are obvious conflicts between the aims of market 'liberalisation' which involves giving free rein to market forces, and the objectives of supply security, pollution control and energy conservation which require long-term planning and some degree of intervention. The diversity of energy supply conditions also makes it difficult to devise policies which are acceptable to all member countries. For example, the Netherlands has large gas reserves relative to its own needs; Britain has considerable oil, gas and coal; Germany has considerable nuclear power, coal and some domestic gas; France and Belgium have little fossil fuel and large nuclear capacity; Spain has some coal and some nuclear capacity. In contrast the remaining EC countries – Denmark, Greece, Ireland, Italy, Luxemburg and Portugal – are heavily import dependent for all their commercial energy supplies.

The long time-scales in this sector – for inter-fuel substitution, major investments, developing and diffusing new technology and resource depletion – make it necessary to take a long-term view. Today there is no conventional wisdom of the kind that gave policymakers confidence when the oil shocks of the 1970s raised the 'marker' price of light Arabian crude from just over $2 a barrel in early 1973 to a peak of nearly $40 in early 1980. The international agencies, governments and fuel industries assumed that OPEC would retain its grip on the oil market, that the crude oil price would continue to rise ($80 a barrel by the year 2000 was a common estimate), and that energy demand was closely coupled to GDP. The resulting forecasts were widely seen as a threat to future prosperity. The response of the OECD countries was to develop non-OPEC oil resources, launch nuclear power programmes and introduce energy conservation measures. These policies, it should be noted, reinforced the price mechanism. The low energy prices and high interest rates of the late 1980s discouraged investment in conservation and in all domestic energy supplies which are high cost or capital intensive.

The Commission has little experience of policy-making in this field and has few of the levers traditionally available to national governments. Despite the 1957 Treaty of Rome which created the Common Market, responsibility for energy policy remained with governments, and they exercised it in the interests of national economic growth, supply security, the balance of payments and employment. Owing to the economic importance of this sector many state corporations were created, and governments often exercised considerable control over private-sector energy companies. Despite the dominance of the international companies, the oil sector was no exception.

After the 1973–74 oil shock, France adopted its '*toute nucléaire*' policy; West Germany and Spain accelerated their nuclear programmes and promoted domestic coal. Britain concentrated on North Sea oil development, the rapid build up of North Sea gas supplies and modernising the coal industry. Responsibility for coordinating an international response fell to the newly created International Energy Agency (IEA), whose members included the United States, Japan and other non-European OECD countries. The Commission therefore tended to focus on supplementing national efforts in energy conservation and R & D, and coordinating an EC position within the IEA oil-sharing arrangements.

In 1975 the EC put a restriction upon the use of oil and gas in power stations. Since these fuels were expensive and in short supply, member countries already had a strong incentive to replace oil in power stations where possible (for example, with nuclear power and coal) and to use whatever natural gas was available to replace oil in direct heating uses. The restriction had to be sufficiently flexible to allow for the use of oil and gas in power generation where it could not be avoided.

The best indication of policy thinking at the EC level before the Single European Act was a Council of Ministers' Resolution in 1986 which set out energy policy objectives for the EC up to 1995. The aims were to increase the efficiency of final energy use by 20 per cent; limit oil's share to 40 per cent of total energy consumption; reduce the share of oil and gas in electricity generation to under 15 per cent; increase coal's share of EC energy consumption; and for natural gas, to maintain the level of domestic exploration and production and to diversify import sources. Oil prices had recently collapsed and the resolution stated that if oil prices remained low the policies might have to be strengthened – a clear indication that the Council of Ministers in 1986 thought that long-term policy considerations should have precedence over short-run market forces.

Following agreement to complete the Common Market, the EC's policy agenda underwent a fundamental change. The Commission published its Single Energy Market (SEM) proposals to eliminate market distortions in the supply of energy owing to monopoly and state assistance (CEC, 1988g). The most controversial proposals relate to the supply of electricity and gas to large industrial users and open procurement for capital projects. Electric and gas utilities were required to provide open access and common carriage on their transmission and distribution systems for third parties and to charge 'transparent' prices. Energy companies were required to operate open bidding lists despite the higher transaction costs involved and the technical diffe-rentiation between both suppliers and products. These raise many detailed questions which are outside the scope of this chapter. Suffice it to say that the SEM proposals concern market distortions on the supply side, not distortions on the demand side which are widely held to warrant state assistance to improve consumer choice and the efficient use of energy; they say little about nuclear power, despite the prevalence of state support at various stages of the fuel cycle, and nothing about the concentrated structure of oil supply in the EC; nor do they conclude that if electricity and gas prices for large industrial users are to be transparent, the same should apply to oil prices in the same market.

The SEM proposals are very different from the policy concerns of the 1986 Resolution and of national governments over many of the previous 40 years. This was possible only because of a major improvement in the global energy outlook. At times of energy glut it is easier to believe that energy is no different from other commodities and products, especially as it then seems so much more costly to provide for long-term supply security.

I The new global context

The crucial change was a large improvement in the efficiency of energy use among the OECD countries. By 1987 they were on average using 24 per cent less energy per $1,000 of GDP than in 1973 (OECD, 1989c). This was due to far-reaching changes in product and process technology and in industrial structure as well as to price effects. Energy conservation in the industrial and residential sectors had been so pronounced and estimates of the further potential were so large (20 per cent by 2000 was a common figure) that it was now widely thought that OECD energy demand would grow very slowly, even though transport fuel demand continued to rise.

Major improvements in the supply outlook had also occurred. By the mid-1980s the OECD countries imported far less oil from OPEC sources than a decade earlier and the major Middle East exporters now had much idle production capacity. In 1988 OPEC oil export revenues were $88 billion, one third of their peak level in 1980 (Shell, 1989). This was due to energy conservation combined with large growth of non-OPEC oil production in the period of high oil prices (especially in the Soviet Union, Mexico, Alaska, Norway, the United Kingdom and Brazil) and the substitution of gas, coal and nuclear power for oil. Non-OPEC oil exports rose from 4.4 million barrels a day in 1980 to 9.6 million in 1988.

The crude oil price collapsed in early 1986. Despite subsequent price fluctuations between $10 and $18 a barrel, it seemed likely that surplus production capacity would persist and that the crude oil price would probably not go much above $20 a barrel in the 1990s. The end of the war between Iran and Iraq brought the hope that these countries would be too preoccupied with reconstruction to withhold their own oil supplies or disrupt those of the Gulf producers. This was a remarkable change in expectations compared with the forecasts made in the decade after the 1973–74 oil shock. However the new optimism was an OECD-centred view. OECD oil demand in 1988 was still below its 1973 level; but global oil demand was considerably higher – mainly because developing countries' oil demand, despite the oil shocks of the 1970s and the debt burden of the 1980s, had continued growing at almost 4 per cent a year since 1973.

The gas supply outlook had also been transformed due to new discoveries, the proving of reserves in earlier discoveries in various parts of the world and greater confidence in the ability to develop them commercially. For the EC countries the main factors were the large discoveries in the North Sea, particularly on the Norwegian Continental Shelf; the expansion of Dutch onshore reserves; and that the Soviet Union, despite having increased its gas production by 250 per cent from 1973 to 1988, has 40 per cent of world proven gas reserves and plenty of scope to raise production further (Shell, 1989). Europe's large imports of Siberian gas, far from being the grave risk painted earlier, could now be seen to be of considerable benefit in the period of *perestroika* and building trade bridges with the Soviet Union. At the same time Algeria was proving to be a keen gas exporter to southern Europe. Should gas prices rise sufficiently to cover the transport cost, large gas reserves were waiting to be developed in Nigeria and the Middle East.

Owing to very large expansion in China, the United States, India, South Africa and Australia, world coal production in 1988 was 50 per cent above the 1973 level (Shell, 1989). In the mid-1980s considerable overcapacity had developed which caused coal export prices and ocean freight rates to slump. Very cheap coal was now available for Europe's power stations and industries.

Given that the bulk of international energy trade is denominated in US dollars, the large American trade deficit encouraged the view that the dollar would remain weak and that this would help to perpetuate Europe's cheap energy imports.

Besides reducing concern with supply security, the changed supply outlook had three further consequences. First, nuclear power suffered a big setback owing to cost increases and the public opposition engendered by Chernobyl and a variety of health and safety concerns. Energy abundance made this far less of a problem even for governments hitherto intent on developing nuclear power programmes, and it encouraged some prominent political parties in Europe to adopt non-nuclear energy policies. Second, lower crude oil prices had reduced the profitability of high-cost oil production and brought sharp cuts in exploration. However, fiscal adjustments (to take account of changes in the economic rent in production) combined with cost-cutting technological change enabled sufficient new fields in the British sector of the North Sea to be developed to avoid a sharp rundown in the 1990s. Third, in the new global energy context the energy R & D priorities of the 1970s were now seen to be high-cost technologies without customers for the foreseeable future. Priorities for the fast breeder reactor, coal liquefaction and coal gasification were greatly scaled down. Renewables came under the same cost discipline imposed by lower energy prices.

Although the energy outlook had improved considerably when the Single Market rose to the top of the EC agenda, there were some blots on the horizon. OPEC still had considerable excess production capacity, but global oil consumption was above the 1973 level and rising. Lower fossil fuel prices were a further setback to nuclear power and put European coal producers under great pressure to reduce costs and output. Above all, the greenhouse effect had emerged as the most difficult issue of all for the consumer society (House of Commons, 1989; Keepin and Katz, 1988). The time-scale of climatic change is such that action has to precede full scientific proof. Unlike acid rain emissions (SO_2 and NO_x), there are no technical fixes to reduce greenhouse gas emissions, especially CO_2 and methane. To be effective, any action would need to be coordinated internationally and include not only the industrial countries but also industrialising countries such as India, China

Table 17.1 EC primary energy consumption, 1989

	Oil	Natural gas	Solid fuels	Nuclear & hydro	Total
Million tonnes oil equivalent	480	200	230	174	1,084
Per cent	44	18	21	16	100

Source: EUROSTAT, 1990

Table 17.2 EC energy consumption: changes since 1980 (Million tonnes of oil equivalent)

Final energy by sector	1980	1988	Final energy by fuel	1980	1988	Primary fuel	1980	1989
Industry	245	219	Solid fuel	63	58	Coal	238	229
Transport	170	211	Oil products	397	358	Oil	551	480
Homes, shops & offices	278	277	Gas	127	163	Natural gas	171	200
			Electricity	102	124	Nuclear	46	157
			Heat	4	4	Hydro	17	13
						Others	1	4
Total	693	707		693	707		1,024	1,084

Source: EUROSTAT, 1989

and Brazil which contain over half the human race. The 1988 Toronto Conference recommended targets to reduce CO_2 emissions by 20 per cent by 2005 and 50 per cent by 2020. These would require sustained effort to raise energy use efficiency in all sectors, to change the fuel mix towards low- and zero-carbon fuels, and to develop and apply less-polluting combustion and conversion processes. More efficient energy use is essential because it can reduce CO_2 emissions in all sectors and take effect comparatively quickly. Fuel switching tends to be more limited in application (mainly electricity generation) and slower in taking effect if new power stations have to be built and new technology developed. If, after more study, these seem insufficient, changes in transport and other policies would be needed to reduce energy demand. European politicians were quick to call for a coordinated international response, but international talks so far have exposed wide differences in what the EC member countries are actually prepared to do.

Policy issues

Let us now examine EC energy issues in relation to the objectives of maintaining supply security, reducing environmental pollution and increasing market efficiency. It is useful to begin with the broad picture.

As shown in Table 17.1, EC primary energy consumption is currently around 1100 million tonnes of oil equivalent and oil's share is above the 40 per cent ceiling called for in the 1986 Council Resolution.

Table 17.2 shows that EC total energy consumption grew only slightly in

Table 17.3 The EC's energy imports, 1989 (%)

	Import dependence	Main foreign suppliers & shares of EC imports
Crude oil and and feedstocks	80	OPEC 62 , Soviet Union 12, Norway 12
Oil products	20	OPEC 29, Soviet Union 18, Eastern Europe 14 Norway and Sweden 13
Natural gas	40	Soviet Union 39, Norway 31, Algeria 28
Coal	33	United States 41, South Africa 20, Australia 13, Colombia 8

Source: EUROSTAT, 1990

Table 17.4 World primary energy consumption, 1989 (million tonnes of oil equivalent)

	West Europe	United States	Japan	Soviet Union	Rest of world	World
Oil	592	793	233	435	1,045	3,098
Solid Fuels	258	473	75	289	1,136	2,231
Natural gas	208	488	42	563	406	1,707
Nuclear & hydro	256	214	67	102	338	977
	1,314	1,968	417	1,389	2,925	8,013

Source: BP Statistical Review of World Energy, 1990

the 1980s; but there were big differences between the various sectors. Owing to the decline of the former 'smokestack' industries, the rise of electronics and service industries and a stream of product and process innovations, industrial energy demand fell by 11 per cent from 1980 to 1988. In contrast, transport fuel demand rose by 24 per cent owing to the unabated growth of road and air traffic; much of the increase was therefore for light oil distillates. The growth of nuclear power, especially in France, combined with the growth of natural gas supplies and more coal in power generation in some countries, led to a considerable reduction in demand for heating oils, especially heavy fuel in power generation. Together with the increased demand for light distillates, this led to a considerable change in refinery operation and economics.

Despite the large growth of nuclear power in France, in Dutch and British gas production and in British North Sea oil production, the EC remains dependent upon imports for nearly half of its total energy supplies. Four-fifths of EC crude oil supplies and one-fifth of EC oil product supplies are imported – in both cases principally from OPEC producers in the Middle East (see Table 17.3). Imports from the Soviet Union, Algeria and Norway supply 40 per cent of EC natural gas consumption. One-third of EC coal consumption is now supplied by low-cost overseas suppliers.

From Table 17.4 it can be seen that Western Europe (which is predominantly the EC) accounts for 20 per cent of world oil consumption, 12 per cent of world gas consumption and 11 per cent of world coal consumption. The

EC already accounts for a substantial share of total international trade in these fuels.

The EC has only 1 per cent of world proven oil reserves and 3 per cent of world proven gas reserves. Although it has substantial coal reserves (sufficient to last 300 years at the present rate of output), they are expensive to produce relative to the current price of imported coal and exchange rates, and only small quantities of domestic output are suitable for blast furnace coke or sufficiently low in sulphur to meet current environmental regulations without the use of expensive clean-up processes (chiefly flue gas desulphurisation). Since there is little reason to expect over the next 20 years a repetition of the two big changes that occurred in the period 1975–90 (the historically high rate of energy conservation and the big growth in domestic production of oil, natural gas and nuclear power), it is likely that growth in EC energy consumption in the future will lead to even higher EC dependence on imported fuels. The continuing growth in EC transport fuel demand would almost certainly involve even heavier dependence on OPEC sources, especially in the Middle East. Higher EC gas demand would likewise involve increased dependence on imported gas. A significant reduction in domestic coal production would further increase EC imports of coal and/or natural gas. A nuclear moratorium of any kind would have broadly similar effects.

Given the very unequal distribution of fuel resources across the globe, the growing energy needs of the developing countries and the rising energy import dependence of other major industrial countries, energy supply security will be a recurrent concern for the EC. The need to reduce greenhouse gas and acid rain emissions will make the problems of EC energy policy even more difficult. This is because of the changes likely to be required on the demand side, the difficulty in increasing nuclear power owing to widespread public opposition and the fact that the substitution of natural gas on a large scale for coal would involve a very large increase in EC gas imports. This provides the broad picture against which to consider the following issues.

Energy Conservation

Energy intensity per unit of output for the EC countries as a whole fell by 20 per cent from 1973 to 1983; but the rate of energy saving then declined as industrial production picked up and after oil prices had slumped (House of Lords, 1989). Numerous studies have indicated further scope for cost-effective energy saving but also showed that lack of information and other factors inhibit their take up by consumers, the building trades and appliance manufacturers. Governments in most EC countries have policies to promote energy conservation. These vary from passive and small-scale measures (for example, publicity and grants for loft and draught insulation) to more active ones such as mandatory appliance labelling, high insulation standards for new buildings, compulsory certification of the energy efficiency of existing dwellings before they can be sold and high taxes on petrol and big cars.

Increasing the efficiency of energy use is not part of the SEM proposals even though inefficient use of energy constitutes a prime cause of market failure. Energy efficiency, however, is widely seen to be essential for reducing environmental pollution from energy production, conversion and use and especially for reducing greenhouse gases. To achieve the large reduction in world greenhouse gas emissions called for at the 1988 Toronto Conference would require concerted action to improve energy efficiency across the board, especially in road transport and power generation since these account for a large proportion of the emissions. Reducing road vehicle emissions would require a concerted switch to public transport and small cars. In power generation, whatever combination of technical fixes and change in fuel mix is sought, it would also be necessary to promote electricity conservation with considerable vigour. An approach which is being widely canvassed, though probably difficult to implement, is that electric utilities should be allowed to invest in new supply capacity only if they can show that this is more economic than schemes to promote the more efficient use of electricity.

Oil

EC imports of crude oil come mainly from the principal OPEC suppliers in the Middle East. At January 1989 world proven oil reserves were 907 billion barrels, 30 per cent higher than the level from 1973 to 1987; 75 per cent are located in the OPEC states and over 60 per cent are in the Middle East where production costs are only $1–5 a barrel. Domestic oil production consists largely of production from the British Continental Shelf. Most new fields in the British sector of the North Sea are likely to be small and tend to be only moderately profitable at an oil price of $15–18 a barrel. Growth in EC oil consumption would almost certainly increase imports from OPEC sources.

Owing to the unabated growth in demand for light oil products, reduced demand for heavy oil products and reduced refinery distillation capacity in the OECD countries, a fifth of the EC's oil products is now imported. This gives non-EC suppliers, particularly those in the Middle East, a stronger commercial stake in the EC market, but a supply disruption which closed refineries as well as oil wells in the Middle East could cause serious economic disruption in Europe. The Gulf Crisis which was triggered by Iraq's annexation of Kuwait in the summer of 1990 was yet another reminder of the basic instability of the region which will account for an increasing proportion of total oil supplies as long as world demand keeps rising.

The EC oil market is dominated by the major international companies, especially BP, Shell and Exxon. Being European-based, BP and Shell tend to be regarded more favourably in Europe than other oil multinationals. Since the 1960s Exxon and Shell have operated as partners in exploration and production throughout Europe. The oil majors also have a large presence in the European gas market.

The Commission appears to regard oil as a reasonably competitive business and to assume that any departures can be rectified through competition law. The SEM proposals call for the removal of protection of national refineries in Spain, Portugal and Greece, the harmonisation of specifications of oil products throughout the EC and the equalisation of VAT and excise duties across the 12 member countries. Harmonising specifications appear not to present great difficulties. Equalising VAT and excise duties would cause substantial price changes and is opposed by member countries which are unwilling to relinquish sovereignty over taxation in advance of any agreement on monetary and political integration.

There are clear conflicts between the SEM proposals and the ceiling levels for oil in the 1986 Council Resolution and between the SEM and the aim of reducing environmental pollution. A policy to reduce greenhouse gases would need to reduce transport fuel consumption which is almost entirely oil specific. Burning oil produces 20 per cent less CO_2 than burning an equivalent amount of coal, but there are strong arguments against encouraging a switch to oil in power generation. Burning gas produces 40 per cent less CO_2 than an equivalent amount of coal; gas is even more attractive if used in combined cycle plant to replace coal burned in older, relatively inefficient stations (the CO_2 reduction would be 75 per cent using a combined cycle gas turbine with a thermal efficiency of 42 per cent instead of an old coal plant with only 27 per cent efficiency). Switching to oil in power generation could involve substantial imports of heavy fuel oil which might well increase the price – as happened dramatically owing to the extra power station demand in Britain during the 1984–85 miners' strike.

Natural gas

The chief domestic gas sources are Dutch onshore gas and British offshore gas. There are small deposits in several other countries but production from Lacq in France and the Po Valley in Italy is running down. The principal imports are from the Soviet Union and Norway into Western Europe and from Algeria into Southern Europe. The development of the large Troll and Sleipner fields will permit bigger imports from Norway, which is regarded as a reliable trading partner and a potential member of the EC. Whether the Norwegians would be prepared to meet a markedly higher level of EC gas demand is unclear since they have always wanted to avoid oil and gas production unduly distorting their economy.

Perestroika and the easing of super-power tensions have reduced the strategic risks associated with Siberian gas imports, and the large gas trade gives the EC and the Soviet Union further reason to improve political relations. Gas exports constitute a large proportion of the Soviet Union's hard-currency income available to pay for technology imports. The main uncertainty for the future is not the size of the Siberian reserves but the extent to which the Soviet Union could step up gas exports given the need to increase domestic gas consumption west of the Urals in order to make up for the post-Chernobyl problems with nuclear power and to reduce pollution from the large-scale, inefficient use of coal. Supply security could become a problem again if the present unrest in the Soviet Republics leads to turmoil in the vast territory traversed by the gas pipelines *en route* to the EC.

The substitution of gas in factories, offices and homes has been a major factor in the energy efficiency improvement and the cleaner air achieved in some member countries, notably the United Kingdom. So far no gas has been used in power stations in the United Kingdom and France and only a little in Spain (see Table 17.5). Significant amounts are used in power stations in the Netherlands, Italy, Germany and Ireland owing to large 'take or pay' contracts and insufficient major markets other than power generation to absorb the surplus gas. Gas is increasingly attractive, environmentally and

Table 17.5 Natural gas in the EC, 1989

	United King.	West Germany	Italy	Netherlands	France	Spain	Others[a]
Consumption (TJ)	2,102	2,124	1,717	1,447	1,130	218	455
Used in power stations	0	13%	16%	21%	0	3%	21%
Production (TJ)	1,689	536	640	2,517	107	64	210
Import dependence	20%	75%	63%	0	91%	71%	54%

[a] Belgium, Denmark, Greece, Ireland, Luxemburg and Portugal

Source: EUROSTAT, 1990

economically, for power generation – especially at the high efficiencies obtainable with combined cycle plant. There is large potential for using more gas in power stations in various parts of the EC, especially as existing stations fall due for replacement or refurbishment. However, the availability of natural gas varies widely within the EC, and large areas are not connected to gas transmission and distribution pipelines. The difference is most pronounced between the well-developed gas systems of the Netherlands, Belgium, the United Kingdom and Germany on the one hand and Southern Europe on the other.

Owing to the combination of stricter pollution regulations and lower gas prices there has been strong pressure to lift the 1975 restriction on gas use in power generation. The Council of Ministers acceded to this in 1990. Although there is obvious commercial advantage in using combined cycle gas turbine plant with high thermal efficiency (rather than equipping all coal-fired plant with flue gas desulphurisation), it is to be hoped that the result is not a switch to gas firing on a scale which would lead to higher gas prices and faster depletion of Europe's own gas resources.

These issues do not feature in the SEM proposals for the gas market which focus on access for new entrants, common carriage terms and price transparency for industrial contracts. These matters are relevant for those parts of the EC which are already connected to gas grids, but it is difficult to see how gas grids will be extended to the other parts of the EC if it is uneconomic to do so. Large-scale funding by the Commission of gas 'infrastructure' projects seems unlikely as it could easily cause dissension among the core members of the EC who are afraid of being paymasters for the development of the periphery.

There are monopoly questions in the European gas market owing to the dominant position of a small number of oil companies and gas utilities. These are not addressed directly by the SEM proposals. For the time being at least, the Commission appears to prefer to chip away at the monopoly through its common carriage and price transparency proposals. How effective these will be, especially in the absence of a common regulatory framework throughout the EC, remains to be seen. So far, the Commission has not proposed any regulatory institutions.

Coal

For 150 years the mainstay of Europe's industrial economy, coal production is now concentrated mainly in the United Kingdom, Germany and Spain (see Table 17.6). Despite considerable government support of various kinds, output capacity has been cut back in the 1980s owing to falling demand and imports. Coal mining is extinct in the Netherlands and going that way in Belgium and France. Coking coal for steel making is now mainly imported. Consumption of steam coal is heavily concentrated on power generation, industrial steam raising and cement production.

Though not yet on the scale predicted after the 1978–9 oil shock (World Coal Study Report, 1980), EC coal imports have risen substantially and now account for a third of EC coal consumption. Delivered prices at Rotterdam slumped in the mid-1980s owing to large surplus capacity overseas (MacKerron, 1989; Prior, 1989). In national currency terms coal imports were made considerably cheaper by the low dollar exchange rate. The main sources have been the United States, Australia and South Africa. Surface mining is still profitable, but reserves of hard coal accessible to surface mining in the EC are comparatively small, and local environmental opposition tends to limit the number of new sites approved for development each year.

The main problem of coal policy involves the underground coal industry. The features which make it difficult are that once mines are closed it is generally uneconomic to reopen them later (enforced 'mothballing' during the 1984–5 British miners' strike led to the permanent loss of 73 coalfaces and high costs of restoring many others); large scale closures can lead to severe regional unemployment; and investment in major new capacity is discouraged if there is a strong risk that it will be made uneconomic through distress pricing by overseas producers in the future.

In the absence of protection a great deal of EC underground capacity would have had to close at the import prices of 1986–8. Not much of the German coal industry could have survived, for its problems were worsened in the 1980s by the rise in the Deutschmark relative to the US dollar and in 1989 it was receiving subsidies totalling DM 10bn (£3.6bn). The British industry would have shrunk to those mines which were insulated by the cost of trans-shipment at Rotterdam and rail transport from British ports. However, by mid-1989 coal import prices at Rotterdam were $41–5 a tonne – around 50 per cent higher than the nadir reached two years earlier; by late-1989 prices of up to $50 a tonne were reported for imported low sulphur coal.

Combined with currency movements this made coal imports dearer than previously in some EC countries. Meanwhile, strong productivity growth had been occurring, especially in the United Kingdom where the mining labour force has been cut drastically since 1985. By phasing out state assistance the SEM proposals would hit the German coal industry particularly hard, and they are therefore strongly resisted in the mining areas of North Rhine Westphalia which would suffer most. Without the *Kohlepfennig* (a tax on electricity which is used to subsidise coal) a great deal of German hard coal output would be uneconomic. The British coal industry will be less directly affected by the SEM proposals because electricity privatisation in the United Kingdom will remove any remaining protection after 1993, when the present coal contracts expire. Up until the early 1980s British coal output in the

Table 17.6 Hard coal in the EC, 1980 and 1989*

		United Kingd.	West Germany	Spain	France	Belg.	Others	EUR 12
	million tonnes							
Deliveries	1980	123.8	91.8	16.8	46.4	16.7	35.4	330.9
	1989	108.2	78.6	29.0	28.8	13.9	52.4	310.9
Production	1980	128.2	94.5	12.9	18.1	6.3	0.3	260.3
	1989	98.3	77.8	19.3	11.5	1.9	0.2	208.7
Imports[a]	1980	7.3	9.3	5.7	29.5	10.1	35.9	97.8
	1989	12.1	6.3	10.6	16.0	10.9	56.0	111.9
	kg/man hour							
Productivity[b]	1980	382	539	219	352	279	n.a.	426
	1989	680	645	329	589	328	n.a.	603
	thousands							
Employees[b]	1980	184.4	121.6	33.2	30.4	16.4	1.1	387.1
	1989	52.0	87.9	35.0	9.6	2.0	0.9	187.4

*This table excludes brown coal which is mined mainly in the FRG
[a] EUR 12 figures are imports from non-EC countries; figures for individual countries include imports from EC and non-EC countries
[b] Relates to underground mining; 1980 figures are whole year averages; 1988 figures are for August 1989

Source: EUROSTAT, 1990

1990s was expected to be 130–50 m tonnes, but the current expectation is that it may fall to 40–60 m tonnes after 1995, depending on exchange rates.

Yet coal import prices are likely to remain volatile. Even if the major coal exporters had no excess capacity, there are various producers (for example, Colombia, Poland, Indonesia, the Soviet Union and China) which urgently need hard currency and have, or could develop, considerable export capacity.

In contrast with the aim in the 1986 Council Resolution of retaining a sizeable coal industry, the SEM proposals allow no protection against imports; this could lead to the rundown of the remaining coal industry in the EC, especially if there are further big temporary falls in import prices. In world terms the EC is a large importer of steam coal: EC coal imports in 1988 represented 40-45% of all seaborne traded steam coal (*Financial Times*, 1989: May). The central policy issue is the extent to which a considerable increase in EC coal imports would increase the price of all imported coal. A *dirigiste* policy against coal imports would be costly and would not work: member countries without domestic coal industries have no interest in subsidising coal production; they want access to imports, as do increasingly the larger customers in the United Kingdom, Germany and Spain. To retain a sizeable coal industry requires a policy to protect against big temporary falls in import prices and encourage further productivity growth so as to retain those mines which would be economic at the long-run or trend price of imported coal.

Technical fixes to reduce acid rain emissions (for example, flue gas desulphurisation and fluidised bed combustion) are required under the Commission's 1988 Large Combustion Plant Directive. More far-reaching changes would be needed to meet the Toronto targets for reducing

Table 17.7 Nuclear power in the EC, 1989 (%)

	France	Belgium	West Germany	Spain	United Kingdom	Nether-lands	Others
Nuclear share of electricity generated (%)	75	61	34	39	22	5	0

Source: Energy Statistics, EUROSTAT, Monthly; EUROSTAT, 1989

greenhouse gases. These might include using more gas in combined cycle gas turbine plant and closing older coal-fired stations with low thermal efficiency, using coal more efficiently by means of combined heat and power schemes, and perhaps installing more nuclear plant.

Nuclear power

Outside France, however, nuclear programmes in the EC countries are at a virtual standstill. The nuclear share of electricity generation varies widely. It is high in France and Belgium, medium in Germany, Spain and the United Kingdom, negligible in the Netherlands, and nil in Denmark, Greece, Ireland, Italy, Portugal and Luxemburg (see Table 17.7). Within Germany and Spain the nuclear share varies widely among utilities and regions.

Public opposition and economics cloud nuclear power's future. Outside France that future will be small unless there is a marked change in public attitudes and improvement in the economics. With 50GWe of nuclear reactors installed, France accounts for 54 per cent of EC nuclear capacity; this proportion will rise as the French reactors under construction enter service. France has a large surplus of nuclear plant in relation to its base load electricity demand (but not its peak demand) and is therefore exporting as much as other EC Member countries will take during the periods of the year when France has considerable surplus capacity.

The oldest reactors in the EC – the early gas-cooled reactors in Britain, France and Italy – will probably all be closed by 2000. A further 30 GWe of the EC's existing nuclear capacity would be over 30 years old by 2010 and all 90 GWe of currently installed EC nuclear capacity would be over 30 years old by 2020. Without major new ordering (and/or refurbishment of existing plants if that proves possible), nuclear capacity would start to decline around 2005.

After Chernobyl another serious nuclear accident might well result in the premature closure of some or all of Europe's reactors. The majority of EC countries have no nuclear plant and would be little affected, except to the extent that they import French nuclear electricity. Germany, Spain and the United Kingdom would have difficulty for perhaps 10 years and would have to operate existing coal-fired plant more intensively, convert some of them to gas or use imported low sulphur coal to comply with SO_2 emission regulations. Owing to their heavy dependence on nuclear power, however, it could take France and Belgium all of 20 years to adapt their electric power systems to a nuclear moratorium.

Would a return to energy scarcity and high fossil fuel prices revive nuclear ordering? France's large overcapacity makes it unlikely that it would order more nuclear plant than necessary to maintain its existing level of capacity; to do more would stretch out the period of overcapacity even longer and increase Electricité de France's capital debt even more than at present (over £20bn). The ordering would be correspondingly smaller if it proves possible to life-extend France's existing reactors from 40 years to 80 years through refurbishment. Nuclear ordering might be revived in Germany, Britain and Spain, but this would require a fundamental change in public attitudes and probably also the development of an inherently safer reactor design and acceptable solutions of the problems of spent fuel and radioactive waste. The smaller member countries with no previous nuclear ambitions would additionally require the development of smaller as well as safer reactors.

Nuclear power could have a role in a strategy to reduce greenhouse gas emissions; but this, too, would be possible only with changes in public attitudes and if the public could see that everything possible was being done to reduce greenhouse gas emissions through policies to improve the efficiency of energy (especially electricity) use. The 'not in my backyard' syndrome is particularly strong if a minority are asked to bear risks and social costs on behalf of a profligate majority.

At the present time the EC can do little to promote nuclear power in the interest of the environment or fuel supply security. Even if public attitudes and economics do change in favour of nuclear power, it would probably be necessary to develop improved reactor designs with more inherent safety features before nuclear power could be re-launched on a substantial scale. Not only is it unclear who would finance the necessary development and design work to produce a new reactor system, but the lead times suggest that if such work were undertaken it would not be reflected in additional nuclear capacity in under 15–20 years.

The SEM proposals wisely say little about nuclear power. Yet two politically contentious nuclear issues are likely to be important in the SEM process. One is that of ensuring access for French nuclear power to the German electricity market in the face of entrenched opposition from domestic coal and electricity interests and from politicians nervous about votes in the coal mining region of North Rhine–Westphalia. The SEM requires free access across internal borders and the abolition of state support. In this case open access is being sought for a product which derives from unstinted state support over a period of 30 years. The other issue is the protected status of nuclear power in Britain (an ironical result of electricity privatisation) where the Non-Fossil Fuel Obligation and the Fossil Fuel Levy are prime examples of what the SEM proposals were intended to phase out. The EC's approval of these two instruments of protection extends only to 1998, making it unlikely that any new reactor will be built in Britain unless gas and coal prices increase to levels that hardly anyone now considers credible.

II Conclusions and implications

The conflicts noted at the beginning of this chapter suggest that 1992 will be the beginning rather than the end of a long process of regulatory harmonisation and policy development in the energy sector. Owing to the conflicts and the uncertainties surrounding the energy outlook, the EC is having to grope its way forward in the uncharted, muddy waters of second best; a series of running battles over its more controversial SEM proposals is likely. Even if the principles are finally accepted (for example, open access to transmission systems and transparent pricing), disputes are likely to arise over matters of practical implementation.

The SEM proposals are really statements of principle about market efficiency on the supply side of the energy sector. Their application will involve considerable difficulties and they may not get very far unless they are backed up with an EC-wide regulatory system to enforce open access, common carriage and transparent pricing which currently is not proposed by the Commission; unless the Commission is allowed to invest heavily in electricity and gas interconnections to link the periphery with the core which currently seems unlikely; nor if major countries remain implacably opposed to important principles, as West Germany seems to be on the removal of coal subsidies, importing French nuclear power and giving third-party access on its gas transmission and distribution system.

Supply efficiency is important but not the only objective in this sector which is subject to periodic upheavals and long periods of adjustment in demand and supply. The SEM proposals show no recognition of the need for policies to deal with situations where the market works badly – notably the economic costs of shortages and the need to insure against the risks of supply disruptions and high import prices. If the apparatus of national energy policy were dismantled owing to the phasing out of state assistance and intervention, no one would be responsible for ensuring adequate energy supplies and insuring against the risks of high import dependence. Unless the need for these responsibilities is given formal recognition, the single energy market is unlikely to function effectively in the long run. It would also be difficult to move forward coherently on energy and environmental policy issues and necessary to deal with each problem through *ad hoc* derogations from the strict SEM principles. The resultant policies would be neither coherent nor coordinated.

Whatever form political integration may eventually take, responsibility for energy policy should be shared between the EC institutions and member countries (much like the division between federal and state responsibilities in the United States). This appears necessary to give the flexibility to cater for the differences in national energy conditions as well as avoiding a bigger Euro-bureaucracy. The Commission (or whatever may replace it after political integration) would need to be responsible for economic, environmental and safety regulation; for reviewing oil stocks and gas storage capacity and maintaining emergency oil-sharing among member countries; and for aid and trade agreements which help to improve fuel supply security. It should also promote energy conservation, long-term R & D and the diffusion of existing innovations and should be prepared to protect domestic coal producers against short-term 'distress' pricing by overseas producers. If a carbon or energy tax is introduced it should be done at the EC level and be buttressed with more direct measures; the tax revenue should be used to finance energy conservation. This is because large areas of energy demand are price inelastic. Without some form of assistance many energy consumers cannot respond to high energy prices for long periods. For example, it remains difficult for many domestic consumers even to identify, let alone install the cheapest forms of space and water heating for their particular requirements or even the most energy-efficient refrigerator; many pensioners and low-income families cannot afford to install efficient heating systems and insulation; owing to the high replacement cost, many consumers are effectively tied for years to their existing plant, equipment and major appliances; and many motorists are shielded from market forces by the prevalence of company-owned cars and expense-account petrol.

This is not to pretend that the Commission could operate a 'Common Energy Policy' of centrally ordained market shares for the various fuels or high prices to protect domestic producers. Nor, given the differences among member countries, could the Commission effectively promote domestic coal and nuclear power throughout the EC.

The need for flexibility to cope with the diversity suggests that national governments should retain the responsibility to ensure that energy supplies will be adequate to meet future demand and insure against the risks of supply disruptions and higher prices in the future. Clearly, they would need to do this in consultation with the Commission if it acquired the type of responsibilities suggested above.

Were the SEM proposals to be pushed to the extreme at which governments could do nothing, the risk would be that of no policy at all. The greenhouse effect and long-term supply security both require long-term coordinated policies to achieve results which are unlikely to be achieved through market forces alone. Equally, the European Parliament should be aware of the concentration of market power among the large corporations in this field and the increasing decision-making power of the Commission as it acquires ever wider responsibilities for policy. The Parliament must somehow ensure democratic accountability over policy and public expenditure in this sector, perhaps using the investigating committees of the United States Congress and the British Parliament as models.

Meanwhile, the latest Gulf crisis following the annexation of Kuwait by Iraq sent oil prices up to levels which many commentators in the West have refused to contemplate since the early 1980s. It renewed concern among the OECD countries over the instability of the region which possesses so much of the world's oil resources and it has also led to increased official interest in the neglected subject of energy conservation. Fortunately, the diversity which precludes a common European energy policy provides good examples of approaches which could usefully be copied by the rest of the EC – most notably Denmark in the case of energy conservation and West Germany in the case of pollution control. A good starting point for the EC in grappling with the problems of energy and the environment is to encourage the spread of best-practice technologies and policies and thus to build on the advantages of diversity.

18 Regulating industrial wastes in Europe: the limits of harmonisation

Frans Berkhout

Since matter is indestructible, all raw materials and manufactured products, when they have no further value for use or recovery, become waste.

[House of Lords, 1981]

I The finite world

The discussion of nuclear power in Chapter 17 raises the acute problem of radioactive waste disposal. But this is not unique: wastes are ubiquitous. Each human activity produces them either directly or indirectly, immediately or in the future. Some are easily absorbed and recycled by the environment, others pose persistent hazards to nature and to human health.

Wastes are such a basic product of human endeavour that they bear redescription. The word 'waste' strongly suggests redundancy, something to be thrown away and forgotten about, an 'externality'. But wastes are a condition of life, the most visible or odourous of which have been suppressed since the beginning of history. What has changed is the perception that wastes should not be discarded without an understanding of the hazards they may pose; that we live in fragile ecologies, on a planet whose grand cycles and stabilities are now being altered by the wastes of industrialised society. There is a fear that what is thrown away will come back to haunt us as climatic change, the poisoning of children or the devastation of the animal world. One reaction is to think again about the way things are made and used and to consider the pollution impacts of all the technological–organisational commitments implicit in a new product. As the value of goods and services is adjusted to account for their environmental effects, patterns of production and consumption will inevitably change. Developments in industrial waste disposal are only the first signs of the pervasive influence which environmental protection is likely to have on industrial and technological transformation over the next decades.

If we imagine a boundary to exist between human activity and the environment, then we could describe pollution control as one of the main determinants of the shape of that boundary and the flows across it. In the long term the aim would be to establish some form of equilibrium along it (Pearce *et al.*, 1989). But we labour in ignorance, and no equilibrium can yet be shown to exist. Until it can, the regulation of polluting activities will continue to inspire compelling social disputes.

States have traditionally taken on the role of defining and enforcing the environmental regulation of industry. From rudimentary rule-of-thumb methods science has progressively come to play a role in identifying the rules of regulation, in defining the performance of pollution control regimes and in providing the justification for regulatory practices. This science is often partial. It serves particular interests: the regulated industry, the regulatory bureaucracy, the state in pursuing wider goals, and, less commonly, the groups in whose name regulation was being carried out. Furthermore, science is incomplete. A politics of regulation evolves which can never abolish technical uncertainty about the measures which it generates and organises. Unlike a new lawnmower, which can be seen to work, a repository for radioactive wastes has no such opportunity for *complete* and long-term validation; it stands or falls by a socially generated agreement about its performance. All Environmental Protection Technologies (EPT) are therefore implicitly the embodiment of a social contract, and the general tendency we are witnessing is for this contract to be haggled over more fiercely, as shown in Chapter 19 with the example of acid rain.

II Harmonisation and environmental policy

The drive to harmonisation is reflected in the Single European Act which gave environmental policy the same status as other Community policies. Its broad aim is to equalise the quality and the costs of environmental protection across the Community. Like many aspects of 1992 this objective evokes grand themes. The hand of the regulator will apply the same pressure in Aarhus as in Bilbao, and the water drunk in Londonderry will be as clean as that in Thessaloniki. A clean environment will be the right of all Europeans. As the boundary between human activities and the environment becomes less permeable, the need to draw it more evenly increases. Harmonisation is an attempt to draw the boundary for the mutual benefit of producers and consumers alike.

The waste industry is already enormous. In 1982 it was estimated that some 2–2.5 million people were employed in about 350,000 firms in a broadly defined waste industry in the European Community. Spending on Environmental Protection (EP) in Europe amounted to some DM84bn (£28 bn) in 1987, of which a little over 20 per cent was allocated to waste disposal and material recovery. Expenditures are set to rise sharply over the next decade (Figure 18.1). Although these figures have to be treated with some care, the costs of EP are now clearly substantial and rising. In the United States, for instance, the waste industry is the fourth largest industrial sector (Bennett, 1989). As the costs of environmental protection increase, questions of competitive advantage within Europe, and in relation to Japan and the United States become more insistent.

However, we are not only interested in the scale or nature of economic effects. Harmonisation has come of age for other reasons as well. It aims at an integrated or optimised control of pollution. Through the 1980s the objective of matching production and distribution infrastructures to some ideal of minimised environmental harm has come to define much environmental policy.

Figure 18.1 European environmental spending, 1987 and 2000

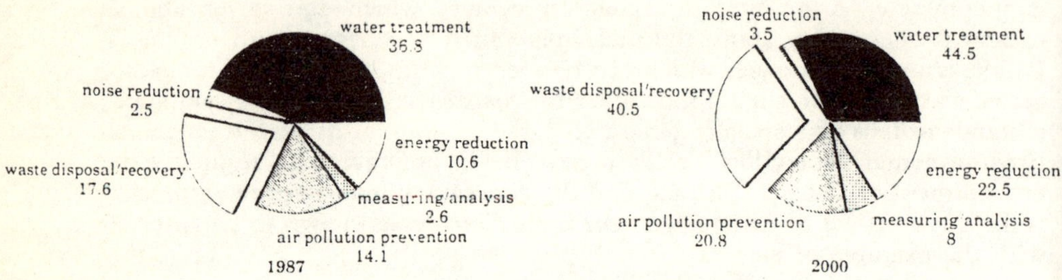

Western European environmental spending (DM '000 million)

Source: Haznews, 1989c: 10

The first effect of progressive regulatory change is on the production of wastes. For example, the reduction of liquid waste discharges can only be achieved through the production of more solid wastes requiring disposal. As regulations for waste treatment and disposal get tighter and costs increase, new technological opportunities will be presented both for ways of dealing with wastes and for designing products and processes which produce less waste – so-called 'clean' technologies. All of these interactions between emission abatement, waste disposal practice, waste reduction and avoidance and clean technologies are highly dynamic and quite imperfectly understood. But, however uneven this change is, the integration of pollution control policies will have profound consequences for technological innovation.

Harmonisation also presents the problem of how, in a liberalised economic climate, a policy of integrated control can be enforced. Regulators have traditionally been concerned with the control of emissions whereas greater effort will now need to be expended on licensing the different steps of the waste life cycle. All the material flows through industrial processes now become subject, in principle, to regulatory control. Integration therefore appears to imply a more interventionist role for governments, or the EC, in the design of waste-generating products and processes.

Lastly there is a question of how a policy of progressive optimisation remains dynamic: that is, how regulatory change can be made progressive over the longer term. Environmental policy is always meshed with other goals of industrial, energy and resource conservation policy. The European Commission may well be an appropriate centre for progressive improvements to environmental protection, but its success will depend on the broader integration of environmental policy with other economic and strategic goals. This may already be happening to a small extent as in the case of the PROMETHEUS Project discussed in Chapter 16, although the pattern of these linkages is not yet clear.

The following discussion begins with radioactive waste (radwaste) management, one of the prime constraints on the nuclear industry since its birth; the question of the disposal of its waste continues to haunt the industry. It can be seen, in Häfele's term, as a 'pathfinder' in encoutering and adapting to the pressures of environmental protection (Häfele, 1974). There are conceptual, policy-making and technological lessons to be drawn from this experience which can be applied to hazardous waste management and regulation more generally.

III Radioactive and toxic wastes: some facts and figures

Radioactive wastes

Radioactive wastes are produced by three groups of activities: the production of nuclear electricity; the production of nuclear weapons; and in much smaller quantities in nuclear research, medical practice and certain industrial activities. Most radwaste management is associated with civil nuclear power production, although military, industrial and medical uses of radioactive materials also produce diverse waste streams.

Nuclear reactor operation is supported by a range of industrial activities collectively known as the nuclear fuel cycle (NFC). This includes uranium extraction, uranium enrichment and fuel fabrication (the 'front-end' of the fuel cycle); reactor operation itself; and spent fuel storage, spent fuel reprocessing and other aspects of waste management (the 'back-end' of the fuel cycle).

Radioactive wastes may be classified according to several criteria: phase, origin, general characteristics (fission products, actinides – heat-producing or non-heat-producing), radioactivity, lifetime, disposal route. No labelling system exists which can encompass all of these characteristics and, depending on the situation, the same waste stream may be classified differently. For example, an ion exchange sludge (origin) will eventually be transformed into a packaged intermediate-level waste form designated to a specific disposal route.

In the United Kingdom a hybrid of the last four criteria: low-level waste (LLW), intermediate-level (ILW), plutonium-contaminated material (PCM) and high-level (HLW) is normally used. In Germany a basic classification of non-heat-generating and heat-generating wastes is used; in Sweden wastes are defined according to whether they arise in reactor operation, decommissioning or as spent nuclear fuel; in France distinctions are drawn among low-level, alpha-contaminated and high-level wastes. Considering that the basic characteristic which these classifications define is radioactivity, this variety is perhaps surprising. But the hazard of a radioactive waste is, in fact, not so simple. Its activity will be made up by a number of different isotopes, each of which is metabolised differently by the human body, each emitting a variety of types of ionising radiations with different energies. The hazard

Table 18.1 Nuclear fuel cycle and radioactive waste statistics in Europe, United States and Japan, 1990

Country	Number of reactors (civil)	Spent fuel (tonnes/yr) (approx)	Number of reactor sites	Solid waste (m^3/yr) ILW*	Solid waste (m^3/yr) LLW	Number of waste streams
United States	110	3,600	91	n.a.	n.a.	n.a.
Japan	38	1,000	26	2,000	10,000	n.a.
Germany	23	700	28	7,400+		300
United Kingdom	40	1,200	20	1,900	27,500	500
France	54	2,100	28	3,000	40,000	700
Italy	2	40	4	150	500	n.a.
Netherlands	2	20	3	150	500	n.a.
Spain	10	200	8	700	2,500	n.a.

* LWR waste taken to be 70 m^3 ILW/yr (conditioned) and 250 m^3 LLW/yr (conditioned)

+ Packaged non-heat generating wastes.

Source: *World Nuclear Industry Handbook 1990*, 1989.

of a waste is related to the risk that its radioactivity will lead to the exposure of human beings (other parts of the ecosystem are ignored) and how this cumulative radiotoxicity then affects this population, usually a small so-called 'critical group'. Calculating the hazard is usually simplified because only a small number of these isotopes will dominate the radiological risks. Attempts have been made in the past to set international standards for radwaste classification, but have come to nothing primarily because a waste's hazard flows not just from its inherent characteristics but is indivisibly linked to the way it is managed (IAEA, 1970).

Most radioactivity is produced during the 'burning' of uranium (and plutonium) nuclear fuel in power reactors. Most of this activity is retained within the irradiated nuclear fuel elements. Some activity escapes from the fuel to form waste streams of low- and intermediate-level activity. After about three years in an operating reactor the nuclear fuel becomes depleted and is discharged as spent fuel. Spent fuel may be regarded as a waste, as in Sweden, or as a resource, as in France, Germany and the United Kingdom. Uranium extracted from spent fuel may be recycled as fuel in thermal reactors and plutonium in both thermal and fast reactors. Extraction of these materials is achieved through the chemical reprocessing of spent fuel. Over 95 per cent of the total activity in the nuclear fuel cycle is contained in high-level waste residues from reprocessing. A small proportion of fission and actinide products (less than 5 per cent) is carried away in a diluted form in mainly liquid waste streams. In terms of volume and environmental impact it is these radwastes which impose the main present-day burden on the regulation of NFC wastes. If the fuel is not reprocessed it may eventually be disposed of directly as waste.

Two indicators can be used to give an impression of the magnitude of radwaste management in Europe. First, the gross figures for accumulations of waste in each country can be estimated. Second, it is possible to give some idea of the complexity of the management task. The number of waste streams produced can be used as an indicator of system complexity. A waste stream is a discrete source of waste which in practice may be combined with other wastes of a similar type to ease management. The scale and diversity of radwaste arisings in Europe, Japan and the United States is shown in Table 18.1.

Within the continuum of waste management and regulation exists a panoply of technical and administrative interventions. This includes everything from the direct discharge to the environment of fluid radwastes to the storage, conditioning, packaging (in cement, glass or bitumen), transport and disposal of solid wastes to an engineered repository. Above all, radwaste management policy is concerned with the organisation and operation of complex technological systems of control. These end in the storage or disposal of the waste by 'dilution and dispersion' or 'concentration and containment'. All disposal techniques will include an element of dilute and disperse in their design because nothing is indestructible; the main distinction exists in the degree of engineered restraint imposed between the wastes and their environment.

What precisely this environment will be depends on the disposal route chosen. Many gaseous wastes are discharged straight into the atmosphere; low-level liquid effluents are discharged to the sea or fresh-water bodies. For solid radwastes land-based disposal (at or below the surface) has become the technological standard, though some argue for indefinite monitored and retrievable storage instead.

The objectives of radiation protection

At the simplest level the objective of radwaste management is to ensure into the long-term future the protection of human beings from unacceptable harm associated with man-made ionizing radiations. A large body of scientific competence has developed around this basic objective which has been formalised into three basic principles:

(i) justification: no practice involving the use of ionising radiation shall be adopted unless its introduction is judged to produce a net positive benefit;
(ii) optimisation: all exposures to ionizing radiation shall be kept as low as reasonably achievable, economic and social factors being taken into account; and
(iii) dose limitation: the dose to individuals shall not exceed the limits recommended by the International Commission on Radiation Protection (ICRP) (ICRP, 1977).

Of these, optimisation is now regarded as paramount. Narrowly interpreted, optimisation means no more than finding the cheapest, most convenient disposal option for a waste stream which is estimated to give radiation doses below ICRP recommended limits. Interpreted more generously, optimisation can be seen as a more thorough style of control in which waste production and management is a primary factor in the design, construction and operation of a reactor or fuel cycle plant. In an ideal world the management, packaging and disposal path of each waste stream would be plotted as the plant was conceived. This is seldom what happens, either because disposal routes are

Table 18.2 Toxic waste production in the EC (excl. Luxemburg and Portugal)

Country	Toxic waste production ('000s MT/yr, mid-1980s)	GDP Total (ECUbn, 1988)
Belgium	1,000	127
Denmark	70	91
Spain	1,800	288
France	4,300	804
Greece	—	44
Ireland	80	27
Italy	2–5,000	701
Netherlands	1,000	193
Former West Germany	4,500	1,018
United Kingdom	3,700	697
Total	18,450-21,450	3,990

Sources: Mettelet, 1989; EUROSTAT, 1990.

not available or because management practices change over the course of time.

Nevertheless, a wealth of optimisation concepts exist: as low as reasonably achievable (ALARA), best practicable environmental option (BPEO), best practicable means (BPM), best available technology not entailing excessive cost (BATNEEC), best demonstrated available technology (BDAT), the state of science and technology (SST). All of these have meanings specific to the legal and administrative context in which they are used. Sometimes their effect is transparent, at others completely opaque. It should be noted that all are *relative* concepts, not absolutes, since the risk of harm can never be eliminated. Second, the list shows a development from *science-based* to *technology-based* concepts which mirrors the tendency within radwaste management. The tendency is away from justifying control systems on scientific grounds to installing abatement and disposal technologies as they become available. The pretence of benefit–costs assessments has been shelved in many countries because these often turned out to be politically contentious. Furthermore, there has until recently been a widely held assumption that waste management accounted for only a small proportion of the cost of nuclear electricity, even if very tight pollution control measures were applied. In recent years, it has become clear that waste management is a significant component in nuclear costs, especially if spent fuel is reprocessed. As electricity markets in Europe become more competitive, radwaste management costs could become an important factor of comparative advantage.

The hazards of exposure to radiation became apparent to the earliest pioneers in nuclear physics. Systematic scientific work on radiation effects began with growing clinical concern over the radiation effects caused to radiographers and their patients during X-ray therapy and diagnosis. In 1925 the International Congress on Radiology was set up to standardize units and propose maximum doses of radiation. Twenty-five years later, following research begun at Hiroshima and Nagasaki, the ICRP was established in its current form to formulate principles for radiation protection and to review and revise these as appropriate. With the growth of civilian and military applications of nuclear power, and the need for these installations to be furnished with defensible operating standards, ICRP principles and dose limits recommendations were adopted around the world. Today these limits (which are transformed into risk limits in waste disposal assessments) are designed to reduce the risk of random somatic (corporeal) health effects caused by exposure to low levels of radiation to acceptable levels (ICRP, 1985).

What distinguishes the nuclear industry is that an agreed frame of reference was established early on through international scientific consensus which became the basis for control regimes the world over. Standardisation of the theoretical basis for control was seen as one of the conditions of the new industry. Of course, standardising principles and dose limits are very far from uniform enforcement; nor is the ICRP consensus a stranger to controversy. But the long-standing quantitative statement of objectives makes radwaste regulation a good exemplar of the effects and limits of standardisation in environmental policy.

Toxic wastes

The origin of toxic wastes cannot be described neatly. They arise in a number of major industrial sectors: chemicals, plastics, pharmaceuticals, paper, leather products, metal products, but also in the domestic and service sectors. Their diversity is immense. Today, for instance there are over 50,000 chemicals on the market, many of which may be toxic in some contexts. It is difficult even to guess at the number of toxic waste streams in Europe. The United Kingdom Department of Environment estimated in 1985 that there were about 100,000 special waste producers in Britain (DoE, 1985). Table 18.2 gives estimates of toxic waste arisings in European countries. There are serious doubts about the accuracy and comparability of the data; their function here is merely indicative.

The classification of toxic wastes has become one of the main objects of standardisation. Classifications define the subject of control. Some materials can easily be defined as hazardous and as wastes, but there is a much larger group of materials for which this label is less clearly suited. No 'base-line' classification exists, as with radioactive wastes, of whether a material is hazardous. But any adequate programme of waste management needs an agreed definition scheme to separate wastes controlled by management from those for which no special precautions are necessary.

There are several aspects of this problem which differ from radioactive wastes and which create problems for standardisation. First, waste has many characteristics which make it hazardous: flammability, toxicity, corrosivity, carcinogenicity and so on. This heterogeneity is matched by a corresponding lack of information and scientific uncertainty about the specific hazard of different substances in different circumstances. Defining a potential hazard normally requires instrumentation, expertise and standardised test methods; hazard is not usually obvious. Radioactivity can be analysed using a radiation counter or spectrometer; toxic wastes in their diversity require a more wide-ranging approach to detection and assessment. Such assessment techniques may not always be available, or may be too cumbersome for day-to-day operation (Lehmann, 1981).

Second, the relative hazard of a waste may change through its lifecycle owing to chemical or physical alteration. Moreover, the same material may in some contexts be relatively benign while in others toxic: milk in sewers is the classic example. Unlike radioactive materials whose relative hazard can be altered only by dilution or concentration, some toxic wastes can be destroyed altogether (PCBs during high temperature incineration). Such mutability means that hazard is in large part situation specific. This deepens the problem of the indeterminacy of classification already present with radwastes. While there is a strong linearity in radwaste life cycles, toxic waste cycles display strong backward linkages with recycling and treatment measures which may dramatically alter the handling of the material. Indeed, Laurence and Wynne (1989) claim that the boundary between wastes and products is a continuum and therefore cannot be defined. Clearly boundaries do exist, but they are set differently in different places, depending on the infrastructure for recycling, the demand for recycled materials like solvents, oils and resins, and whether a profit can be made.

Classifications are instruments of control, and where the parameters of control are wider (theoretically infinite in this case), the classification scheme must follow suit. It must either become more detailed to cover more diverse situations or it must be reconciled to its generality while retaining flexibility and inclusiveness. The preferred route will be the one which conforms to the legal and administrative context of the classification's usage. Classification approaches differ greatly. The American system employs a *constituent list* system. If a waste contains one of 380 substances it is further assessed according to ll factors – of toxicity, concentration, etc., and then defined as hazardous or not. An additional *characteristics* approach is used to catch materials not covered by the list system. In Germany a basic *inclusive list*, containing 86 waste types, is in force. Some states have, in addition, classification categories which define the permitted disposal route for waste streams. In the United Kingdom both a characteristics and an inclusive list approach is used (Dowling and Linerooth, 1987).

Third, the relationship between classification and enforcement is different for toxic wastes. Unlike the nuclear fuel cycle which is comparatively centralised, and within which basic industrial processes do not differ much, the very diversity of the industrial base which gives rise to toxic wastes has tended to lead to decentralised regulatory structures in Europe. Diverse institutional settings have developed which act as barriers to standardisation because classification and regulatory enforcement are intimately related; changing one will usually require transforming the other. Not only are toxic wastes recalcitrant to definition, but their sheer diversity has generated institutional forms of control which are resistant to standard classification schemes.

Despite these difficulties it is generally recognised that some form of 'degree-of-hazard' measure will be necessary for the efficient allocation of resources and more evenly spread control. Such a measure would allow a more transparent comparison of the environmental consequences of different management options and be the first step in a benefit–cost assessment of them. According to the crude model, these hazardousness measures would be matched to standards of environmental quality. Present list approaches are entirely unqualitative and so preclude comparative analysis across the population of Treatment and Disposal (T&D) options for different waste streams.

Degree-of-hazard measures have long been available to regulators of radioactive wastes, but their impact on rationalising management and disposal policy has been piecemeal and often controversial. This does not alter the benefits, in principle, of similar measures for toxic wastes. In the longer term stable recycling markets and T&D infrastructures can only develop with the structuration of waste life cycles which degree-of-hazard measures imply. The new TA Abfall ordinance in Germany plots the future obliteration of large segments of indeterminacy by giving regulators the power to set disposal methods for each waste stream (BMU, 1989). One of the main objections to this move has been that it may serve as a disincentive to innovative recycling strategies by fixing waste life cycles. A balance must be struck between this stultification and the benefits of control and accountability arising from more determinate systems of regulation.

The objectives of toxic waste management

The objectives of toxic waste management are usually stated less precisely than for radwastes because the dimensions of hazard are multiple; because they affect not only humans but the environment also; and because qualitative data on hazards is so poor. Radiological protection measures are typically only concerned with the prevention of health risks to man, by assuming that if these can be kept acceptably low then the environmental vectors of activity will also be unaffected. No suprastructure comparable with radiation protection exists with its simple system of dose and risk limits, although similar concepts are applied. Again it is the sheer diversity of hazardous waste which disrupts these rather cosy principles since 'no single best technology exists even for similar processes.' (Barbour, 1989).

The adoption of technology-based definitions raises another problem specific to non-radioactive wastes. Abatement technologies, as they become more costly, will also become a more integral part of process design. Abatement will not normally be achieved by end-of-pipe retrofits with bought-in plant. In other words, the means of controlling pollution will itself become proprietorial. The role of the regulator in mediating this kind of contradiction is still unclear.

In the design of nuclear technology, particularly for waste conditioning and disposal, the regulator has been highly intrusive, with a tendency towards deeper involvement. National systems vary however. British nuclear licensing includes a high level of private interaction between the operator and a regulator exercising considerable discretion in detailed decision-making. Nuclear licensing in West Germany is formally decentralised and conflictual, and, as a consequence, far more structured by legally defined objectives and prohibitions. Licensing also has to contend with the extensive scope for judicial review open to protesting citizens. In Sweden the licensing process has many similarities with the cooperative, discretionary British model, but differs in devoting more resources to gaining public acceptance for decisions.

IV The aims of harmonisation

What is the meaning of harmonisation in the context of toxic wastes? The European Commission is interested in three aspects: the equalisation of environmental standards; the evening out of environmental costs to industry; and the free trade in wastes. Put simply, harmonisation is the process of making industrial production accountable for its environmental impacts by creating a more uniform framework for the control of its wastes. This includes the naming of wastes, setting standards for EP, such as for example, groundwater (CEC, 1980) and in some cases specifying management practice, as in the case of waste and disposal (CEC, 1987a). If, as has been suggested above, there is a tendency towards greater structuration of waste streams, then the problem of institutional reform also arises. The European Commission has ruled out the possibility of it sponsoring such change.

There are three basic elements to enforcement: the licensing by the state of waste handling, storage, transport and disposal; record-keeping by waste producers, handlers and disposers; and a system of inspections of waste facilities again by the state. Waste life cycles are legally connected through the liability of producers, handlers and disposers of wastes to deal with them properly. 'Polluter pays' defines the basic object of liability arrangements (CEC, 1988a). This is the lynchpin of regulatory regimes because producers, handlers and disposers are usually separated, with different agencies or firms dealing with specific segments of a waste stream's life cycle. Under a well-functioning polluter pays system, the waste producer would be liable to ensure that wastes followed non-hazardous life cycles and hence liable to penalties if they did not. In many cases, this places completely new demands for information and expertise on the waste producer. Standardisation could, in principle, encroach here by creating progressively similar definitions of liability, authorisation and administrative procedures.

Today, national legal definitions of liability vary greatly. Stringent 'cradle-to-grave' liability underpinned by a consignment note procedure which traces the path of waste from generation to disposal has existed in the United States and Germany since the mid-1970s. Similarly in Belgium, toxic waste producers are 'liable for any damage whatsoever which may be caused at any time during transport, destruction, neutralisation or disposal'. In Britain, liability has traditionally been transferred along the waste life cycle, although the government has now adopted the concept of a 'duty to care' which requires waste producers 'to take reasonable care to ensure its lawful disposal' (HMSO, 1989).

What does it mean to equalise the costs of EP across Europe? As an ideal for harmonisation it means that waste producers should pay an equivalent amount for waste management and should have the same access to means of disposing of wastes. Through higher costs of T&D, the object is to create equivalent pressures across the community of waste producers to reduce and reclaim wastes and to develop or install 'clean' technologies under zero-waste scenarios. This is a visionary menu of aims which takes the 'managerialist' line that waste facilities should be situated independently of national boundaries so as to minimise overall risk. It is clear that the information currently available on waste arisings, life cycles and hazards is at least an order of magnitude less than would be necessary before European best

Table 18.3 The 'dirtiness' of industry in parts of Europe, the United States and Japan

Country	Share of high-polluting branch of manufacturing sector (%)[b]		High-polluting industries as share of GDP (%)	Toxic waste productivity[a] (ECUm/mt waste)
	1973-75	1983-84		
France	37.3	30.6 (13.3)[c]	7.65	0.142
Germany[d]	37.4	33.5 (13.7)	10.39	0.085
Greece	46.8	43.5 (11.4)	6.96	—
Italy	44.1	40.9 (15.2)	—	—
Japan	40.6	35.6 (13.6)	10.68	n.a.
Netherlands	38.9	34.6 (15.6)	6.23	0.143
Spain	49.1	39.3 (13.8)	11.76	0.106
United Kingdom	38.5	35.4 (14.5)	7.79	0.124
United States	36.6	34.4 (13.4)	7.22	n.a.

[a]Relation of the value of high-polluting industries to the total production of hazardous wastes (according to definitions used in Table 18.2)
[b]High-polluting industries are taken to be: textiles (SIC 321), leather and fur (323), paper (341), industrial chemicals (351), other chemicals (352), rubber products (355), plastic products (356), iron and steel (371), Non-ferrous metals (372), metal products (381)
[c]Figure in brackets is the proportion of the manufacturing sector accounted for by the chemicals, rubber and plastics subsectors
[d]Note that these are pre-unification figures; with the incorporation of the GDR, the German figures have, no doubt, been shifted towards the dirty end of the spectrum

Sources: UNIDO, *Handbook of Industrial Statistics, 1986.* Vienna, 1986, UN, *National Accounts Statistics: Main Aggregates and Detailed Tables, 1985* New York, 1987

practicable environmental options could be implemented. The real significance is that waste management is now no longer left to individual and local solutions, but is beginning to be thought of in a global perspective which seeks to match the industrial, economic and environmental dimensions of waste management.

The first obstacle to this version of harmonisation is the historical development of industry in European countries. Sectors of high-polluting industries clearly vary in size so that the relative costs of EP to economies in aggregate must also vary. A representation of this is given in Table 18.3 which shows the share of high-polluting branches of the manufacturing sector and their share of the GDP in a range of countries, including the United States and Japan. The last column gives an indicator for the relative intensity of waste production in different countries.

Table 18.3 shows that in all the countries listed 'dirty' industries have declined in relative importance. There are basically two groups of countries, high-polluters with over 10 per cent of GDP accounted for by dirty industries, and low-polluters with 7–8 per cent of national product coming

Table 18.4 Waste disposal and incineration costs in the former West Germany, France and the United Kingdom, 1989

	(ECU per tonne)		
	W. Germany	France	United Kingdom
Landfill			
(municipal wastes)	55	12	10
Incineration			
(chemical wastes)	200-500	200-900	300-900
(PCBs)	1,500	800	2,500
ECU1 = \$1.137			

Source: Haznews, 1989c: 20

from these sectors. Japan, Germany and Spain are high-polluters. Waste productivity varies widely, with Germany showing, on this definition, the lowest intensity of waste.

Second, waste management infrastructures vary greatly across the Community. In 1988 in Britain, for instance, there were 4,156 landfill sites licensed to accept hazardous wastes, while in the Netherlands there are fewer than 20. As a result, the costs of disposal vary accordingly. Table 18.4 gives comparative costs for waste disposal and incineration in Germany, France and the United Kingdom. The pattern of ownership of the waste industry also differs greatly. In the United Kingdom over 90 per cent of waste is handled within a highly competitive market, whereas in Germany and Denmark publically owned monopolies operate disposal facilities (Laurence and Wynne, 1989).

Third, harmonisation must take account in some way of the natural environmental advantages which some countries might have for landfill or underground disposal of wastes. Large parts of the Netherlands may be unsuitable for landfilling because of the threat posed to drinking water supplies, for instance. But making an inventory of environmental suitability for waste management is not a simple matter. Long-term safety assessments of radwaste repositories all emphasize the importance of *site-specific* factors. Each repository is a unique structure, adapted to its hydrogeological environment and containing wastes which have in turn been adapted to the repository. One of the unresolved contradictions of repository siting is over whether there can be shown to exist a 'best' site for such a structure (hence extensive site-search programmes) or whether equally safe repositories can be built in different rock masses. Experience shows that expediency usually wins the day, and the ideology of best sites is quietly forgotten (compare the example of the siting of the world's first deep repository at the Forsmark reactor site in Sweden).

The difficulties of harmonisation have most recently been demonstrated by the general problem of toxic waste trade in Europe. Trade has been an established feature of toxic waste management for many years. France, for instance, has exported certain waste types which it could not treat for storage to the German disposal facility at Herfa-Neurode (European Study Service,

1985). Waste export is chosen as a management option for two reasons – either because wastes cannot be disposed of domestically or because it is cheaper abroad. In aggregate, Denmark, the former West Germany and the Netherlands have been net exporters of wastes, while France and the United Kingdom have been net importers (Mettelet, 1989).

From the mid-1980s volumes of waste trade began to increase, with Britain, the former East Germany, Yugoslavia, West Africa and France becoming favoured destinations. Between 1981 and 1988 waste exports to Britain rose from 3,800 tonnes to about 80,000 tonnes (HoC, 1989). It was an attractive option for exporters because surplus landfill capacity was available at low prices. This competitive edge was not owing to a clear-cut natural advantage, nor to a decisive technological lead on the part of British disposal companies. It was caused by differences between statutory and enforcement regimes. In the Netherlands, for instance, the disposal of certain chemical wastes in the ground is banned, so that they must either be disposed of at the site of production, sent for incineration or exported. Landfills in Britain have accepted these wastes.

Several things have acted against European waste trade. Not only has it become politically contentious for exporters and importers alike, but there has been a need in some countries to ensure that expensive waste treatment facilities (as in Germany) are utilized. Gradually there has been an attempt by governments to restrict exports, with the Dutch government, for instance, insisting that they will be permitted only if domestic treatment and disposal standards are met (Wynne and Hortensius, 1987). More recently the question of waste trade has forced the European Commission to adopt the contradictory concept of 'self-sufficiency' in waste management which appears to fly in the face of the notion of a free trade of wastes. The British have given notice of their intention to reduce the scale of waste imports, and for countries like the Netherlands and Germany it has become more difficult to defend large-scale exports. The Dutch were by the late 1980s exporting some 60 per cent of their chemical wastes. Rather than unification, a Balkanisation of waste management is occurring.

All trade cannot be stifled because there are serious shortfalls and overcapacities for T&D scattered across Europe. Nor does the Commission currently have the power to influence directly the prices charged in different countries because it goes against the grain of market liberalisation and would be resisted fiercely by member states. Powerful institutional and political obstacles will have to be overcome if the waste industry is to be liberalised in a way which encourages waste reduction and recycling, while at the same time its regulation is being standardised.

V The orientation towards standardisation

Despite these evident difficulties a strong movement towards harmonisation is afoot. Eight European Commission Directives directly related to hazardous wastes have been adopted since 1975, and the Community's role in environmental policy was strengthened by article 130R of the Single European Act which made the needs of the environment a factor in the creation of the Internal Market. There is a general perception that since 'the environment' is one of the key political topics of the age, the Commission has a good opportunity to make its mark.

The drive to harmonisation has so far been unsteady, and the record of implementation by member countries has on the whole been poor. Directives have concentrated on standardising waste classifications (unsuccessfully), the central collection of information about waste management plans (unsuccessfully), intra-European trade in wastes (effect as yet unclear) and general exhortations about the need to encourage waste minimisation and recycling. Waste trade has been an obvious target for the Commission's attention, and the 1984 and 1986 Directives on transfrontier shipments (CEC, 1984, 1986a) marked its most deliberate intervention in the field.

Lack of success has not reduced the Commission's zeal in pursuing reform. In a communication to a discussion of environment ministers in September 1989 it argued that harmonisation of waste management costs across the community was the objective of EC policy; that low-cost, sub-standard alternatives should be eliminated so that uniform incentives to reduce waste production at source would be created. This was not to be achieved through the 'rigid imposition of administrative procedures' because the Commission is sensitive to being type-cast as a socialist super-state. Instead, it would be left to voluntary action by individual countries.

The first need of the project of harmonisation is information: about waste arisings, handling and disposal; about the pattern of waste life cycles in different regions; about price structures; about the standards and modes of regulation. Information in this field is notoriously poor and variable, but the advantages of their centralised collection, together with other industrial and economic data has been obvious since the mid-1970s. The 1978 directive on hazardous wastes contained a requirement that member states submit to the Commission waste disposal plans. Germany is on the point of becoming the first country to submit such a document, 11 years later. At first the concern was with environmental quality in a descriptive way; more recently interest has focused on costs.

It is unrealistic to think that all arisings and disposals of toxic wastes could ever be monitored centrally, and neither is this the intention. What then would all this data allow the EC to do? First, it would enable the Commission to make better comparative judgements about the real extent of problems in different countries. The Commission has traditionally been loath to make such comparisons for fear of antagonising governments or of inflaming national disputes. Nevertheless, on subjects like contaminated land the EC could make a very useful contribution. The costs of land reclamation are potentially enormous. Estimates in Germany (before unification) for expenditures on clean-ups ranged up to DM50bn over the next 10 years. The American EPA clean-up programme looks likely to cost at least $30bn (Haznews, 1988a,1989b).

Second, the Commission would finally be able to prescribe improvements in practice through knowledge and technology transfers. Again this may be a problem because practices differ so radically, for instance in relation to landfilling or co-disposal. Third, it could foster EPT markets within Europe through standardisation and information flow and connect these markets with those outside Europe. A comprehensive view of the situation would in any case make clearer market opportunities both for waste management and for recycling technologies.

VI The costs to industry

Industrial pressure for harmonisation is intense. It is estimated that European chemical firms currently spend 10–15 per cent of their annual capital investment on environmental measures (about ECU1.2 bn per year), and that this proportion could rise to about 25 per cent during the 1990s (Marsh, 1989). In Germany operating costs of environmental protection plant were over four times the scale of investment in 1987 (DM4,430m compared to DM1,034m) (Haznews, 1989a). New EC legislation on environment impact assessment is likely to have a major impact on the chemical industry (CEC, 1985).

All of these developments affect the competitiveness of industry. Interpreted crudely they mean that the return on capital must be raised significantly for European, especially German, industry to compete with companies operating under slacker regulatory regimes. Even within Europe, spending on environmental protection is distributed very unevenly. A German study has estimated that in 1987 some 40 per cent of non-energy-conservation EP investments were made in West Germany, about 15 per cent in France, 11 per cent in the United Kingdom, 8.5 per cent in Italy and so on. Harmonisation of standards and practices would be a way of spreading the burden more evenly.

But the argument about competitiveness is far from clear-cut. Higher regulatory standards bring two opposite effects: they lead to an increase in the production of wastes and they produce incentives for waste reduction, recycling and eventually to the adoption of cleaner technologies. The net effect is usually that less waste is produced and that costs are saved. Both the OECD and the United States Office of Technology Assessment have found that waste reduction was a tool to improve industrial efficiency, growth and industrial competitiveness (OECD, 1985a; OTA, 1986).

Stricter standards therefore produce a cascade of adjustments in industrial processes, some of which may be cost savers. All the design reference points are shaken up, innovations take a new direction, old systems are replaced by new ones. In this regard it is interesting to note that the German chemical industry which labours under some of the toughest regulations in the world is also the most dynamic (Chapter 12). The moral must be that if you run a tight ship then you remain competitive almost regardless of the climate of environmental regulation.

Industrial opportunities

Tighter regulation also opens up new markets for EPTs. As has been argued in other chapters, the leaders in environmental regulation may derive great benefits from the export of technology, plant and equipment as standards rise elsewhere in the world. A good example is the case study in Chapter 19. The German electricity industry initially incurred heavy costs in leading the way in de-sulphurisation equipment but German plant manufacturers are now greatly benefiting from exports to Eastern Europe and elsewhere. Total spending on pollution control was estimated to be worth £50bn in the OECD in 1988. In January 1989 the Deutsche Bank recommended greater commitment in the commercial field of environmental protection as a means of achieving competitive advantage (Haznews, 1989c; Toepfer, 1989). The bank estimated that heightened environmental awareness would open up secure future markets of at least DM20bn per annum. And markets are growing not only in the industrialised world, West and East, but also in developing countries. The costs of repairing the damage of previous generations of waste disposal are also high and provide the scope for new technologies. For example, *in situ* bioremediation offers good prospects for future contaminated land clean-ups (Haznews, 1988b). Environmental protection has been recognised as a huge market opportunity. But the market suffers from a paucity of information and information exchange. Market conditions, however, are complex and unpredictable. A harmonisation of standards would internationalise markets for EPTs.

VII Conclusion: the limits of standardisation

There are strong pressures for redefining industrial economies as green, but there remain deep conceptual and practical difficulties with creating a standardised framework for such redefinition. This chapter has revealed several of these contradictions. We have argued, following Wynne, that the structuration of waste life cycles in any comprehensive way may be beyond the capacity of all but the most intrusive regulatory regimes. The line between what is a product and a waste is often indeterminate. Moreover, structuration might act as a disincentive to waste avoidance and the development of clean technologies by creating inflexible and bureaucratic procedures. Such standardised systems would also be incapable of taking into account the special site-specific features of waste handling and disposal sites, just as command economies fail to achieve high efficiencies in their ignorance of local conditions. There would be the risk that global, standardised systems of control could become obstacles to change, even if at first they were created by progressive optimisers.

Centralised, intrusive regulatory systems are disavowed by everyone, and it is clear that the European Commission is in no position to take such a regulatory role. Nothing has so far come of proposals for a unified environmental protection agency; nor is it likely to in the near future. In particular, its precise role *vis-à-vis* the national regulators has still to be decided. Instead, the talk is of loosening the grip of regulations and substituting structures of taxes and incentives for waste producers. But what will be the basis of these measures? Costing environmental damage through scientific assessment may not even be possible in principle and is unlikely to gain political acceptance because of the deep uncertainty which still exists about the toxicity, persistence and bioaccumulation of the awesome variety of hazardous materials currently being dumped and discharged into the environment. It also fails to answer the question of how the introduction of fiscal measures will allow the state to withdraw from its traditional monitoring role of regulation. A 'light touch' regulation quickly loses sight of the broader objective of converting industrial capital to cleanliness. Much therefore will depend, as it always has in the analogous case of law enforcement, on informal public opinion and on public attitudes and standards at all levels of society.

It is likely that this problem of how to strike the balance between intervention and liberalisation will dominate environmental policy from now on. In many ways the debate around harmonisation is a model for the future, whose wider significance may be felt as global regimes are created to ameliorate the major environmental problems like global warming. The limits of such regulatory frameworks should be clearly understood now.

19 The environmental challenge in Europe: the policy framework

Sonja Boehmer-Christiansen and Jim Skea

Since the publication of Rachel Carson's seminal book *Silent Spring* in 1962, 'environmentalism' has emerged as a powerful philosophical and ethical outlook in its own right. Environmentalism now aspires to a universality which has been attained, outside the religious world, only by concern for human rights. Like human rights groups environmental groups operate on the assumption that the ideals which they pursue transcend national boundaries.

On the surface there appear to be fundamental conflicts between the environmental world view and technological development. For the environmentalist science and technology may appear as forces which threaten nature and disrupt social organisation (for example, Amery, 1985; Goldsmith, 1988). Technologies that are thought to produce irreversible ecological impacts are particularly resented.

On the other hand, the application of technology has brought economic benefits and has improved the quality of life enjoyed by the majority of the population in the developed world; many in the developing world have yet to realise these benefits. Economists have also pointed to the capacity of technology to solve environmental problems and the stimulus which such attempts can give to innovative activity (for example, van Tulden and Junne, 1987; OECD, 1987; *Economist*, 1990).

The romantic and technological strands of thought are deeply embedded in Western society and have remained in creative tension for 150 years. Each has played its role in the evolution of technologically advanced societies with their subtle and extensive effects on the natural environment (Collingridge, 1980). While William Blake took arms against the 'dark Satanic mills', Karl Marx called upon the proletariat to throw off its chains and subdue nature, as well as its masters, in its own interest.

Over the last century the nature of perceived environmental problems has changed. Traditionally the effect of pollution on human health was the major concern. More recently effects on landscape, wildlife, the cultural heritage and eco-systems have become political issues. Thus, the scale and complexity of environmental problems has grown. Older environmental problems were either strictly local in nature or remained unrecognised because of their long-term, incremental nature. Now truly global problems such as pollution of the sea, transfrontier air pollution and climate change have become the subject of international diplomacy (Caroll, 1988). The negotiation of a convention on global climate change within the United Nations Environmental Programme (UNEP)is perhaps the most ambitious exercise in international environmental diplomacy ever mounted.

Many factors influence the development and effectiveness of environmental policy. Within individual countries economic interests, public opinion and deeper-seated cultural values influence the priority attached to particular problems. Technological capabilities and the strength of institutional arrangements for pollution control affect a country's readiness to deal with new issues promptly and effectively. Furthermore, environmental issues may be integrated with imperatives in related policy fields, such as agriculture, transport, trade and, of course, energy (Chapter 17).

In what follows, the policy-making frameworks in the EC and the wider international arena are described. The processes by which new environmental issues develop and are tackled are outlined with particular attention to the role of governments, Non-Governmental Organisations (NGOs) and affected industries. To illustrate how a major international environmental issue may be handled, the case of acid rain in Europe is presented, focusing on policies in the United Kingdom and Germany as well as the EC.

I The responsibility for environmental policy

One of the growing problems of environmental policy is the widening regional and global impacts of problems which have their origins at local or national level. These impacts may be direct – through the transfrontier movement of pollutants – or indirect – through alterations in the terms of trade between countries setting differential environmental standards. Such impacts are the engine of international environmental policy-making, leading to a growing number of internationally derived obligations on national authorities.

The internationalisation of environmental issues may appear to reduce the responsibility of the nation state. But individual states play a vital role in determining the course of negotiations which result in international agreements and they are responsible for implementation which is, after all, the major, though often neglected, component of environmental policy. In implementing internationally agreed policies there is ample scope for the exercise of considerable national discretion and initiative. International agreements may even strengthen the role of the nation state; countries may need to assert control over industries operating within their boundaries which were previously unregulated. As large-scale environmental problems are addressed, domestic and intergovernmental policy processes thus become increasingly intertwined (as the acid rain case study presented below demonstrates).

As well as increasing the internal authority of nation states, international agreements also increase their responsibility. Agreements, such as those on dumping at sea or transfrontier air pollution, will ultimately stop the use of the global commons as a dumping ground for unwanted wastes (see Chapter 18). As this happens, countries will be forced to seek solutions to environmental problems within their own boundaries. For example, the drive to reduce air pollution at the international level is leading to the creation of new solid and liquid waste streams from abatement facilities. The associated problems often fall to local authorities to resolve.

Environmental policy in the European Community

The EC effectively entered the environmental policy domain in 1972, just before Britain's accession to the Community. Following the 1972 Stockholm Conference on the Human Environment, the EC formulated a 'Programme of Action Concerning the Environment' which was based on the following principles: prevention of pollution at source; the polluter pays; activities carried out in one state should not degrade the environment in another; and the long-term harmonisation of Community and member state environmental policies. In 1974 a separate directorate-general (DG-XI) on environmental protection, nuclear safety and consumer protection was established.

Like technology policy (see Chapter 4) until 1986, the legal basis for EC environmental policy was weak, relying heavily on Article 2 of the original Treaty of Rome which requires the Community to 'promote ... a harmonious development of economic activities, a continuous and balanced expansion'. Consequently, most environmental measures adopted in the past have been justified with reference to removing barriers to trade and needed to be approved unanimously in the Council of Ministers.

The 1986 Single European Act (SEA) placed environmental policy on a firm legal footing. The objectives are 'to protect and improve the quality of the environment, contribute to the protection of human health, and ensure a prudent and rational utilisation of natural resources'. In addition, member states, as already confirmed by the European Court, are entitled to set environmental standards higher than the Community norm, while the EC itself aims to establish common standards as high as politically possible. These two provisions, taken in tandem, establish considerable institutional momentum behind the Commission's ambition to expand its competence in the environmental field.

The SEA allows qualified majority voting in the Council of Ministers whenever environmental standards can be justified with reference to the internal market. The SEA also permits the European Parliament, which takes great interest in environmental affairs, to refer legislation back to the Council and the Commission. Both of these new features were instrumental in securing a directive which will bring emission limits for all new motor vehicles closer to present American standards.

The principal legislative device used in the environmental field has been the directive, the provisions of which are binding on member states which then have discretion over implementation at the national level. Over 200 environmental directives have so far been issued. EC environment ministers are close to agreeing the formation of a European Environment Agency, the remit of which would initially be restricted to information gathering. The development of such an agency may help to overcome British criticisms that some of the Commission's legislative proposals have been inadequately based on scientific understanding.

The process of negotiating directives within the Council of Ministers can be extraordinarily protracted given the diverse interests and environmental concerns of member states. Some draft directives are never agreed while negotiations on others have taken five years or more. Inevitably, after this length of time, the original Commission proposals may be watered down considerably. Gaining agreement in some areas has become even more difficult since recent accessions to the Community have increased the range of economic development of member states.

The EC is now attempting to reach a uniform position on global environmental issues in wider international fora. This was particularly evident during negotiations on the Montreal Protocol restricting production of CFCs. This development is attracting criticism, notably from the United States (International Environmental Affairs, 1989). While it often takes a considerable length of time for the EC to agree a uniform position, member states subsequently wield 12 separate votes in international meetings in favour of their previously agreed position, allegedly exerting an undue influence over final agreements.

The wider international framework

International treaties on environmental issues are often arranged at the regional or global level under OECD or United Nations auspices. While legally binding, such agreements are essentially unenforceable and rely on the goodwill of contracting states.

The daughter programmes and commissions of the United Nations, notably the Economic Commission for Europe (UNECE) and the Environment Programme (UNEP) have been particularly active. In spite of being under-resourced and subject to political in-fighting, United Nations bodies are now increasingly called upon to act as clearing houses for information, to coordinate international responses and to provide the umbrellas under which new laws and conventions can be negotiated and monitored.

During the 1970s agreements on marine pollution problems, notably on oil pollution and the dumping of radioactive and other wastes at sea, were reached by treaty at the regional level as well as within the UN framework. Progress was accelerated by pressure to define national competences over marine resources for non-environmental reasons. Vessel source pollution is now regulated globally under the MARPOL Convention of 1974 and its four protocols. Dumping is subject to the London Dumping Convention of 1972, but is over-ridden by more stringent EC measures. Slower progress has been made under the Paris Convention of 1976 which covers land-derived marine pollutants. Progress on transfrontier air pollution has been slower, partly because, unlike marine pollution, the issue has had to be negotiated entirely on its own merits. The pre-existence of diverse national regulatory regimes has also slowed progress.

The establishment of the 1979 Convention on the Long Range Transport of Air Pollutants (LRTAP) and the development of subsequent protocols is described later in relation to acid rain. It was agreed under the auspices of the UNECE. For truly global problems, such as climate change and ozone depletion, other United Nations bodies are clearly more appropriate, although developments reflect regional initiatives, with the EC playing an active role.

The idea of an all-embracing global convention for the atmosphere similar to that for the ocean has recently been rejected as too cumbersome to negotiate. Instead, individual pollutants are becoming the subject of treaties. The 1986 Vienna Treaty on the Protection of the Ozone Layer is enhanced by the Montreal Protocol which regulates CFCs, incorporating trade restrictions and production cut backs of 85 per cent.

The biggest challenge to the international system comes from the problem of global climate change, caused largely by emissions of carbon dioxide (CO_2). A major international conference, organised by the Canadian government in 1988, recommended that developed countries should cut their national output of CO_2 by 20 per cent between 1988 and the year 2005. A rapid move to such a measure has been resisted by the United Kingdom and the United States.

The equity implications of such simple measures for countries with very dissimilar energy resources are now widely debated internationally. In addition, the use of fiscal mechanisms to reduce CO_2 emissions and the operational meaning of concepts such as sustainable economic growth, proposed by the Brundtland Commission on Environment and Development in 1986, are also the subject of controversy.

Under UNEP auspices, an Intergovernmental Panel for Climate Change (IPCC), acting as the focus for the international research effort on climate change, was established. The IPCC set up three working groups covering climate effects, environmental impacts and possible responses. There is now more hope that relatively rapid progress can be made.

International conferences provide important landmarks in the process of environmental policy-making. Much of the current effort is focused on making adequate progress for the 1992 Stockholm Conference on Environment and Development. This will, among other things, celebrate achievements since the 1972 Stockholm Conference effectively initiated the international environmental policy-making process.

II The emergence and resolution of environmental issues

New environmental issues can emerge very slowly. While, for example, acid rain was recognised as early as 1872, effects of acid emissions on eco-systems became apparent only gradually. Even after the acidification of fresh waters was understood it took 20 years for a political solution to be found at the European level. It will take another 20 years before regulatory responses have their full effect on emissions, let alone the acidification of soils.

Many groups, including the public, NGOs, scientists, industry and governments have a stake in the resolution of environmental issues. The influence of the public and NGOs is often greatest during the early stages of the environmental policy cycle. It is at this point that new knowledge is identified and agendas are set. As issues move towards possible solutions, the roles of industry and government are emphasised because regulatory instruments are sought which are compatible with wider economic interests.

Indeed, few environmental problems are ever laid to rest completely. Often the 'resolution' of a problem is little more than an uneasy truce between irreconcilable interests. The environmental issue may also resolve itself by being forgotten. Large tracts of nature have already disappeared, vast quantities of soil blown away and species lost without apparent regret. Doing nothing is clearly one possible response.

Wider economic and social implications of environmental controls may affect the stringency of environmental protection either positively or negatively, depending on economic culture and particular circumstances. As well as meeting political needs, the strengthening of environmental controls may also foster the environmental protection industry and improve export opportunities. Some governments, notably that of Germany, have recognised this more quickly than others (Chapter 18).

Public concern about environmental issues is a latent political force. Without direction and articulation it cannot have a direct policy impact, but direction is provided by the diverse range of NGOs currently active and increasingly organised transnationally. Environmental groups are not simply channels for public concern, but play a much more pro-active role which increasingly includes action aimed at industry and consumers as well as government. Groups such as Greenpeace, Friends of the Earth and the World-Wide Fund for Nature have grown greatly in influence and membership.

The staff of the larger environmental groups now display a degree of professional skill matching that found in government and industrial organisations. As a result, NGOs may enjoy regular access to government through personal contacts and various formal and informal meetings. Increasingly bureaucrats concerned with environmental protection find informal alliances with such groups helpful in promoting their own influence, status and resource base within the larger government machine. NGOs now frequently participate as observers at workshop meetings of international conventions.

In some countries, notably Germany and Belgium, the green movement has spawned political parties which have had a notable impact on parliamentary politics. In the last European election green parties in most EC countries attracted a significant share of the vote, making it imperative for established parties to give greater priority to environmental issues.

In spite of the 'green' image projected by major corporations, industry seldom sees environmental regulation as being in its interest. However, once regulation is perceived to be inevitable because of political demands, industry tries to influence the type of regulation adopted. Regulations can have major implications for production costs, technological choice and the allocation of capital and skills.

The selection of regulatory instruments – ambient air quality standards, emission limits, pollution taxes or the specification of operational practices – itself has profound implications for technical choice. Hence, governments may be able to force technological development through the specification, for example, of emission limits which are beyond the reach of current technology. Flue gas desulphurisation (FGD) developed this way in the United States and in Japan.

Ironically, companies may urge government to regulate more stringently. For example, in the 1970s American coal companies with high sulphur reserves urged the federal government to promulgate tough regulations which would force the use of abatement equipment at all other sources, regardless of the quality of fuel used. In addition, companies with good R & D bases and developed technological capacities may urge tougher controls in the hope of expanding market share by offering more environmentally benign products.

On the other hand, conflict among members of a single industry may delay decisions for years. This happened in the European motor industry because of the direct impact of exhaust emission standards on technology choice and the relative cost of cars (Chapter 16).

III Acid rain – a case study in environmental policy-making

By 1968 Swedish scientists felt that they had conclusively identified the basic cause of acidification of their country's lakes as sulphur dioxide (SO_2) emissions originating from the United Kingdom and other parts of Europe. At the 1972 United Nations Conference on the Human Environment they presented their case, which has by now been universally accepted, to a still sceptical world.

The opportunity for the Scandinavians to press their case came in 1975 when President Brezhnev suggested that the environment was an issue upon which East and West ought to cooperate. Developments in East–West relations in the late 1980s have created a similar opportunity to seek broadly based solutions to the problem of global climate change.

The Scandinavians pursued their concerns through the framework of the UNECE. The result, after two years of negotiation, was the 1979 Geneva LRTAP Convention (HMSO, 1980). Swedish and Norwegian delegates proposed strict standstill clauses (preventing SO_2 emissions from increasing) and rollback clauses (specifying fixed percentage reductions for aggregate national SO_2 emissions), but this proved not to be possible given opposition from the United States, the United Kingdom and West Germany (Wetstone and Rosencranz, 1987). The final treaty was a compromise, lacking bite, but was nevertheless an important symbol of progress towards tighter emission controls.

The Convention members promised to 'endeavour to limit and, as far as possible, gradually reduce and prevent air pollution' and to use the 'best available control technology economically feasible'. In addition, an Executive Body was set up by the Convention to allow consultation between countries and to initiate relevant research. The LRTAP Convention has been ratified by 31 of its 34 signatories.

Unsatisfied with the watered down LRTAP Convention, the Scandinavian countries made a specific proposal in 1982 that countries should undertake to reduce SO_2 emissions 30 per cent below 1980 levels by the year 1993. This was taken to the Executive Body of the LRTAP Convention in June 1983, receiving support from Canada, Austria, Switzerland and the former West Germany (which had by then embarked on an ambitious emission reduction plan on its own). It was blocked by the United Kingdom, the United States, France and the Eastern bloc countries.

In March 1984, 10 UNECE countries met in Ottawa to initiate an informal '30 per cent Club' (ENDS, 1984). At a major international conference on acid rain hosted by Germany in Munich a few months later, the 30 per cent Club grew to 18 members, including the Eastern bloc countries which undertook to reduce transboundary flows of pollution. A formal 30 per cent protocol was attached to the LRTAP Convention in July 1985, coming into force in September 1987 after 16 countries had ratified. Neither the United Kingdom nor the United States signed this protocol.

In 1988, a further protocol requiring a standstill in nitrogen oxide (NO_x) emissions was signed in Sofia by a number of countries, including the United Kingdom. However, 12 countries, among them Germany and Sweden, have pressed for a 30 per cent reduction in NO_x emissions. The implications of this for emission abatement for both stationary and mobile sources are considerable, potentially requiring changes to transport policies as well as the adoption of expensive control technologies.

Forest dieback in Germany

Although the international community was taken by surprise when in the summer of 1982 Germany promised to join the SO_2 30 per cent club, this step was merely the outward manifestation of domestic policy. In September 1982, the federal government agreed in principle the Large Combustion Plant Ordinance (Grossfeuerungs-Anlagenverordnung–GFAVo) which was to result in a reduction in emissions of SO_2 and NO_x far greater than 30 per cent.

The GFAVo was enabled by the 1974 Federal Air Quality Protection Law (Bundes Immissionen Schutzgesetz–BImSchG) and was, in fact, the 13th in a series of ordinances issued during the 1970s and 1980s. The BImSchG bases German environmental policy on the principles of anticipation of problems (*Vorsorge*), the use of best available technology and economic proportionality, that is, having regard for cost/benefit considerations. Early drafts of the GFAVo had been prepared by the Federal Environment Office (Umweltbundesamt–UBA) as early as 1978.

Widespread political support for major emission controls developed quickly, however, between 1981 and 1983 during a time of great political turbulence. In 1981 Germany became aware of the problem of *Waldsterben* (forest dieback), which affected large areas. As well as being a potent cultural symbol, forests remain an important source of economic activity and employment, particularly in Bavaria. Early scientific hypotheses linked forest damage to the modification of soil chemistry caused by deposition of SO_2, deriving largely from coal-fired power stations based in northern Germany.

At the same time, the Green Party was posing a major electoral challenge to the established political parties. Since 1969, the Free Democratic Party (FDP) has held the balance of power. The Greens threatened to displace the FDP as the SPD's coalition partner in the run-up to the 1983 general election. This provided a great incentive for all established parties, and the FDP in particular, to 'green' their policies. As it happened, the FDP had control over the interior and economics ministries, critical for environmental policy-making.

For the CDU/CSU, *Waldsterben* also provided a credible argument in support of nuclear power which does not contribute to acid emissions. The Greens were demanding the immediate abandonment of Germany's nuclear programme. Licensing approval for new nuclear reactors was urgently required by the utilities in 1981 (Thomas, 1988) and was reluctantly given in the summer of 1982 by the same FDP interior minister who was responsible for air pollution.

The GFAVo finally approved in June 1983 has had far-reaching effects. Between 1983 and 1988, 37,000 megawatts of coal-fired power station capacity was retrofitted with flue gas desulphurisation (FGD) equipment and 12,000 megawatts was closed down. As a result, SO_2 emissions from the power sector have declined by 80 per cent. In addition, many denitrification facilities have also been installed. Such a result could be achieved only with the express consent of the federal economics ministry, which shortly afterwards also allowed the development of strict exhaust emission limits for motor vehicles.

Political demands for environmental controls coincided with a period when the federal government was interested in stimulating capital investment as part of its economic counter-cyclical policy. The rapid implementation of the GFAVo allowed government to stimulate investment in the electricity sector; to stimulate activity in the engineering sectors which were operating well below capacity; and to promote the development of vehicle manufacture and the environmental protection equipment industry, as major export opportunities. For vehicles, existing catalytic convertor technology had to be adapted for German conditions, in particular the high speeds characteristic of the autobahns. The widespread use of German-made information technology was also promoted as a result of the greatly enhanced need for continuous pollution-monitoring equipment. However, the cost of the programme was considerable and the short-term economic impacts on German industry in its role as an electricity consumer were severe. Consequently, strong efforts were made to spread the German domestic legislation to trading partners via the EC so as to re-establish 'fair competition'.

Ironically, the spread of German forest damage predicted in the early 1980s has not occurred. Since the mid-1980s the proportion of German forest in the moderately damaged category has stabilised at around 15 per cent. A growing, though as yet far from complete, understanding of the *Waldsterben* phenomenon shows that there are in fact several categories of tree damage. Sulphur deposition is only one of a range of factors implicated.

The British response

During the 1970s, Britain had resisted pressures by the Scandinavians to reduce sulphur dioxide emissions by fitting FGD equipment to major power stations. Complaints had been directed against the Central Electricity Generating Board (CEGB), the nationalised electricity company responsible in 1980 for more SO_2 emissions than any single country in Western Europe. In response, the CEGB had initiated major in-house research on the long-range transport of air pollutants and their impacts on Scandinavian ecosystems. Initially, the CEGB argued that the deposition of sulphur in Scandinavia had not been proved to originate from Britain. Later, the extent to which sulphur deposition led to alleged environmental damage was contested.

Until 1984, Scandinavian complaints had not been politically compelling. However, in December 1983, the European Commission issued its draft LCP Directive proposing strict emission limits for new power stations and other large industrial sources, as well as a 60 per cent reduction in SO_2 emissions from existing installations. The United Kingdom was also embarrassed by the emergence of a wide range of support for the 30 per cent Club which the United Kingdom could not join without being prepared to undertake major SO_2 abatement measures.

This international pressure triggered an escalation in domestic attacks on the government's environmental policies. British pressure groups such as Friends of the Earth were aided by a 'Stop Acid Rain' Campaign. However, the most politically damaging attack on the government came from the House of Commons Environment Select Committee which criticised both the government and the CEGB in forthright terms and called for British membership of the 30 per cent Club as well as acceptance of the proposed LCP directive.

In May 1984 the prime minister, who had taken a personal interest in the acid rain issue, called a weekend seminar at Chequers where eminent scientists and representatives of relevant government departments debated acid rain. The Department of the Environment is reported to have 'argued forcefully' for membership of the 30 per cent Club though not for agreeing to the LCP Directive. However, the department lost the argument to, among others, Sir Walter (now Lord) Marshall, then Chairman of the CEGB. The Industrial Air Pollution Inspectorate (IAPI), formally charged with air pollution control, played no part in these discussions. This informal forum, outside the normal cabinet committee system, and including individuals from outside the government, set the seal on Britain's acid rain policy for the following two years.

The location of official British technological expertise on pollution control within an unassertive and politically isolated inspectorate, separated from the policy divisions of the Department of the Environment, can be argued to have contributed to the United Kingdom's unsatisfactory response to environmental problems. Technological expertise, where it exists, has been remote from the policy-making process and information provided to decision-takers has often been generated by those very industries which have an economic interest in the result of policy decisions.

The government's resistance to acid rain controls was largely owing to the high cost of the controls, estimated at £2bn in 1983 money, but also to strong resistance from the CEGB which was unprepared for major engineering work required at its power stations, which are among the largest in the Western world.

During 1987, Community pressure on the United Kingdom intensified; the United Kingdom had been isolated from its partners and stood alone in resisting a compromise LCP directive. Faced with this situation, and the desire to reduce uncertainty about the electricity supply industry's investment programme after privatisation, the British government conceded an agreement which would mean retrofitting FGD to a total of 12,000 megawatts of power station capacity by the year 2003. This compares with the 37,000-megawatt German programme completed between 1983 and 1988. The first British FGD unit will not begin to operate until 1993.

The role of the European Community

Although the most convincing evidence about damage caused by acid rain has come from Scandinavia, German pressure through EC institutions was the most potent force for a common European approach. However, the path to this was long and tortuous. The EC Commission introduced its draft LCP directive in December 1983, only six months after the German GFAVo had become law. The proposal for a 60 per cent reduction in SO_2 emissions from existing plant by 1995 proved very hard for countries such as the United Kingdom, Greece and Ireland to accept. Britain led the opposition to the

Commission's proposal which was strongly supported by the 'green' member states, Denmark and the Netherlands. France with its large nuclear programme was also content to support the Commission.

However, agreement on the LCP directive had to be unanimous. During the early negotiations, the Commission's strategy was to isolate Britain by offering derogations to the smaller countries which it saw as less central in the fight against acid rain. This policy had not succeeded by 1986 when the Netherlands proposed extending the timetable for emissions reductions into the next century by using a two-stage compliance schedule that considerably reduced abatement objectives in the mid-1990s. It was the Belgian proposals of early 1987 that led to an agreement. An ingenious juggling of emission figures took into account the principal valid objections to the European Commission's original proposal: it did not take account of emission reductions prior to 1980, nor of the fact that some countries had major fossil power station construction programmes under way during the 1980s. The Belgian proposal flushed out, for the first time, the views of the smaller and newer EC countries which had been hiding behind the United Kingdom and its resistance. Spain announced, for the first time, that it had reservations about the entire principle of the proposed directive.

The need to make derogations for countries such as Spain was recognised; Denmark and Germany brought the negotiations to a close by June 1988. Denmark proposed a three-stage emission reduction schedule for 1993, 1998 and 2010 which came into the broad range of acceptability for the United Kingdom. Britain conceded SO_2 emission reductions which would require an FGD retrofit programme about double the size of that originally intended. Overall, EC emissions of SO_2 from existing power stations and other large plant must decline by 23 per cent, 42 per cent and 57 per cent by the years 1993, 1998 and 2003 respectively. Reflecting economic realities, the negotiations evolved thus from a Commission-led pursuit of harmonisation to a more familiar bartering between nations. Nevertheless, the supra-national authority of the EC added urgency to the negotiations and, arguably, compelled greater concessions from the United Kingdom.

The case study of acid rain displays many of the features which characterise debate on other pressing environmental issues. Policy-making processes at the national level, at the EC level and on the wider international stage are intimately intertwined. While science was a major ingredient in the policy process, technological and economic considerations had equal, or even greater weights in political deliberations.

Comparing the British with the German case is instructive. Broadly speaking, the pluralism in the German system, expressed through federal government and coalition politics, coupled with an established and well-integrated environmental bureaucracy, allowed a rapid response to perceived damage. Centralised decision-making in Britain based on government by a political party with an unassailable majority, coupled with powerful vested interests and an unassertive pollution control system, resulted in constant procrastination and delay. EC activity was instrumental in bringing a semblance of a harmonised European response to the acid rain problem.

Environmental concerns will occupy a major position in the policy debates of the coming years, both at national and international levels. By its very nature, environmental policy-making is intimately linked to questions of science and technology. EC R & D policies therefore have an important role to play in promoting the accumulation of scientific evidence and technological alternatives.

Technological capabilities play an increasing role in determining the resolution of environmental problems. Environmental protection is coming to be regarded as an opportunity as well as a cost. Technologically advanced economies may find considerable advantages in promoting control technologies at home in order to create markets for products which will ultimately provide major export opportunities. This has been the case with German policy both in the energy and in the automobile industries.

Scientific knowledge does not, of course, necessarily provide imperatives for any particular course of action. In the case of acid rain the scale and complexity of recent environmental problems, coupled with the incompleteness of scientific knowledge, led to the adoption of anticipatory policy-making, notably in countries such as Germany and at the EC level. The British approach during much of the 1980s, heavily dominated by the interests of major polluters, has been manifestly incompatible with the principle of anticipation.

In view of the need to anticipate possible environmental catastrophies, policy decisions must be made with incomplete information in an increasingly politicised context, in which public participation plays an increasing role. The complexity of the issues involved is increased by their impact on other policies, notably for energy, industry, transport and agriculture. This makes it imperative to strive for the widest possible public debate, fed by the dissemination of relevant information.

IV In conclusion

The problem of acid rain is not the most difficult policy issue. The greenhouse effect is likely to dominate the environmental policy agenda for some time to come. The sustainability as well as the quality of life found in the developed world is now under challenge. If solutions are actively sought, they may need to go beyond technical fixes and begin to address the fundamental issues of energy use in everyday life – how people live, work, travel etc.

Within Europe the EC will in future clearly play a critical role. Under the SEA with qualified majority voting, it is most unlikely that a single member state would in future be able to hold up actions by others in, for example, establishing a policy response to climate change. Whether solutions are sought through directive (for example, setting minimum standards or new product regulations) or through the 'market-based mechanism' of a pollution tax, their implementation, although the responsibility of individual member states, will perforce be EC-wide. Of critical importance also is Europe's role on the world stage. Only at the truly global level can solutions to climate change be found. The participation of the EC in global negotiations will be all the more effective when it has demonstrated that harmonisation of policies for the protection of the environment is possible within the diversity of its own sphere.

20 Consumer protection policies in the EC: the quality of food

Erik Millstone

Responsibility for ensuring an adequate level of consumer protection has historically fallen to national governments, but in the process of completing the internal market it has become a Community-wide responsibility of the Commission, the Council and the European Parliament. The public costs of inadequate protection for consumer interests can be readily discerned in circumstances where public health has been avoidably damaged by the actions and products of commerce and industry. This chapter seeks to describe and explain what is happening to levels of consumer protection in the process of completing the Single European Market.

Since, as Foster (1990) has observed, the completion of the internal market was never meant to be a consumer protection measure, there is a danger that if consumer protection is weakened in the course of completing the market, the political consent needed for European stability could be jeopardised. Sceptics might think that this view exaggerates the importance of consumer protection policies, but traditional centrifugal tendencies in Europe are still so powerful that it is important not to play gratuitously into their hands.

The potential for conflict between consumer protection and the pursuit of free trade was highlighted in May 1990 when the governments of France, the former West Germany and Italy banned imports of British beef. Their ostensible reason was to protect their consumers from Bovine Spongiform Encephalopathy (also known as BSE or Mad Cow Disease). Clearly, however, these governments were also concerned to protect their beef producers.

The pursuit of European free trade and the completion of the EC's internal market will require the elimination of an enormous range of non-tariff barriers to the free movement of goods and services. Free trade within the food sector, for example, is hampered by complex non-tariff barriers, most of which are related to national consumer protection policies. The completion of the internal market will not require the elimination of these consumer protection measures; all it requires is uniformity and harmonisation of the regulations. It is the differences between national regulations that constitute the barriers, not the existence of those regulations as such.

The process of harmonising regulations requires in effect establishing new Community-wide standards. For both consumers and producers the key question is: at which level will those standards be set? If they are too high they may inhibit innovation and weaken the competitive position of manufactures in those member states with lower standards. If they are too low then there is a serious risk that the welfare of consumers could be damaged. The recent history of British food policy indicates that the political costs of getting this wrong can be very high.

Rather than attempting to review the entire field of consumer legislation, this chapter will concentrate specifically on the food sector, and within that relatively narrow area will focus on two topics: compositional standards and food additives.

I The benefits of harmonisation

In theory the completion of the internal market is supposed to benefit both producers and consumers. A free market in raw materials, labour, services and manufactures should enable the overall efficiency of commerce and industry to rise. Competition should mean that the savings due to productivity gains will be passed to consumers as lower prices. Consumers should also gain access to a wider range of products. In relation to food this seems particularly attractive because all consumers welcome price reductions that are not accompanied by any deterioration in quality, and Europe has such a rich range of cuisines that almost everyone would welcome increased access to each other's foods. The question is, can this be accomplished without a general decline in standards? A decline is neither necessary nor inevitable, but the evidence clearly indicates that it is occurring.

The most optimistic claims about the scale of the benefits from 1992 can be found in *The Cecchini Report* where it is claimed that potential savings are available to the EC of between ECU500m and 1,000m (Cecchini, 1988). Cecchini's treatment of the food sector is based upon a study conducted by the Paris-based Groupe Mac. Their study has received some well-deserved criticism. Swinbank points out that 'On occasions the presentation is so glib and superficial that serious doubts are bound to be raised about the quality of the underlying research' (Swinbank, 1990b:176). Groupe Mac's analysis is partial and selective. For example, if - as seems likely - the compositional restrictions covering ice cream are abolished and vegetable fat is allowed to replace some or all of the dairy fat, then the production costs of what will doubtless still be called ice cream will decline, and those savings may be passed on to consumers, but since the EC has a stored surplus of butter fat, consumers may have to pay increased taxes to cope with the consequent decline in sales of dairy fat. That fact, and others like it, were neglected by Groupe Mac.

Similarly, some Italian municipalities concerned about air pollution from the incineration of plastic bottles have banned their use for sales of mineral water. Harmonisation may force them to accept the sale of mineral water in plastics, and that may reduce the price of some bottled waters, but Italian tax payers will have to meet the extra costs of waste disposal, and/or suffer a deterioration in air quality. Groupe Mac and Cecchini have ignored many of the repercussions and externalities of harmonisation.

The degree of concentration in food retailing in all the north European member states already means that it is very difficult for food products to find space on the supermarket shelves unless they sell in large volume. The degree of concentration in food manufacturing, especally when combined with that in retailing, means that economies of scale are already there. As Swinbank says, it is 'unlikely that significant scale economies in food manufacture would be captured as a result of further liberalisation of intra-Community trade (Swinbank, 1990a:42). The only foods which will become easier to market after harmonisation are those which are currently excluded because they fall below prevailing consumer protection standards.

II The development of regulatory regimes

In the period before the signature in 1957 of the Treaty of Rome, countries had developed extensive statutes with which to control numerous aspects of their food supply. Food regulation became a problem for the EC because there are complex differences between the various national regulatory systems. Different countries variously permit and forbid different ingredients and practices. Consequently products which can lawfully be manufactured and sold in some countries are not permitted for sale in all of the others. Understandable though this might be, it does not provide for a free and open market, and that is a problem which the EC has sought to resolve.

The bright red artificial colouring called Amaranth (E123) provides a simple illustration. This compound is permitted and widely used in the United Kingdom and Ireland while in France and Italy it was banned until the early 1980s, and now is permitted in those countries only in a handful of products and only at very low levels. Only in Britain and Ireland are meat products allowed to contain the synthetic dye Red 2G (E128). Other countries have historically rejected the practice of colouring their sausages on the grounds that it may mislead consumers because it allows mixtures of rusk and fat to be dyed so as to resemble meat. Similarly, in the Federal Republic of Germany there are regulations which specify maximum residue levels in most groups of products, while there are few corresponding regulations in most other member states.

Non-tariff barriers to intra-Community trade are occasioned by such regulatory differences in all sectors, and such problems are not, of course, confined to foods. The task of harmonising all these regulatory regimes is vastly complicated; there is a great deal of political pressure for it to be completed in time for full implementation by the end of 1992. Before reviewing attempts to resolve these differences it may be useful to understand why they arose in the first place.

III Accounting for the differences

Many factors have contributed to producing these differences, some are scientific and technical while others are social, institutional and political. There are massive lacunae and uncertainties in the relevant scientific disciplines. There is, for example, some agreement (though it falls short of a complete consensus) that dietary factors significantly affect vulnerability to coronary heart disease (CHD) and cancer, but there is far less agreement about precisely which dietary factors are most crucial. Many scientists have concluded that a reduction in the consumption of fats can reduce the risks of CHD, but the putative benefits of mono- and poly-unsaturated fats are hotly debated (British Nutrition Foundation, 1988; Walker and Cannon, 1985). Some argue that what is most important is a reduction in total fat consumption, while others argue that what matters far more is a shift from saturated fats towards those which are less saturated. Similarly, in relation to the artificial sweetener saccharin, there is almost complete agreement that at high doses saccharin can increase the incidence of bladder cancer in male rats, but the significance of this finding for humans is the subject of an enduring controversy (Arnold *et al.*, 1983; Millstone and Abraham, 1988). The scale of the uncertainties is so huge that a team of American scientists has calculated that over the next 70 years, the expected number of cases of human bladder cancer in the United States resulting from a daily exposure to 120 milligrams of saccharin might range from as few as 0.22 cases in the entire American population to as many as 1,144,000! (Wilkinson, C,1983) Given uncertainties on such a massive scale it is easy to appreciate why different countries set conflicting regulations to cover compounds such as saccharin.

Scientific and technical factors are, however, only part of the story. Cultural, political and institutional differences contribute significantly to differences in systems of regulation. French and German regulatory systems are structured so that they are inherently more cautious and prudent than, for example, the British system, and this factor is perhaps most important in accounting for regulatory differences given the toxicological uncertainties (Abraham and Millstone, 1989).

The key difference between the British system on the one hand and the French and German on the other is that while the governments of all three countries both regulate and sponsor their food industries, only in Britain are these two functions located within a single ministry, thereby generating unavoidable conflicts of interest. This seems to explain why industrial interests so predominate over those of consumers in the United Kingdom, but do so to a markedly lesser extent in France and Germany. In contrast to Britain the majority of French and German consumers trust their governments to be prudent and cautious on matters of food safety (Millstone, 1988). If one compares the structure and operation of the EC's regulatory system to those in France, Germany and the United Kingdom one must conclude that the EC's system more closely resembles the British than either the French or the German (Abraham and Millstone, 1989). This is one reason why it is possible that harmonisation will involve a significant decline in standards of consumer protection.

Article 100 of the Treaty of Rome empowers the EC to take steps to protect public health and consumers. In the Commission the directorate-general with responsibility for food regulation is DG-III. DG-III is primarily responsible for the operation and completion of the internal market. The fact that it is DG-III which is responsible for food regulation has occasioned some comment because consumer representatives have argued that this responsibility could more properly be located in (what was until the summer of 1989) DG-XI with responsibility for environment, consumer protection and nuclear safety. (During the summer of 1989 these responsibilities were reorganised, and a distinct Consumer Policy Service (CPS) was created.) By virtue of their consumer protection brief, if DG-XI or the CPS were responsible for food regulation they might be expected to try to set EC-wide standards as close as possible to the highest prevailing in the Community, while DG-III's responsibility is to facilitate trade and commerce and this partly explains why they are trying to harmonise at significantly lower levels.

This tendency is powerfully reinforced by the fact that the Commission is under instructions to provide the Council of Ministers with a comprehensive set of draft regulations in time for full implementation at the end of 1992. In those circumstances, when the Commission is negotiating with the representatives of member states and seeking to reach agreement on the wording of draft regulations, there is an almost overwhelming temptation for the Commission to offer proposals which will meet least resistance, and those will be the ones which will cause the least disruption to production, allowing industry to continue with all existing practices. Article 100A(3) of the Single European Act states that 'The Commission in its proposals ... concerning health, environmental protection and consumer protection, will take as a base a high level of protection.' But in spite of that clause, there is a tendency in the entire system for harmonised regulations to move towards the level of the lowest common denominator.

IV The history of harmonisation

Harmonisation is not a new objective, it is at least as old as the Treaty of Rome. One of the problems, however, is that while the Treaty set the objective it failed to provide adequate means. Consequently, the Community has had a long and unhappy experience trying to harmonise regulations covering food products. The progress which they had achieved prior to the signature of the Single European Act was painfully slow. The first instrument of harmonisation in the food sector was a 1962 directive covering food colourings. Subsequently it took the Council 10 years to agree the fruit juice directive, and 14 years were spent on the jams directive. After more than 16 years of relatively fruitless negotiations a spokesman for the foodstuffs division of DG-III said that 'Free movement of goods is neither a practical reality nor is it necessary for products which are of but local importance and whose producers have no aspirations to wider sales' (Haigh, 1978:491). That remark should be interpreted as an expression of frustration rather than a statement of the Commission's genuine aspirations.

Many of the regulations which prevailed historically in member states set minimum compositional or recipe standards. The Commission consequently set out initially to establish a comprehensive set of agreements on what came to be called 'Euro-recipes'. After more than 16 years' work, the process of pursuing comprehensive agreements had accomplished only a minute fraction of what had originally been planned. In 1985, as part of the policy shift that culminated in the Single European Act, it was decided that instead of trying to reach agreements on further compositional standards the EC would in future only set maximum levels of use for particular ingredients across all product groups. In the vocabulary of the Commission they would set no more vertical regulations but only horizontal ones. The first stage was to abandon attempts to set any further compositional standards, but the final stage requires their abolition before the end of 1992 (European Commission, 1985)

The meaning of this retreat can readily be illustrated by reference to food additives. Instead of starting with a list of products and then specifying what each must or may contain, one starts from a list of permissible ingredients and compounds, and specifies the levels at which they may be used across a range of product types. This approach has the advantage that it allows for a far greater variety of products and is less likely to inhibit innovation. It has, however, the potential disadvantage that it requires dismantling compositional standards covering, for example, bread and meat products which historically have protected consumers against adulteration and debasement. While the number of compositional standards currently in force in the United Kingdom is low, in West Germany, Belgium and France there are innumerable detailed product standards (Hauser, 1988; Temmerman, 1989; Millstone, 1989).

The shift from vertical to horizontal regulations needs to be understood as one element in a broader political and regulatory process. The two main landmarks of that process were the European Court's judgement in the Cassis de Dijon case in 1979 and the signature in 1986 of the Single European Act. The Cassis de Dijon case concerned a traditional French alcoholic blackcurrent drink which the German authorities tried to exclude because it violated their compositional standards. The court ruled against the German position and in favour of free trade. The case was complex and the decision must be interpreted carefully. The court did not rule simply that any product which can be legally produced and sold in any one country must be acceptable for sale in all member states. The court ruled, however, that governments must mutually recognise and accept each others product standards except where their regulations protect public health and safety, in accordance with Article 36 of the Treaty of Rome.

The Cassis de Dijon decision has mistakenly been supposed by some commentators to provide a judicial basis for regulatory harmonisation without the need to negotiate detailed agreements in the Council of Ministers (Mattera, 1980). If that interpretation were correct the court would merely have endorsed what is termed 'mutual recognition'. Under conditions of full mutual recognition whatever is permissible in any one country must be acceptable in all. The problem with implementing full reciprocal recognition is that standards would rapidly decline to the lowest common denominator. Few producers would choose to conform to high standards if lower ones are sufficient to gain universal access to EC markets.

The Cassis de Dijon judgement stopped short of endorsing the full mutual recognition of standards. The court ruled, in effect, that all member states are obliged mutually to recognise and accept each others' product standards except where a government has scientific grounds for believing that the use of a product might endanger human health and safety. Under those circumstances governments are empowered to restrict or prohibit the sale of suspect products, though they are also obliged to inform the Commission and all other member states so that an EC-wide decision can be reached. Any steps must also be in proportion to the scale of the problem, and be designed to restrict trade as little as possible. Rather than rely on case law and judicial arguments about whether or not mutual recognition should apply, the Commission and Council are enjoined by the Single European Act to set Community-wide standards which must take into account relevant decisions of the European court. In 1981, the European court upheld the Dutch government's ban on the use of nisin as a preservative in cheese despite its acceptability elsewhere in the EC. They accepted that the scientific uncertainties concerning its safety were serious enough to justify the Dutch regulation (Weatherill, 1989). That judgment, and similar case laws, do not prevent the Commission from setting harmonised regulations approaching the lowest common denominator, but they set an implicit floor below which the court will not allow them to go.

V Compositional standards and food labelling

There was no dissent when the EC gave up trying to establish any new vertical standards, but as we approach the point at which all remaining compositional standards are to be abolished disquiet is growing. The key question is whether the abolition of compositional standards will entail a decline in product quality. Both the British government and the European Commission have argued that it should not result in any decline in quality, as long as consumers are provided with adequate labelling (Cockbill, 1990; Gray, 1984). Whether or not that argument is valid, it is clear that labelling of food products in a major issue.

The British government argues, furthermore, that quality is a purely subjective notion. The Commission, however, has acknowledged that the most obvious respect in which quality can be 'objectified' is in terms of the proportion of the ingredients. A beef sausage containing 65 per cent of beef is poorer in quality than one containing 85 per cent of beef. A strawberry jam comprising 65 per cent of strawberries is a better quality product than one containing only 45 per cent of the fruit. The use of bulking agents such as air, water and starch can constitute adulteration if the consequent reduction in the proportion by weight or volume of characterising ingredients is not self-evident to the purchaser as it is when one knowingly buys, for example, a low-fat margarine.

The position of the British Ministry of Agriculture, Fisheries and Food (MAFF) is that there is no evidence that the abolition of compositional standards leads to a decline in product quality. To adopt that view amounts to turning a blind eye to the obvious facts. Evidence does exist, and MAFF is familiar with it. As a representative of the British National Consumer Council has said, 'there is a clear and well known risk that standards may fall if compositional standards are abolished' (Foster, 1990:25). She summarises the evidence concisely:

> 'In the UK in 1987, the standards for some meat products were removed and percentage meat declarations were introduced. A study of 22 meat products found that in every case the meat content had dropped. Under the old standards the average minimum meat content would have been 46.85%. Under the new provisions, it fell to 31.3%. Furthermore, consumer choice decreased as the range of products with a higher meat content was no longer on offer.' [Ibid].

The report, from Shropshire Trading Standards officers to which Foster refers, was sent to MAFF in 1986 and is widely known and discussed. This demonstrable decline in standards occurred in circumstances where the abolition of compositional regulations went hand-in-hand with improvements in labelling. Representatives of both MAFF and the foodstuffs division of DG-III have argued that Community-wide improvements in labelling will be sufficient to counteract the potentially negative consequences of the abolition of all compositional standards. The evidence cited above casts doubt on that claim, but is sufficient to establish that if compositional standards are abolished without sufficient improvements to food labelling then standards will decline. Labelling is crucial, and it is therefore useful to compare the decision on labelling which was taken early in 1990 with the corresponding proposal which preceeded it, and with the labelling improvements which would be necessary to protect product standards.

When there are compositional standards consumers do not necessarily know what goes into the food products which they are buying, but they could find out if they chose to, and there are clear standards against which enforcement officers can judge products and processes. When compositional standards are abolished, there will be no standards to enforce and Trading Standards and Environmental Health Officers will be less able to act on behalf of consumers. In those circumstances, if consumers are to be in a position to know what they are buying they will require what is known in the trade as QUIDs or 'quantitative ingredient declarations'. When buying a pair of socks the label states that they contain, for example, 65 per cent cotton and 35 per cent nylon. It is not clear why consumers should be entitled to more information about what goes into their socks than what goes into their sausages. Where a full QUID labelling system is in place, all ingredients are listed, and their proportions are given in relation to unit size, weight or volume. A necessary, but not necessarily sufficient, condition therefore of the maintenance of consumer protection standards is that comprehensive QUID labelling be introduced for the food sector.

In 1988 the Commission published a proposal to amend Article 7 of the Labelling Directive (79/112/EEC), so as to require '... a statement of the quantity of an ingredient where that ingredient is likely to form an essential item of information.' Foster has pointed out that 'these proposals received a somewhat chilly reception from some parts of the food industry' (Foster, 1990:27). They were, however, warmly welcomed by consumer representatives who were concerned only with the vagueness concerning precisely what information would be deemed to be essential.

The processes of consultation and negotiation which took place in the intervening period remain unclear but on 22 February 1990 the Council of Ministers agreed not to require a full QUID system, but opted for a system which will be predominantly voluntary and optional, except where a nutritional claim is made on the label or in advertising. Where QUID statements are required, they will either show figures for what are known as the 'Big Four': energy, protein, carbohydrate and fat, or they may list the Big Four plus fibre and sodium. Figures may also be provided on starch, polyols, mono-unsaturates, polyunsaturates, cholesterol and minerals and vitamins if they are on a positive list (Council of Ministers, 1990).

The pattern is therefore clear. The decision was first taken to abolish all remaining compositional standards, and consumers were told not to worry because labelling changes would provide complete protection. By the time the labelling decision was announced a substantial retreat had occurred, and consumers had been deprived of the means by which food quality standards could be protected.

In the negotiations about regulatory harmonisation one of the main issues has been which institution, the Council of Ministers or the Commission, should make the key decisions. The Commission initially proposed that the Council should comprehensively delegate to the Commission their authority to establish regulations covering foodstuffs. Their reasoning was curious. The principle governing delegation stipulates that delegation is appropriate for issues which are purely technical rather than for those which are political. The Commission attributed the painfully slow progress in reaching harmonised regulations to the fact that the issues were highly political or at any rate they had become highly politicised. They argued therefore that far greater progress could be made by removing decision-making from the political arena and treating it as if it were a technical matter; but the implicit logic is curious indeed.

Issues which are irredeemably political cannot be depoliticised merely by handing responsibility for them to the bureaucracy. That fact may be administratively inconvenient, but the Council's decision to retain responsibility for regulating the use of food additives was consistent with their established policy and politically astute. The Council's decision-making will almost certainly be slower than might have been achieved if authority had been delegated, but in the long run their decisions will be easier to defend politically. One important consequence of the Council's retention of control is that proposed regulations must go to the European Parliament for scrutiny and comment. If the Commission had had their way, the Parliament's potential contribution would have been substantially diminished.

VI The example of food additives

In relation to additives, the Council has the responsibility for deciding which particular compounds are permitted, but they act on advice from an expert committee which is, in turn, serviced by and acts in close consultation with the Commission. The expert group is called The Scientific Committee for Food (SCF). It was established in 1974 with a brief to advise the Commission and the Council on matters relating to food and health. The role of the SCF is crucial and so is its relationship with corresponding national expert advisory agencies. Those relationships are important because the SCF and its secretariat are very poorly resourced, and it has had to rely substantially on material, administrative and intellectual support from national bodies. As 1992 approaches, the role and significance of the national agencies becomes increasingly problematic because they soon may lose their role and thus no longer be of use to the SCF. If the SCF is properly to fulfil its role it will require significantly greater resources than are currently available.

There are essentially just two ways of regulating food additives. One can either take the relatively weaker approach of establishing a list of those compounds which are prohibited and allow all others. Alternatively one can list only those which are permitted, banning all others. The former is called a negative list, the latter a positive one. The crucial difference between the two approaches is that with a negative list industry is free to use an unlimited range of chemicals and is not required to provide evidence of their safety, while the burden of proof falls on governments to establish evidence that a compound may be hazardous. With a positive list the burden of proof falls upon industry. The government refuses to permit compounds unless and until industry provides data sufficient to satisfy them that the chemicals can safely be used. In the EC, all countries now have positive lists for many groups of additives, but no country has positive lists for all groups.

A fully harmonised regulatory regime covering food additives would consist of two components: there must be a unified and comprehensive positive list, and that must be complemented with a set of detailed regulations defining conditions of permissible use for all positively listed compounds. Unless positive lists are supplemented by regulations which define how much of each compound may be used in specific kinds of products there would be nothing to prevent levels of consumption exceeding those considered toxicologically acceptable.

To establish the impact of harmonisation on regulatory standards it is necessary, therefore, to review two aspects, the length of the positive list and the severity of the conditions of use. In 1988, when the Commission was starting to give serious attention to the problem of determining the Community-wide positive list, the United Kingdom had the longest list in the EC consisting of 299 compounds in the regulated categories (which exclude flavourings, processing aids and enzymes). All other member states had shorter positive lists. By July 1989 when a Draft Directive was published, it contained approximately 500 compounds (European Commission, 1989). That list had been generated essentially by a simple process of consolidating the positive lists of the 12 member states. To that extent, the proposed Community-wide regulatory system is not at the level of the lowest common denominator, but is set to sink even lower.

The proposals in respect of conditions of use cannot be characterised quite so simply. To cut a very long and complicated story as short as possible, they are not set to sink to the very lowest common denominator because that would effectively entail setting no restrictions on the conditions in which they can be used. There are, in the United Kingdom for example, no restrictions on the levels at which artificial sweeteners can be used in low-calorie carbonated drinks, although quantitative restrictions do apply in most other member states. The proposals which have been made by the Commission for sweeteners involve quantitative restrictions, but the proposed levels are not based on any estimates of actual patterns of consumption nor on the cautious 'Danish Budget Method' but have been chosen as most likely to be agreeable to the governments of the member states (Hansen, 1979).

At the time of writing, the Commission is proposing to set quantitative restrictions only on the use of a small minority of additives, namely the preservatives, a few of the antioxidants and the sweeteners, while for all the remainder no quantitative restrictions are proposed. The proposals which the Commission issued in July 1989 merely enjoined maufacturers to confine their use of chemical additives to those levels consistent with 'Good Manufacturing Practice' (GMP).

When I first encountered the expression 'good manufacturing practice' I contacted the Ministry of Agriculture Fisheries and Food to discover what it might mean. No clear answer was forthcoming. I asked therefore: who decides whether or not any particular manufacturing practice is a good one? 'The manufacturer' I was told. Since manufacturers are unlikely to volunteer the suggestion that their manufacturing practice is poor, a regulation interpreted in this way is effectively non-existent. Unless a level is specified which must not be exceeded, no conviction could ever be obtained, however gross the product. A manufacturer could always claim that the level which it used was necessary to produce the kind of products in which it specialised.

Since those proposals were issued the Commission has superficially modified its approach. The problem with the expression 'good manufacturing practice' is that it implies that not all manufacturing practices are 'good'. The Commission has subsequently reduced its reliance on the expression 'GMP' and introduced in its place the latin phrase *Quantum Satis* (QS). This expression means a level of use which is sufficient, but that is no less vague than GMP, and the use of Latin hardly adds to the comprehensibility of the system. It also does not enable enforcement officers to establish whether or not any particular pattern of use conforms to or is in violation of the regulations.

It is clear that Community-wide additive standards are declining not to the level of the lowest common denominator but to one below it. A positive list of 500 compounds, the vast majority of which are restricted only by reference to GMP or QS, entails a striking liberalisation of controls. The proposals are substantially less rigorous than any of those currently applying in any member state. This will almost certainly facilitate the free trade of some products which are currently excluded because they fail to meet existing standards and thus entail a decline in the quality of some food products.

VII Decision-making and consumer consultation

How has this happened? On matters of food safety the Commission is supposed to consult with the Advisory Committee on Foodstuffs (ACF), which is a statutory body that includes consumer representatives. In the 18 months prior to the autumn of 1989, during which time almost all the key decisions were taken, the Commission failed to call a single meeting of the ACF. They did, however, hold a lengthy series of meetings with an *ad hoc* Commission Working Party on Food Additives which included representatives of governments and industry but excluded consumer representatives. On 12 January 1990 an MEP asked the Commission a Parliamentary Question to establish why this had occurred. Between that date and the time when the question was answered on 21 February 1990, a meeting of the ACF was summoned, and some working papers of the *ad hoc* group were sent to consumer organisations, nominally for consultation, but by then they had reached a stage at which there was little left to decide (Ford, 1990; Bangermann, 1990). One might reasonably speculate that if the entire matter had been in the hands of DG-XI or the Consumer Policy Service rather than DG-III, the procedures and outcome would have been rather different.

Even if the ACF had met, it is not clear that it could have accomplished a great deal. As Crampton, the Secretary of Consumers in the European Community Group, has said: 'The Commission calls its few meetings and decides the agenda; the ACF makes no attempt to come to a common view; and there are no reports and not really any minutes' (Crampton, 1989:10).

Another route for involving consumers would have been to discuss these issues at the Consumers Consultative Committee (CCC) which was established by the Commission in 1973. But as Crampton has said:

> The CCC meets in theory about four times a year in Brussels ... [but] ... Consumer organisations form less than a quarter of its total membership, and there are sometimes conflicts of interest. Because of its lack of independence, it fails to act effectively in disseminating its views. The CCC is often consulted late, only after the Commission has already decided its policies. Its resources are totally inadequate and meetings are cancelled, often at short notice, because of lack of funds and meeting rooms. The CCC is increasingly ignored by the Commission. Subjects on which it has not been asked to give an opinion include ... food policy [Crampton, 1989:9]

VIII Food standards in general

At a meeting with the Guild of Food Writers in November 1989 the Secretary of State for Agriculture, Fisheries and Foods, John Selwyn Gummer said in relation to harmonisation in the run up to 1992 '... we shall be fighting to get the standards which we have in this country accepted where they are higher than other people's, and where they are not we shall be seeking to make our own standards higher'... In the event, it is difficult to take those words at face value.

It was pointed out earlier that additives were being discussed as an illustrative example but one for which the negotiations had reached a relatively advanced stage. It is reasonable to inquire, therefore, whether a similar picture would emerge if other aspects of food policy were scrutinised and other non-food consumer protection issues. The example of bacterial standards presents some contradictory indications. On the one hand, some very tough proposals emerged in February 1990 from the Commission concerning minced meat. Those proposals have generally been interpreted as having confused the requirements for minced meat intended to be eaten raw (for example, in steak tartare) with those for minced meat which will be cooked before being eaten. For this reason they are not expected to survive in their present form (MAFF, 1990; Meat & Livestock Commission, 1990). On the other hand, if all border controls are eliminated as currently proposed, then this would seriously weaken protection against bacterial spoilage. A consequence of abolishing border controls is that meat, for example, only has to meet bacterial standards when loaded for transport in order to be allowed to travel freely throughout the Community. It is not difficult to imagine that a consignment of perishable foods could deteriorate substantially on a journey, for example, from the Mediterranean coast to Denmark.

The case of pesticides is particularly tricky for the Commission. Some commentators have argued that because geographic and climatic conditions and entymological populations vary so considerably within the Community, it will not be possible to have a unified set of rules governing the use of chemical pesticides. Some progress has been made, however, in drafting proposals concerning maximum permitted residues in particular food groups.

In this case it is important to acknowledge that standards will not be falling exactly to the lowest common denominator. Prior to the moves towards harmonisation the United Kingdom had hardly any quantitative limits on pesticide residues in foods, and the credit for the introduction of a limited set of such standards in the United Kingdom must go to EC integration. The Community-wide standards which are being proposed, however, are in many respects less restrictive than those which currently apply in Germany, France and Italy.

The technology of food irradiation poses a severe problem for the Community. The head of the foodstuffs division in DG-III was formerly a senior manager at the Winfrith nuclear power station and is an enthusiastic supporter of food irradiation. He has argued that this technology is already permitted in the Netherlands and France, so has to be accepted throughout the Community if all non-tariff barriers are to be abolished. At the time of writing the European Parliament is considering a proposed amendment to a joint position of the Council and the Commission that, if passed, would effectively outlaw irradiated food from intra-Community trade. It is not yet possible to forecast the outcome, but it is possible that a sufficiently large majority against food irradiation may be achieved to over-rule both the Commission and the Council. If that were to occur, it would be the first occasion when the Parliament would exercise that power acquired under the Single European Act.

IX Prospect and alternatives

It is difficult to be certain how representative the food sector is of other consumer protection issues in the process of completing the EC's internal market. The London-based group, Consumers in the European Community Group (or CECG), are responsible for monitoring these matters and for representing British consumers in Community institutions. They were asked if they could identify a single sector in which Community-wide standards are expected to rise above the level of the lowest common denominator. They were unable to do so.

The evidence clearly indicates that the tendency is for regulatory harmonisation to be accomplished in ways which will weaken the protection afforded to consumers. This may facilitate the completion of the internal market in time for the end of 1992, but in the longer term the Community is giving hostages to fortune. Community institutions and public support for European integration could be seriously damaged as well as public health and safety.

In principle, it is not too late for policies to be changed in time for the 1992 deadline, but the timetable is very tight. If the market is to be completed and maintained without damaging the interests of consumers a range of reforms would be required. First and foremost, it would be necessary to ensure that the responsibility for consumer protection policies within the Commission was clearly separated from responsibilities for promoting industry and commerce. In relation to the food sector this has crystalised into the idea that it may well be desirable to establish a European Community version of the United States Food and Drug Administration.

That proposal has surfaced in several contexts, ranging from the Bureau of European Consumer Unions to the European Parliament and at high levels in the Commission (Foster, 1990:29; van Miert, 1989:5; Bureau of European Consumer Unions, 1989:2; Bangermann, 1989/90:1). Its implementation might entail a considerable realignment of forces and considerable revision of EC regulatory policies, but for those reasons it will be strongly resisted by those with a vested interest in the status quo.

One of the greatest causes of disquiet among consumer organisations is that the information, on the basis of which the governments of member states and the Commission make policies, is kept secret; in the United Kingdom most aspects of regulatory policy-making are covered by the Official Secrets Act. The Scientific Committee for Food, for example, is as willing to accept unpublished data as it is to review information in the public domain, and they respect the wishes of firms to keep their toxicological data secret. In such circumstances it is not possible for the scientific community to subject the work to any peer review, and consumer groups are prone to speculate that the industry, and maybe also the regulators, have something to hide. Public confidence in the regulatory system is undermined by secrecy and a major step towards its enhancement could be achieved by adopting a freedom of information policy. At a meeting at Westminster of the Parliamentary Food and Health Forum in June 1988, the head of the foodstuffs division at the Commission, Paul Gray, stated that '... he felt that it was essential to have greater disclosure of information for greater public confidence when there was freer circulation for foodstuffs. The Commission was moving in that direction and those submitting data were now warned that this could eventually be published' (*Transcript*, p. 6). However reassuring those words may have been, there has been no indication that any subsequent action has been taken along those lines.

Even if all the toxicological data were published this would not address the problems concerning the dubious relevance of animal studies to human consumption. In order to address this problem, it would be necessary to embark on long-term research projects designed to estimate the validity of existing toxicological methods and to develop methods which are demonstrably more relevant to human health. The uncertainties are severe, and steps need to be taken to reduce them. Action on this front could most effectively be taken on an international scale, and the EC is well placed to take a lead in this field.

Another related step in the right direction would be to ensure that expert advisory committees were properly resourced. It is necessary and desirable to improve both the quality and the independence of scientific advisers. Committees should be staffed only by individuals who have no commercial links with firms whose products they are evaluating. From the consumers' point of view it is important that their interests are properly represented in the decision-making processes within the Commission and the member states.

Part V
The International Context and Policy Agenda

Preface to Part V

Every chapter in Part IV has served to emphasise that the problems of regulation and harmonisation within the EC cannot be resolved except within a broader international context. Some of the most acute pollution hazards are inherently global in nature, while the movement of goods, people and capital means that many others require international regulatory arrangements.

The end of the Cold War in the 1980s offers an unusually favourable opportunity for promoting the necessary degree of international cooperation and also puts the relationship between Eastern and Western Europe on an entirely new basis. While ultimately some Eastern European countries may gain admission to the EC there are urgent problems of economic collapse, unemployment and social tensions which require more immediate attention. Mary Kaldor argues in her chapter that the scale of the problems calls for a major new initiative from the EC, comparable to the Marshall Plan in the early post-war period.

There are also complex new defence and security issues which are the main concern of the following chapter by William Walker. The future evolution of the EC's political institutions as well as the possible redeployment of its scientific and technological resources and its defence industry will alike depend on the way in which these issues are approached.

The fact that we focus mainly on the relationship between Eastern and Western Europe should not be taken in any way as lack of concern for the problems of the Third World. Rather, the problems of Eastern Europe should be seen as one illustration of the more general problem of international economic cooperation and technology transfer to assist the poorer countries to improve their standard of living and quality of life.

In the final chapter we return to the policy agenda which this implies. We do not pretend that there are any simple universal prescriptions, but we do point out some general principles which we believe should guide EC policy-making. In particular we indicate the great importance of overcoming the 'democratic deficit' within the institutions of the EC. Only a combination of well-informed public opinion, good research and continuous critical debate can develop the long-term strategies which are needed.

21 East and West Europe

Mary Kaldor

I Introduction

The democratic revolutions of 1989 have transformed the economic prospects for the whole of Europe. The direction and form of potential transformation is, however, as yet undetermined. It is possible that the military division between East and West will be replaced by a new economic division between a prosperous West and a poorer, less-developed East. It is also possible that both halves of Europe will gain from increased East–West economic cooperation.

The transition to a new socio-economic model in East-Central Europe will be successful only if three conditions are met: first, that real restructuring takes place – namely, that there are fundamental changes in patterns of production and consumption and not just changes in the mechanisms of managing the economy. At present, discussion tends to focus on what Marx called the sphere of circulation and Keynes 'monetary' as opposed to 'real' phenomena – that is, on how to introduce a free market, rather than on the goals of economic development. Up to now, patterns of production have been dominated by energy-intensive, large-scale heavy industry and patterns of consumption by state investment decisions. The problem is how to shift these patterns in such a way as to improve the standard and style of life and the quality of the environment.

Second, there must be major changes in the mechanisms for managing the economy in order to bring about restructuring. It is of course important to free the markets for goods and labour by removing physical or administrative controls and reducing price and wage controls. It is also important to develop new forms of ownership and to reduce drastically centralized state ownership of property. However, if these steps are to stimulate the economy, it is equally important to develop a democratic framework for planning at local, regional and national levels. Even if the role of the state is reduced, what remains is the heritage of the state structures of the past. The state will continue to be a major investor and to be responsible for a range of policies (innovation, education, infrastructure, welfare, environment etc.) which in turn will profoundly influence the operations of the market. It is important *now* to develop democratic procedures for determining these policies.

Third, new forms of economic cooperation between East and West must be developed which are not merely financial and based on commercial criteria. Western Europe has a vital interest in the success of democracy in East-Central Europe and in its long-term development. As argued in Chapter 22, Western Europe has a historic opportunity to reduce military spending and to develop mechanisms for transferring resources from rich to poor regions, both in the Third World and in Europe. Now is the time to think and act in an imaginative way about East European development. Much of this assistance needs to be technical rather than financial, that is to say aimed at certain specific projects. The Marshall Plan offers a useful model in many respects, but in today's circumstances environmental objectives and technology transfer are especially important.

Economists tend to assume that provided Eastern Europe introduces the necessary market reforms, foreign capital will be attracted by low wages. The experience of the Asian Newly Industrialising Countries (NICs) is often heralded as an example to emulate. The success of the NICs, however, depended initially on the availability of workers willing to accept extremely low wages and poor working conditions in a context where there was sufficient political stability to attract multinational companies; later, it depended on a very rapid increase in education, skills and R & D. The East Europeans already have a highly educated, industrialised workforce which, after long years of totalitarian rule, is vociferous in its protests about working conditions. The key to the success of East European economies and democracies will be their ability to generate and absorb new sources of technological innovation so as to be able to manufacture a range of goods which are competitive on world markets. This depends on a wide range of social, economic and political changes and not just on market mechanisms.

This chapter focuses on East-Central Europe and does not include the Soviet Union. Although many of the problems of the Soviet republics are similar, they are even more serious and more intractable and would require separate treatment within a global not a European context.

II The historical background

East European scholars often argue that the East–West division of Europe long predated Yalta (for example, Anderson, 1974; Szücs, 1988; Rev, 1984). Indeed, they point out that the division has cultural, social and economic roots that can be traced back to the decline of the Roman empire. The turning point, however, in the shifting boundaries of East and West is usually considered that moment in the 16th and 17th centuries when capitalism began to emerge in Western Europe while Eastern Europe embarked on what Engels called the 'second stage of feudalism', supplying cereals and livestock to Western Europe and expanding output by reversion to extra-economic forms of coercion.

There were considerable differences from region to region. The Balkan region was occupied by the Turks up to the 19th century; in the 17th century Turkish rule had extended northwards to include large parts of Hungary. The Turkish system can be described as a predatory military–political structure. Resources were extracted from pre-existing social and economic structures to finance the state apparatus in cruel and corrupt ways. This had very destructive consequences.

The Western zones of what is now Eastern Europe – Bohemia, Prussia, Mecklenburg, Brandenburg – and Austria represented mixes of Western and Eastern systems. Poland came under both German and Russian rule. Both the Prussian and Hapsburg states bore some resemblance to the centralised Russian state, relying on a hierarchy of noble officeholders. But the Hapsburg

Table 21.1 Basic indicators

	GNP capita 1987[a] ($)	Population 1987 (m)	Life expectancy 1985 Males	Life expectancy 1985 Females	Infant mortality rate per thousand (0-1 yrs) 1985	Urban population 1985 (% of total) 1985	Energy consumption per capita per unit of GNP 1985 (Kgs of oil equivalent)		Scientists and engineers 1985	% of pop.	Technicians 1985	% of pop.
Albania	(953)	3.1	67	73	43	34	1,267	(1.31)	—	—	—	—
Bulgaria	(7,229)	9.0	68	74	16	68	4,332	(0.61)	48,008	5	13,099	1
Czechoslovakia	(9,709)	15.6	66	74	15	66	4,853	(0.51)	61,046	4	47,337	3
East Germany	(11,860)	16.6	68	75	10	76	5,670	(0.50)	128,502	8	69,828	4
Hungary	(8,260)	10.6	67	74	20	55	2,974	(0.37)	22,479	4	17,869	3
Poland	(6,879)	37.7	67	75	19	59	3,438	(0.49)	56,600	2	54,000	2
Romania	(6,365)	22.9	69	74	24	51	3,453	(0.58)	—	—	—	—
Soviet Union	(8,662)	284.0	65	74	29	66	4,885	(0.58)	1,491,300	5	—	—
For comparison												
Greece	4,677	10.0	72	78	16	65	1,341	0.39	2,441[b]	0.3	1,067[b]	0.1
Spain	7,282	39.0	74	80	10	77	1,932	0.28	15,299[c]	0.4	6,181[c]	0.2
Italy	13,010	57.4	74	79	12	67	2,606	0.21	63,021[b]	1.0	28,694[b]	0.5
Austria	15,440	7.6	70	77	11	56	3,217	0.21	6,712[d]	1.0	6,145[d]	1.0
West Germany	18,450	61.0	72	78	10	86	4,451	0.25	133,115[b]	2.0	119,618[b]	2.0
United Kingdom	11,730	56.8	72	77	9	92	3,603	0.32	86,500[e]	1.5	76,600[e]	1.3
United States	18,570	243.0	72	80	11	74	7,278	0.41	728,600	3.0	—	—

[a] There are no direct measurement of GNP for CMEA Countries. Their statistics use a different measure, Net Material Product. These figures are CIA estimates and are taken from US ACDA (see below). They are almost certainly overestimates and the brackets are used to indicate the fact that they are estimates. The same source is used to estimate energy consumption per unit of output. The bracketed figures in the second column are likely to be underestimates, therefore. Paul Marer in a comparison undertaken for the World Bank suggests that, in 1980, the ratios of GNP to NMP were as follows: Bulgaria (1.32), Czechoslovakia (1.19), West Germany (1.26), Hungary (1.24), Poland (1.28), Romania (1.20), Soviet Union (1.25). (Marer, 1985)

[b] 1983; [c] 1984; [d] 1981; [e] 1978

US Arms Control and Disarmament Agency: *World Military Expenditures and Arms Transfer 1988*, 1988; World Bank,

empire was more decentralised than the Russian and Prussian states.

The industrial revolution in Britain and Western Europe stimulated an enormous expansion in the demand for foodstuffs and raw material from Eastern Europe. The Western zones were best able to meet the challenge of industrialisation, particularly after the first unification of Germany (1871). Austria, Bohemia and the German states of what was East Germany (the GDR) developed a fairly high degree of industrialisation, but the Balkan countries had great difficulty responding to the Western challenge. Western imports destroyed traditional handicrafts, while railways and extractive investments (for example, Romanian oil) failed to foster industrialisation but rather led to indebtedness which in turn reinforced backwardness.

The differences between the sub-regions of Eastern Europe persisted up to World War II. Indeed, they were accentuated by the Bolshevik revolution and the emergence of an entirely new system in the Soviet Union, even though it was a system which drew on many characteristics of Russian history. In the aftermath of World War II, however, with the Soviet occupation and subsequent imposition of Soviet control, the variegated socio-economic pattern of East-Central Europe was partially submerged in the uniform adoption of the Soviet socio-economic and political model.

Two other considerations should be mentioned. First, the East European economies suffered proportionally more war damage than the West European economies. Although at the end of the war they received a certain amount of economic assistance from United Nations agencies, they were not eligible for Marshall Aid. Indeed, it has been estimated that the Russians extracted from these countries in the form of reparations and unfair pricing procedures an amount which was equivalent to that which Western Europe received in the form of Marshall Aid (Marer, 1984).

Second, after the imposition of Soviet control in 1947–8, the economies of East-Central Europe were cut off from the West not just by the nature of the Stalinist system but also by Western hostility. The COCOM list of goods to be embargoed was introduced in 1949 and from that date until 1954 it covered approximately 50 per cent of the goods normally traded with the East. In 1938, trade with Western Europe had accounted for 70 per cent of total trade; in 1988, it was 30 per cent.

III Characteristics of East–Central European economies

Table 21.1 provides some basic indicators for the economies of East-Central Europe, compared with selected West European economies. There are no strictly comparable GNP figures; these are CIA figures which probably overestimate the GNP per capita of East European economies. Nevertheless, it is clear that these countries are on the whole poorer than West European countries. They are less urbanized and industrialised; basic social indicators, such as life expectancy and infant mortality, are worse. On the basis of these indicators they lie somewhere between the First and the Third Worlds. Historical differences have persisted. The poorest countries are still the Balkan countries, and the richest countries were the former East Germany and Czechoslovakia, formerly the western zones of East-Central Europe. Poland and Hungary lie somewhere in between.

One interesting feature of East-Central European countries is the very high levels of scientists, engineers and technicians as a proportion of the population. To some extent, this may be explained by a broader definition of scientists used in East-Central Europe; even so, the figures for technicians suggest that the numbers, even in comparison with Western countries, are high. In general, educational attainment is very high and comparable to educational attainment in Western Europe. East Germany, for example, had the highest levels of post-secondary education in Europe (UNESCO, 1987).

The system that was imposed by the Soviet Union in the years 1947–53, had the following characteristics:

(i) *The economies were centrally planned.* Vertical relationships between the state and enterprises replaced horizontal relationships between buyers and sellers. Orders were transmitted from state to enterprise through administrative directives, normally with instructions based on quantitative indicators. Prices and wages were strictly controlled; autonomous economic activity was eliminated or severely restricted, including the partial collectivisation of agriculture and the elimination of artisan activity and of small enterprises like cafes, restaurants or shops.

(ii) *The emphasis of state planning on heavy industry and military production.* Huge factories were constructed, often many times larger than in Western countries, for mass production which was almost an article of faith. Indeed, Lenin had said, 'Socialism has been created through large-scale mechanical industry. And if the working masses introducing socialism do not understand that they should adapt their institutions to the functioning of large-scale mechanical industry, then one cannot even speak of an introduction to socialism'. (Bentley, 1984). This view was echoed by East European leaders throughout the 1950s and 1960s. As a result, agriculture, private consumption and the environment were relatively neglected.

(iii) *There was virtually full employment and income distribution was relatively egalitarian.* Educational opportunities were enormously extended, even though education was authoritarian and propagandist. Women had to participate in the labour force, and the provision of child-care facilities was extensive if over-institutionalised. Contraception and abortion were available on demand and the birth rate was relatively low.

(iv) *Trade was strictly controlled by the ministries of foreign trade* who negotiated state-to-state bilateral trading agreements. Although East-Central European countries traded with each other through COMECON, their trade patterns were dominated by bilateral agreements with the Soviet Union, mainly of manufactured goods in exchange for oil and other raw materials. Trade was relatively small as a proportion of total output; socialist economies were expected to strive for self-sufficiency.

(v) *The systems were characterised politically by the monopoly of the Communist Party* and the elimination of all autonomous political activities. In the Stalinist years personal dictatorships relied on an extensive network of secret police, show trials, deportations and labour camps. Later, party control was reinstated through the nomenklature system which controlled appointments at all levels of society. It was tightly tied to the Soviet Union, primarily through party links, the secret police and the army.

In the first decade or so after the war, these systems generated very rapid rates of growth. The annual average for the period up to 1970 (based on the estimates of the Polish economist, Wlodimierz Brus, which are probably reliable) was equivalent to West Germany and France at that time – between

5 and 7 per cent, sometimes even higher. This, together with full employment and educational opportunities, helps to explain the confidence of Communist leaders at that time – for example, Krushchev's belief that socialism would overtake capitalism.

The main opposition came from the 'revisionists' who led the Hungarian and Czechoslovak revolutions. They opposed the lack of democracy, but they were within the Communist Party and basically believed in planning and state ownership. The Hungarian philosopher György Lukács stated: 'Even the worst socialism is better than the best capitalism'.

By the early 1960s, economic growth had slowed; it began to be very difficult to restructure the economy and raise productivity. East European economists started to talk about the limits of extensive growth, that is, growth through adding inputs (labour, materials, energy etc,) to industry and about the need for intensive growth, by which they meant growth through technological innovation, using inputs more efficiently. The central planning system seemed to lock the economies into the patterns of production and consumption determined in the early post-war years. At a time when Western Europe was moving towards more flexible, market-orientated, R & D-intensive industries, Eastern Europe was caught in a 1950s' economic time warp.

Some of the best analyses of the functioning of centrally planned economies have been developed by Hungarian economists (Kornai, 1982; 1985; 1986; Bauer, 1978). They point out that although these economies were characterised by vertical relationships, they were not top-down economies. Directives issued from above were determined on the basis of various signals transmitted upwards by layer upon layer of bureaucrats. Kornai describes these economies as systems of bureaucratic regulation. Because prices were controlled, such systems tended to shortages resulting in queues, waiting lists for cars, parts, recruitment bottlenecks etc.

These shortages were not absolute (in the sense that there could still exist reserves of labour or raw materials). Rather they were generated by the nature of the planning system itself. They arose from the fact that actual expenditure always exceeded planned expenditure – even when the planners made allowances – the reason being bureaucratic resistance. It was always possible to start a new investment project, but it was almost impossible to phase out (or in some cases even complete) old projects. New investment projects always cost more than anticipated, either because enterprises deliberately underestimated the cost of an investment project in order to 'hook on' to the plan, or because enterprises experienced what Kornai calls a 'soft budget constraint', that is, it did not matter if they over-ran estimates. The main incentive of managers was to remain within the bureaucratic hierarchy and fulfil (quantitative) plan indicators. If they came up against some unforseen problem, their response was to spend more money, and they would not be penalized for this. (Western defence firms also experience a soft-budget constraint, hence the cost-overrun problem.)

Considerable emphasis was placed by planners on technological change. Indeed the technological competition with the West was viewed as a key component of the overall competition between capitalism and socialism. Moreover, technological change was seen as the main mechanism whereby productive capacity could be expanded and the problem of shortage mitigated.

Why was this technological effort not translated into intensive growth? There are many reasons:

(i) R & D institutions were often separated from production, and R & D staff were centrally allocated. This was partly for political reasons (since more creative people were politically suspect), but it also reflected the compartmentalisation tendencies of the bureaucractic mind;
(ii) R & D work was not based on contracts and hence responsive to need. Rather scientists and engineers tended to pursue their own interests;
(iii) there was, therefore, little or no feed-back from customers to research or design staff; rather orders were transmitted by the planners who did not know or heed consumers. (Only in the military sector did there exist an interaction between customers and R & D staff);
(iv) R & D is very difficult to plan; planners lacked technological knowledge to assess projects, or to match them to skills;
(v) most importantly, even when R & D institutions developed new inventions, or imported technology from abroad, it was often extremely difficult to diffuse innovations or to assimilate them into the production process because of resistance at the level of enterprises. Innovation tended to disrupt plan fulfillment and established supply lines and were therefore avoided if possible.

To overcome these problems there had been, since the 1950s, periodic attempts to reform the system, generally includingsome devolution of decision-making and a greater role for the market mechanism, and a shift of state spending from investment and military spending to agriculture and consumption. There was usually also some liberalisation of agriculture and tolerance towards very small-scale private enterprise and artisan activities; administrative directives were replaced by financial incentives. There were also efforts to integrate R & D and production through the creation of large combines in East Germany and Czechoslovakia and smaller contract-based research institutes in Hungary and Bulgaria.

However, these reforms were extremely difficult to carry out in the absence of political reform which met constant resistance from the nomenklature. Reforms tended therefore to lose momentum. The first wave of reforms took place after Stalin's death, and involved an easing of the worst features of Stalinism: the collectivisation of agriculture was halted; forced deliveries of agricultural produce were ended; artisans, cafes, etc. were permitted; there were some reductions in military spending and investment. This was the period when revisionists regained their positions, especially in Poland and Hungary, and it culminated in the Hungarian revolution of 1956 and the Polish October.

The next wave of reforms took place in the mid-1960s. Particularly important was the Hungarian New Economic Mechanism established on 1 January 1968, which succeeded in introducing a significant market element in the Hungarian economy, and the so-called Second Revolution in Yugoslavia in 1966. As in the 1950s, pressure for political reform mounted, culminating in the Prague Spring of 1968. Thereafter came a renewed period of 'normalisation' and the reforms lost momentum.

The *détente* of the 1970s could be described as an attempt to substitute imports for reform to overcome shortage through imports rather than technical change. However, imports were squandered on consumer goods (as in

Poland) or ill-conceived investment projects. They also created a chain of supplementary demands (for parts, appropriate raw materials, etc). These economies were very hard hit by the post-1973 recession in Western Europe with its new protectionism, and by competition from the newly industrialising countries, which were competing for the same patch of the market in cheap manufactured goods (Köves, 1989). By the end of the 1970s, the East-Central European economies were, therefore, heavily indebted to the West. They began to cut back on Western trade and credit even before Afghanistan and the Western decision to halt credits. The increase in the price of Soviet oil and gas in the early 1980s, however, meant even greater dependency on Soviet trade. Table 21.2 shows that despite the increase in East–West trade in the 1970s, trade with Western Europe as a share of the total trade of East-Central Europe was actually *less* in 1985 than in 1970. Only Hungary, and to a lesser extent Poland and the former East Germany, managed to sustain the westwards trade shift. But in these countries levels of indebtedness remained extremely high as can seen from Table 21.3. In Czechoslovakia, Bulgaria and above all Romania, which managed to repay its debt through a hideous austerity programme, the share of East–West trade actually fell.

Since the early 1970s, the rate of growth of labour and capital productivity has declined yet further (UN ECE, 1990), and East-Central Europe's share of world trade has also declined. Material and energy efficiency may also actually have fallen since the mid-1970s (Rostowski, 1988). By the mid-1980s, it was becoming clear that the economic systems of East-Central Europe were environmentally disastrous. Energy consumption per capita, except in Albania, is actually higher than in most Western countries, and energy consumption per unit of output is approximately 2½ times its Western equivalent. The whole philosophy of scientific socialism implied the necessity of controlling and taming nature. If consumers were disregarded, nature was even more so. There were no avenues for public protest or for demanding environmental safeguards. The consequence is an environmental catastrophe – dying forests, polluted rivers and lakes, and an increase in environmentally linked diseases such as food allergies, respiratory problems and cancers.

Even more disappointing was the technological performance of these economies. The failure to respond to increased imports of Western technology and the great difficulty in generating improved technological performance even before the cutbacks of the late 1970s can only be explained in terms of the inability to restructure patterns of production and consumption away from the emphasis on large-scale heavy industry. Even though they developed their own electronics sectors, they were quite unable to keep pace with the successive new generations of semi-conductor technology or to shift the overall structure of industry. For example, the former East Germany had an ambitious micro-electronic sector and had relatively high rates of installation of CNC, FMS and robotics. The failure to use these new technologies effectively can only be understood in terms of the nature of centrally planned economies, the overwhelming tendency for bureaucratic resistance and inertia at both political and economic levels.

If *détente* failed to solve any economic problems, it had some significant political consequences. The invasion of Czechoslovakia in 1968 marked the end of hopes for reforming communism. Increased East–West trade and communication in the 1970s provided space for the emergence of a new form

Table 21.2 East–West trade, 1970–88

	1970 US$ bn	1970 % share	1975 US$ bn	1975 % share	1980 US$ bn	1980 % share	1985 US$ bn	1985 % share	1988 US$ bn	1988 % share
Bulgaria										
World	1.83		5.40		9.67		13.63		16.58	
ECE–West	0.38	21	1.35	25	1.79	19	2.20	16	2.73	16
ECE–East	1.33	73	3.71	69	7.29	75	10.04	74	12.24	74
Other	0.13	7	0.34	7	0.59	6	1.39	11	1.60	10
Czechoslovakia										
World	3.70		9.09		15.18		17.55		23.24	
ECE–West	1.01	28	2.52	27	4.21	28	3.22	18	4.59	20
ECE–East	2.33	63	5.85	64	9.83	65	13.09	75	17.20	74
Other	0.35	10	0.71	8	1.14	8	1.24	7	1.45	6
East Germany										
World	4.92		11.71		20.29		23.96		30.31	
ECE–West	1.43	29	3.91	33	7.49	37	7.57	31	9.98	33
ECE–East	3.19	65	7.09	61	11.27	56	14.80	62	18.74	62
Other	0.30	6	0.70	6	1.54	8	1.60	7	1.59	5
Hungary										
World	2.51		7.15		9.19		8.18		9.86	
ECE–West	0.72	29	2.03	29	3.88	42	3.41	42	4.23	43
ECE–East	1.56	62	4.48	63	4.31	47	4.04	49	4.66	47
Other	0.22	9	0.64	9	1.01	11	0.73	9	0.96	10
Poland										
World	3.61		12.55		19.12		10.84		10.85	
ECE–West	0.98	27	6.31	50	6.92	36	3.82	35	4.62	43
ECE–East	2.36	65	5.45	43	10.09	53	5.87	54	4.93	45
Other	0.27	7	0.78	6	2.11	11	1.14	11	1.30	12
Romania										
World	1.96		5.34		13.20		8.48		9.50	
ECE–West	0.81	51	2.39	45	4.35	33	1.61	19	1.20[a]	13
ECE–East	0.94	58	1.97	37	4.05	31	3.63	42	5.00[a]	53
Other	0.21	13	0.99	19	4.79	36	3.24	38	3.30[a]	35
East Europe										
World	18.52		51.24		86.65		82.64		103.00	
ECE–West	5.33	29	18.51	36	28.64	33	21.83	26	29.64	29
ECE–East	11.71	63	28.56	56	46.83	54	51.46	63	62.91	61
Other	1.48	8	4.17	8	11.18	13	9.35	6	10.44	10

a) Preliminary data

Economic Survey for Europe 1988–9.

Table 21.3 Convertible currency debt of Eastern Europe and the Soviet Union, 1989

	Net debt $ bn	Net interest payments as % of exports to market economies	Net debt as % of exports to market economies
Bulgaria	7.7	24.2	321
Czechoslovakia	3.1	5.1	51
Former East Germany	11.0	11.2	115
Hungary	19.5	24.6	342
Poland	36.5	41.7	456
Romania	0.1	0.6	–1
East Europe	77.7	17.7	204
Soviet Union	36.49[a]	4.9	72
Eastern Europe[a] and Soviet Union	114.1	10.7	136

[a] Includes debt of CMEA Banks

Source: UN/ECE, 1990.

of political opposition outside the party apparatus – first the dissidents and the Samizdat movements and then the emergence of independent citizens' initiatives and social movements, such as Solidarity in Poland. The reversion to austerity and autarky was an attempt to crack down on these new forms of opposition. By 1989 all avenues of reform, save democracy, had been exhausted. Stagnating economic growth, austerity, indebtedness, and environmental disaster compounded deep, bitter and long-standing political and social grievances. As soon as Gorbachev indicated that he would not support them militarily, the regimes toppled one by one.

Current prospects for East-Central Europe

Except in the former East Germany where West German political parties took over political life, the main political divisions in East-Central Europe cannot easily be translated into Western political language. They have to be understood in terms of the historical differences between East and West and within the East-Central European region.

The main political fault line lies between nationalist conservative tendencies and liberal democratic cosmopolitan tendencies (Michnik, 1990). Nationalist groups are often Christian with exclusivist elements – anti-semitic (Hungary and Poland), anti-Turkish (Bulgaria), anti-Hungarian (Romania and Slovakia), anti-gipsy, etc. Democratic liberal groups often include substantial minority groups.

Poland and Czechoslovakia are characterized by mass liberation movements, Solidarity and Civic Forum, which organised the opposition and represent a broad political coalition of those opposed to the former system. Within Civic Forum, liberal democratic tendencies predominate. Solidarity is now divided between the populist Walesa wing and the new liberal democratic party formed by former Solidarity intellectuals like Michnik and Geremek. In Hungary, the ruling coalition (Democratic Forum, Christian Democrats, Smallholders) is conservative/nationalist/Christian while the opposition, the Free Democrats and FIDESZ, are liberal democrats.

In Romania and Bulgaria the most extreme nationalist/democratic divisions are to be observed. In Romania, Vatra Rumaenesca is a nationalist, semi-fascist party with links to the former Securitate, and an extremist anti-Hungarian policy. In Bulgaria, the nationalist communist state apparatus is still largely in place.

All of these groups favour a transition to a market economy. If anything, democratic liberal groups are more strongly in favour of capitalism and the market economy than national conservative groups, although democratic liberal groups also include social democratic and green tendencies. National, conservative groups often have paternalistic social policies. They argue that women should be encouraged to stay at home and produce more Hungarians, Romanians, Poles etc. This would ease the expected burden of unemployment and reduce the cost of child-care provisions!

So far, the most rapid transition to a market economy is taking place in Poland and the former East Germany. Poland adopted a programme of 'capitalist shock therapy' on 1 January 1990. All prices were freed, the zloty was made partially convertible, and a tight monetary and fiscal policy was introduced. Import quotas were eliminated and import duties reduced. Wage rates were, however, still controlled so as to lessen unemployment and prevent cost-push inflationary tendencies.

An even more rapid transition took place in what was the GDR after the currency union with West Germany, that came into effect on 1 July 1990. Although the former East Germans are receiving considerable assistance, the costs of economic union are much greater than anticipated. By the end of 1990, unemployment was expected to reach 30 per cent.

Hungary and Czechoslovakia seem likely to adopt similar market approaches although at a slower pace. Hungarians have more to lose, and the new Hungarian government does not enjoy widespread popular support. Bulgaria and Romania are likely to proceed even more slowly.

What does marketisation mean? It means:

- The removal of physical or administrative controls over the distribution of goods; the allocation of labour; the supply of capital; and foreign trade.
- The freeing of prices and wages, which were formerly centrally controlled. It also means the introduction of convertibility so that the structure of domestic prices can conform to the structure of world prices. Because the freeing of prices and wages is likely to lead to very high rates of inflation, marketisation, in this sense, has been accompanied by drastic macro-economic stabilisation programmes to reduce budget deficits and to soak up forced savings.
- Privatisation. Privatisation is fraught with difficulties in countries where there is no private capital, except fortunes accumulated in irregular ways often by former nomenklatura or by mafia elements, where there are no financial institutions such as stock exchanges, and where property claims dating back to before World War II have become political *causes célèbres*.

The transition to the market is important for political reasons. It creates political space in societies where the state had controlled every aspect of social life. It would destroy the power of the nomenklatura (provided such people do not succeed in transforming themselves into the new bourgoisie). Also, it is difficult to see how else to deal with the massive problems of shortage, increasingly open inflation which rapidly erodes fixed incomes, monetary overhang (forced savings which result from shortages) and the monopoly power of huge inefficient Soviet-style factories.

Transformation to a market economy can proceed with various measures at varying speed; and some procedures currently employed can be criticised. But whatever techniques are adopted to manage the transition, the central question is whether markets can actually generate technological innovation, new investment and ultimately economic renewal. Obviously the techniques make a difference; they influence how painful the transition will be and the opportunities for new investment and innovation. But whether domestic and international competition will be creative or destructive, whether Eastern Europe will find itself caught in the vicious circle of uneven development that characterises many Third World market economies or whether it will catch up with Western Europe depends on the extent of technological and social transformation.

East-Central European economies already have some experience with markets. The so-called 'Second Economy' or informal sector, which includes both legal and illegal activities, is very large in some countries, especially Hungary and Poland. In these two countries the so-called non-socialised sector employed 44 per cent and 29 per cent of the active population respectively in 1986 even before the major changes (Dallago, 1988). The second economy, however, emerged out of the interstices of the first economy – to fill gaps, correct mistakes, and exploit loopholes in the first economy. Some parts of the second economy were very successful, such as private agricultural plots (although their success was partly dependent on the support of agricultural cooperatives or state farms with which they coexisted), small cafes, restaurants or repair shops or even small independent manufacturers, for example, software companies in Hungary. But often the second economy bred on the shortcomings of the first economy, for example, black-market currency dealers, profiteering cooperatives bypassing low prices (and long queues) in state shops, lorries taking vegetables to market when the state distribution broke down. The second economy has often fostered skills of cunning, deviousness and corruption rather than entrepreneurship, and consequently private enterprise often has a bad name in Eastern Europe. The second economy has also fostered inequality between those who work in it and the first economy.

If markets are to be successful in generating intensive growth, state structures have to be reduced and transformed, other non-governmental organisations – enterprises, trades union, universities, laboratories – have to take responsibility for investment and innovation, and a material infrastructure has to be created which includes market institutions (legal guarantees, banks and lending institutions etc), efficient communication and transport, new systems of technical innovation etc. Up to now, the new East-Central European governments seem to be pinning their hopes on foreign capital. What is important is to generate indigeneous sources of innovation, investment and enterprise.

What could governments do to encourage this process? In addition to macro-economic policies, much more attention has to be paid to other aspects of policy. These include:

- The composition of public expenditure. Even if overall expenditure has to be cut, there is some room for manoeuvre in shifting expenditure from state investment, military spending and debt servicing to infrastucture (telecommunications, transport, etc), education, including retraining, science and technology policy, social policy and environmental spending, particularly energy conservation. The 'push' which results from subjecting traditional industries to competition has to be accompanied by a 'pull' of new demands for appropriate restructuring.
- Privatisation. The term 'privatisation' has a different connotation from that in the West. It means reducing central control over production. What replaces centralised state control is still a matter for debate. Are state assets to be sold off at bargain prices to foreign buyers? Are shares to be issued and if so how? Are citizens to get credit to buy shares? Are state enterprises merely to be decentralised and given more autonomy? Are cooperatives, worker-managed enterprises or municipal enterprises to be encouraged? On the whole, Western commentators are rather dismissive of new forms of enterprise. The experience of Yugoslavia is cited to demonstrate the inadequacy of worker self-management schemes. It is said to be safer to copy the tried and tested methods of the West. The problem is that everything is an experiment. In countries where there is no entrepreneurial tradition, it is not evident that new self-made entrepreneurs or old-style managers will be any more or less efficient in managing factories than workers. Likewise, it is not evident that foreign investors will share local concerns. What is important is to permit diversity and experiment and to find local solutions that satisfy as many of the people involved as possible, through extensive consultation at a local as well as national level.
- Municipal and regional development strategies. Given that state policies remain important, they need to be decentralised. Support to new or existing enterprises in the form of credit, marketing expertise, technical assistance or governmental purchases can be better provided and assessed in the light of community needs at a local level. Furthermore, it is local communities that experience the costs of restructuring.
- Technology policies. East-Central Europe has to utilise its own pool of scientists and technologists in the service of socio-economic and ecological development. This involves the restructuring of R & D institutions so that innovative capabilities are harnessed both to commercial markets and to social and environmental need. Undoubtedly, imported technology will be required, especially information technologies. But if the import of foreign technology is to be effectively assimilated and diffused, it needs to be selected and applied by local technical experts who understand local requirements and local culture. The socio-economic experience of East-Central Europe is so different from that of Western Europe that Western technological solutions can probably not be imported without considerable adaptation.

The kind of policies that are adopted depend primarily but not only on whether social energy can be mobilized for this restructuring process. Given the budgetary constraints and the level of indebtedness, the role of foreign donors, particularly Western Europe, will also be important; so will contacts at all levels of society in order to strengthen democratic internationalist tendencies.

IV The role of Western Europe

In one sense, the 17 million people of the former East Germany enjoy a privileged position within Eastern Europe. Through unification with West Germany they will be rapidly integrated into the European Community. Through monetary union, emergency assistance to buttress the effects of monetary union and eventually through a common taxation and benefit system one can expect a continuing and automatic transfer of resources from West to East Germany. Even if the transition period is extremely painful, and there are likely to be income differentials for many years to come, they will eventually achieve what other East Europeans dream of – a Western consumer society. Whether this fairytale wish will turn out to be what they really wanted is another question. In theory, the cost of transferring resources from West to East has to be borne by West German citizens. In practice, assistance to the former East Germany is likely to stimulate economic growth in West Germany and to eliminate excess capacity provided the German Central Bank does not react with a big rise in interest rates. Germany still has a substantial trade surplus even though the unemployment rate has fallen in the West.

The kind of integration experienced in the two Germanies is unlikely to be repeated in the rest of Europe; it is simply too expensive and lacks the cultural cement. At present Western European governments, including Germany, are arguing that economic assistance should be predicated on IMF stablilisation programmes and assessed on the basis of commercial criteria. The reason why credit in the 1970s merely contributed to indebtedness was because the increased availability of foreign resources was not accompanied by internal reform. The appropriate internal reform, so goes the argument, is the adoption of a market economy. Hence economic assistance should be tied to market orientated reforms.

If this approach persists, the likely outcome is a new economic division of Europe along the Oder–Neisse line. The insistence on IMF-stabilisation policies and on the short term profitability of foreign investment is likely to mean that the market has more destructive than creative effects, however hard governments try to offset these effects. The burden of debt repayment, the consequence of austerity, unemployment and inequalities could well strengthen nationalist elements. In these circumstances East-Central Europe could tend more towards the Balkan pattern than the former East German pattern. Pressure for labour migration from East to West may once again raise barriers to travel. Already it is much harder for East Europeans to obtain visas to visit the West than the other way round.

The failure of Western credit in the 1970s was not just due to the absence of markets (although those countries which had introduced market reforms were somewhat more successful, for example, Hungary). The problem was the structures of production and consumption and the inability to shift these structures owing to political and bureaucratic inertia. To achieve this now requires, given the new political and social openings, policies of the kind described above as well as market oriented reforms. Economic assistance has to provide a social and technical framework for restructuring education, environmental issues, infrastructure, etc and by involving local groups and institutions in the decision-making process concerning choice of projects.

Over the period 1948–51, Marshall Aid provided by the United States amounted to some $12.4 bn, which was equivalent to just over 1 per cent of United States GNP. In 1949, it was actually as high as 2.4 per cent of United States GNP (UN ECE, 1990). Moreover, the establishment of NATO and the deployment of American troops to Europe provided a mechanism for a continuing transfer of resources from the United States to Western Europe. The direct foreign exchange costs of stationing troops in Europe amounted to some $3 bn a year, or approximately 1 per cent of American GNP (USA, 1989). Although Marshall Aid is often regarded as the trigger for Western European recovery, at least as important was the continuing flow of dollars which resulted from the presence of American troops in Europe.

East-Central Europe now needs a continuing mechanism for overcoming the ECU and Yen shortage. Moreover, the foreign exchange shortage is getting more acute with the rise in oil prices and the break-up of the CEMA. It needs assistance, not credit, that is, grants or debt relief. To some extent debt relief proposals are being put forward. Poland has achieved a one-year moratorium on debt repayments as a result of the IMF stabilisation programme. There are also proposals for converting part of the debt to environmental assistance (debt-nature swaps). In addition to debt relief schemes, Jacques Delors has suggested that ECU14bn should be provided over the next five to ten years and a further ECU5bn from the European Investment Bank. This is substantial but not as great as Marshall Aid, since it amounts to about .45 per cent of Community income. It is estimated that the former East Germany will receive in excess of DM100bn a year and other East European countries may also receive additional German assistance either in compensation for lost markets (Czechoslovakia) or for the fears aroused by German membership in NATO (Poland and the Soviet Union).

But even if resources are forthcoming on a substantial scale, will they be sufficient to generate self-sustained (not self-sufficient) development? Western Europe after the war differed from present-day Eastern Europe in two important respects. First of all, wartime economic controls were dismantled rather slowly. The European Payments Union (which was not so very different from the CEMA after the introduction of transferable roubles) continued to operate until the late 1950s. Second, the economies of Western Europe had long experience of successful capitalism – interrupted only for a relatively brief period during the war. Eastern Europe lacks most of the basic infrastructure and institutions of an advanced market economy and indeed, even before the period of communist rule, only Czechoslovakia and the former East Germany were culturally and materially comparable with Western economies. Financial assistance could mitigate the hardships associated with the transition to a market economy. But it will not necessarily stimulate private investment and indigeneous innovation. Indeed there is a danger that financial assistance could conceal the costs of neo-liberal policies while not stimulating the construction of new productive capacity. In much the same way North Sea oil revenues concealed the costs of the deindustrialisation of Britain which occurred in the early years of Mrs Thatcher's government.

What this suggests is that Western financial assistance should not be tied to excessively rapid market-orientated reforms, but to concrete technical projects – telecommunications, transport, environmental projects, new energy conservation technologies, new mining technologies, more efficient coal-fired power stations and so on. A British consultancy firm, for example, proposes that East Europe's scrap metal could be used for the construction of mini-mills – small electric arc furnace steel plants (Leadbeater, 1990).

Technical projects of this kind can be financed under the various programmes proposed by the Group of 24, especially the new European Bank for Reconstruction and Development in London, the European Investment Bank and bilateral agreements with the EC. Existing bilateral agreements with the EC typically call for granting of mutual MFN (Most Favoured Nation) treatment, the abolition of EC quantitative restrictions on certain categories of East European exports, improved access to EC markets, co-operation on the environment, science and technology, industry and transport. The question is how, under existing arrangements, technical projects will be selected. There is a risk that Western preconceptions of the kind of technical assistance required to construct a Western-type consumer economy will be imposed on Eastern Europe. It is important that projects are worked out together with local experts and local communities and that a variety of aid-giving institutions be established so as to encourage experiment, in what is, more or less, uncharted economic territory. For example, a new all-European environmental fund might be established to support environmental projects of benefit to both East and West – cleaner, more efficient engines for Eastern European cars, an all-European programme for eliminating acid rain and so on. Similarly, within the framework of the CSCE process, a fund might be established to support local efforts at conversion, that is, the shift from military to civil production; the aim would be simultaneously to reduce political resistance to cuts in military spending and to make use for productive purposes of the skills and technological capabilities tied up in the arms industry.

Although assistance of this kind could benefit both Eastern and Western Europe and should be determined jointly by both Eastern and Western Europeans, finance would have to be provided mainly by Western Europe, either the European Community or individual Western European countries. Given that it is likely to take years to build competitive industries, to establish market institutions, to clean up Eastern Europe's air, land and water, and given that those vulnerable to the social and economic dislocation involved in the process of transformation need to be protected both to sustain democracy and lessen the pressure of East–West migration, it might be necessary for Western Europe to provide some ECU50bn a year, that is, some 1 per cent of Community income, over a long period. (Proportionate to the population, the cost of German reunification and immediate exposure to international competition is, of course, much higher than this.) Why should Western Europe pay such sums?

The answer is twofold. First, the opening up of East-Central Europe will greatly expand Western European markets and provide the kind of stimulus that is required to correct global imbalances without recession. Second, Western European assistance could be considered a form of security which is much less expensive than military spending. Western Europe has developed within the framework of NATO with a permanent enemy and high levels of military spending especially in Britain and France. It is becoming evident that high levels of military spending do undermine competitiveness because of the way in which military spending absorbs and distorts technological resources (see Chapter 22). Nationalism and political instability could provide a continuing enemy of a sort. But a Europe without an enemy would be very liberating for the West. A Europe without an enemy is likely to be more responsive to social and environmental demands.

22 Defence

William Walker

I Introduction

Warfare has always been part of international conflict and technology a part of warfare. The 20th century is exceptional only in the degree to which dominant states have defined their relationships in military terms and have systematically developed and applied technology so as to gain or prevent others from gaining military supremacy. The territorial expansionism of Germany and Japan in the first half of the century, resistance to it by the established powers, and the confrontation between East and West after 1945 involved the marshalling of technological resources in the service of war on a scale that had not previously been experienced.

This stress on military aspects of international relations now seems to be on the wane within the East–West context. For the time being the international order in Europe is being defined more in terms of economic relations and the relative abilities of nations and groups of nations to apply technology for civilian purposes. It is symptomatic that negotiations between Japan and the United States over trade imbalances, and within Europe over the processes of economic integration, are now imbued with greater significance than the nuclear arms control discussions between the United States and the Soviet Union.

This does not mean that military power has lost its relevance. The huge assembly of forces and subsequent war in the Gulf following Iraq's annexation of Kuwait, the civil wars in Cambodia, Liberia and Sri Lanka, and the dispute between Pakistan and India over Kashmir are reminders that recent happenings in Europe do not signal a general end to military confrontation. Furthermore, the Gulf crisis contained worrying signs that new axes of conflict may be emerging-Christianity against Islam, Arab against non-Arab, North against South – which could again involve the mobilisation of huge military forces. It also highlighted the problems that lie ahead in coming to grips with the proliferation of weapons of mass destruction, linked to the acquisition of long-range missiles armed with chemical, nuclear or other high-explosive warheads.

The resource costs of these extra-European conflicts are nevertheless minor compared to the potential savings coming from the end of the Cold War. The extensive military industrial complexes that took root in the United States, the Soviet Union and Europe in the middle decades of the century therefore find themselves in retreat. They have in fact suffered a double loss of prestige. They can no longer claim so central a role in maintaining the security of nations, nor can they make strong claims to be the engines of economic progress. On the contrary, countries possessing them find themselves disadvantaged in the economic game-play which increasingly determines the structure and behaviour of the international system. Leadership in technology has been usurped by civil industries, and by countries which by choice or necessity have taken a back seat in the development of modern weaponry since World War II.

Changes in the balance of industrial power are in turn upsetting previous assumptions about the future location of military advantage. The United States and the Soviet Union are anxious that their military supremacy will also be undermined by these secular trends, since leadership in military technology could ultimately gravitate towards those possessing mastery of the broad technology base. For this reason, the United States is today evincing more anxiety about Japanese than about Soviet intentions where defence technology is concerned.

In Europe, as in the United States and the Soviet Union, the search for ways of adjusting the industrial base to the new realities has therefore begun. It is not a simple task. The loss of a coherent military threat in the East means that equipment 'needs' are less easy to define or agree upon; shifting resources into competitive civil markets, which in many cases already suffer from industrial over-supply, will not be straightforward; and new institutional mechanisms will have to be found, particularly at the level of the firm, for improving the diffusion of technologies between civil and defence activities – in both directions.

In a number of respects, Europe faces more difficult choices than the United States, given the need to redefine Europe's internal and external political relations and hence to decide the role that military forces will play in the new scheme of things. How this will affect the future size, structure and organisation of the European defence sector is hard to predict.

Among all the industrial areas covered in this book, the defence sector may thus face the greatest upheaval. This does not mean that there are no sources of stability: some trends in technology and market structure are strongly established, and the amount of inertial drag in the system should never be underestimated. Nevertheless, unless there are grave political setbacks in Europe, the 1990s seem bound to be a period of retrenchment and transformation.

II Characteristics of the post-war defence sector

In the modern industrial era, large parts of national economies have from time to time been turned over to war-production. The post-1945 period was historically unusual in that an extensive defence sector was maintained in peacetime. It developed a number of distinctive features:

- rapid technological innovation was a primary objective, causing a high proportion of the total costs of weaponry to be accounted for by R & D. In military terms, this was justified by the doctrine of the 'technology multiplier', whereby the Warsaw Pact's superiority in numbers would be countered by NATO's technological advantage; and by 'worst-case analysis' which required technological counter-measures to every conceivable future threat.
- stress was placed on the large-scale, centralised integration of technologies, from the level of military strategy (for example, the NATO doctrine of Flexible

Response) down to individual weapon systems. Large engineering firms acted as the main technology 'synthesizers', particularly in relation to the development and construction of weapon platforms (notably warships, fighter aircraft and tanks) and their associated weaponry and communication systems.

- transactions occurred within a largely autonomous regulatory framework, involving symbiotic relationships between industrial producers, R & D laboratories, defence ministries and the armed forces. Despite belonging to the same military alliance, countries lent a high degree of protection to their defence industries, thereby placing limits on the international division of labour, multinational investment and the diffusion of technology. Within countries, a high degree of tolerance was shown to monopolistic supply structures. Secrecy was endemic.
- the export market was highly politicised, involving inter-governmental bargaining at the highest level. For both the Soviet Union and the United States, defence exports formed part of their international power play. In Europe, the export proportion of defence output has varied widely across countries, with France much the most active arms trader, although Britain, Germany and Italy also became more prominent in the 1980s (see Table 22.1).

The defence sector was therefore granted special status within both national and international regulatory systems. Procurement policy was usually formulated independent of industrial policy and did not conform to the normal rules of the market economy. Moreover, defence transactions were excluded from the multilateral agreements overseeing international trade (notably the GATT and the Treaty of Rome). Trade in defence equipment and technology was either submitted to extremely tight international regulation (namely, the COCOM rules and London Suppliers' Guidelines governing West–East and nuclear trade respectively), or to no regulation at all, as demonstrated by the scramble to export advanced weaponry to the Middle East in the 1970s and 1980s.

An important consequence was that the major firms that made up the industry tended to develop distinct styles of behaviour (Kaldor, 1982). To an unusual degree, competitiveness was determined by firms' adeptness in handling relations with government bureaucracies. Partly for this reason these firms also often specialised in other public procurement markets such as power plant and telecommunications. By and large they were unaccustomed to competing in open international markets and were slow to adopt the multinational styles of operation that were emerging in other areas of production.

The defence industry is nevertheless very heterogeneous. Unlike most of the other sectors it does not rest upon a distinct technology base (Walker *et al.*, 1988). Instead, it is an amalgam of industrial capabilities, such as shipbuilding, aerospace, electronics and motor vehicles, and draws on the whole range of engineering skills; the firms within it vary widely in size and structure; and it encompasses mass, batch and unit production (for example, ammunition, tanks and aircraft carriers). It is thus characterised more by its market relations than by its technological attributes. This said, the aerospace and electronics industries have increasingly come to form the core of the defence sector, not least because they command the vital technological interfaces within and between land, sea and air systems. Moreover, the characteristic and most resource-intensive defence product is the complex weapon system.

Table 22.1 European defence production, trade and R & D, 1988[a]

	Defence spending (%GDP)	Equipment[b] procurement ($bn)	Defence R & D/ total govt. R&D (%)	Defence trade ($m)[c] Imports	Exports	Import penetration (%)[d]
United Kingdom	5.6	8.8	47.3	681	1,722	8
France	3.8	7.2	37.0	173	4,234	2
Germany	2.9	5.1	13.3	463	1,739	9
Italy	2.4	3.9	10.3	203	880	5
Spain	2.2	1.5	19.1	350	453	23
Belgium	2.8	0.51	0.6	321	254	63
Denmark	2.2	0.34	0.4	110	4	33
Netherlands	3.0	1.4	3.2	584	124	42
Greece	6.7	0.81	1.9	295	38	36
Ireland	1.2	0.04	n.a	13	-	33
Luxemburg	1.3	0.02	n.a	6	-	30
Portugal	3.7	0.14	n.a	75	138	52
EC Total	3.6	30.4		3,501	9,604	11
Norway	3.2	0.62	6.8	227	18	37
Sweden	2.7	0.96	n.a	90	185	9
Switzerland	1.8	0.65	n.a	522	287	80
United States	5.9	70.0	65.4	557	11,698	1

Notes: a Defence expenditure according to NATO definitions to give comparability. Note that they usually differ from expenditure figures recorded in national accounts. Trade statistics are taken from the United States Arms Control and Disarmament Agency (ACDA)
b Figures for France, Ireland, Luxemburg, Sweden and Switzerland are estimates
c Average 1983-87, constant 1987 prices
d Imports divided by procurement expenditure

Sources: IISS, 1990; ACDA, 1989; NATO's Sixteen Nations, 1990: OECD R & D statistics from unpublished sources

III The European defence sector

In contrast to earlier periods Europe has, since World War II, not been the main centre of innovation in defence technology. The pace has instead been set by the United States and the Soviet Union. For those European countries which assumed large military roles after the war, the task has been to keep abreast of the lead set by the dominant military industrial powers and to find ways of compensating for the smaller scale of their defence markets. Much technological development in the European context has been conducted within the frame of reference established by the United States.

Across Europe the scale, structure and technological orientation of defence industries has varied widely (see Table 22.1). Among the larger European countries diversity has reflected positions in the post-war political order rather than any distinctive advantage acquired through superior resources or capabilities. In particular, Britain and France maintained large defence sectors while the defeated powers, Germany and Italy, were initially constrained in what they could develop and produce.

Five rough categories of countries can be recognised:

- *Countries with comprehensive, largely autonomous defence capabilities, including nuclear capabilities* (Britain and France). This shows up in the high commitment of resources to defence R & D. The main differences between them is that Britain accepted dependence on American technology in some areas (for example, nuclear missiles) in order to limit expenditure, whereas France tried to achieve scale economies through a strategy of maximising exports (although Britain has recently tended to follow the same approach). France spends less on conventional weapons than Britain.
- *Countries with large but incomplete defence capabilities, with quite high but diminishing dependence on imports of American technology and equipment* (Germany and Italy, with Spain aspiring to join this group). Although not involved in nuclear weapon technology, Germany and Italy have gradually caught up with Britain and France across a wide area of technology and have become major defence exporters in their own rights. Due to political constraints on exporting directly to some regions, Germany has also placed emphasis on foreign investment and licensing (for example, in Argentina and Turkey) and on trading as a partner in international collaborations (for example, the sales of Tornado fighter aircraft to Saudi Arabia).
- *Small countries with sophisticated but highly specialised defence industries which import or manufacture under license the bulk of their defence equipment* (Belgium, Denmark, the Netherlands, Norway and Switzerland). All have exhibited a strong preference for American equipment; Table 22.1 shows that they have a high import propensity. Although their defence industries are small, they are nevertheless significant producers in niche markets: for instance, Belgium in small arms, the Netherlands in naval communications (Philips), Switzerland in guns (Oerlikon) and Norway in anti-ship missiles (Kongsberg).
- *Countries with weak industrial infrastructures* (Greece, Portugal, Turkey). They have tried to acquire defence industrial capabilities, both to gain greater independence and to encourage technological learning in the wider economy. They nevertheless remained highly dependent on foreign technology. Greece and Turkey are also major importers, and have large per capita military commitments.
- *Sweden, which is sui generis*. It has attempted to retain autonomy in the provision of main weapon systems through a policy of imitation and exploitation of foreign

technology – including the operation of a sophisticated 'technology intelligence system' (Hagelin, 1989). Design and assembly have been kept in Swedish hands, while many components and sub-systems (or parts thereof) have been imported.

For much of the post-war period there has been little trade in defence products among these European countries. The main commodity flows have been inward from the United States and outward to the Third World. During the 1960s and 1970s, however, a form of integration in both R & D and production developed through the mechanism of the collaborative project. This involved the pre-arranged allocation of R & D, production and market shares, in line with the financial contributions of the involved countries. A greater European division of labour was not, by and large, the objective. On the contrary, the collaborative project can be seen as a device for maintaining or developing broad *national* capabilities while trying to spread the costs of independence and to achieve some commonalty in military equipment within NATO. As such, it did little to stem duplication and over-capacity within the European industry.

The hub of these international collaborations has been Germany. Ties between the British and French defence industries have been few and intermittent. Instead, each has sought partnerships with Germany, and where possible with Italy and the other NATO countries, in order to widen their markets and exclude American suppliers. On Germany's part, collaboration has allowed capabilities to be established in a number of areas where independent programmes could not be mounted for political reasons. By the 1980s the German defence industry matched its British and French counterparts in sophistication. Since Germany was becoming as reluctant as Britain and France to cede design leadership, collaborative projects became increasingly difficult to organise.

IV Pressures for change during the 1980s

Even before the dismantling of the Iron Curtain in 1989, policies towards defence procurement in Western Europe had come under criticism. Pressures for change came from four main directions.

First, civilian industries had begun to usurp the lead in technology that had formerly resided in the defence sector. The revolution in information technology which followed the introduction of micro-processors in the 1970s, and the resulting emergence of mass markets for sophisticated electronic products and of new approaches to the organization of production, were phenomena belonging substantially to the civilian domain – even if many of the seeds had initially been sown in the defence sector. Moreover, the greatest mastery of the new technologies was exhibited by Japan which had been little involved in defence production since World War II. Defence industries moved from being leaders to laggards with problems assimilating the new technologies. Instead of reducing costs, technical change was often inflationary, not least due to the difficulty of controlling increases in complexity in an environment where the first instinct was to build yet more performance into weapon systems and where the risks of market failure were largely carried by governments. Somehow, the efficiency of defence R & D and production had to be significantly improved.

Second, the pressures to achieve a more thorough integration of European defence industries grew during the decade, partly for reasons of cost and partly out of a desire to reduce dependence on the United States. In the mid-1980s, defence budgets in all European countries (except France) began to be squeezed. Moreover, the aggressive launching of new technological initiatives by the United States (Star Wars being the most famous example) highlighted the costs of developing what could have turned out to be a new generation of defence technology. Coincidentally, concerns were being expressed about the effects of the American budget deficit. As purse strings tightened in the United States, Western Europe might soon have to assume greater responsibility for its defence against the Eastern threat. At the same time, it could face severe competition from an American defence industry which would become hungry for exports as the Reagan defence boom came to an end. All these anxieties were expressed in the Vredeling Report of 1986 which advocated a closer integration of European defence markets (IEPG, 1986).

Third, the Single European Act (SEA) and the general threat of American and Japanese competition set in motion a process of industrial integration that the defence industries could not ignore. High-technology firms with defence interests were becoming involved in new alliances and mergers across European boundaries. But the SEA also embodied liberal economic principles – it looked forward to open markets, an end to national procurement and the elimination of subsidies. The SEA therefore increasingly made the defence sector look like the last bastion of national protectionism and thus implicitly challenged the legitimacy of the manner in which transactions were handled within it. The SEA also aroused fears that the defence sector would become even more backward if it failed to follow suit by embracing the ideal of an open European market.

And fourth, the world export market for military equipment began to shrink in the second half of the 1980s. While Britain and Germany benefited from Saudi Arabia's huge expenditure on a new air defence system (the Al Yamamah deal), France and the Mediterranean countries suffered from a sharp downturn in Third World ordering, growing competition from new suppliers (notably Brazil, China and Israel) and the end of the Iran–Iraq war. Following the Gulf War, it is expected that much greater constraints will be placed on North–South arms trade, and thus on European defence exports.

V The attempted reform of procurement relations

The second half of the 1980s therefore brought a series of attempts to change industrial structures and market relations. The reform of procurement policies began in Britain in 1984 with the implementation of the 'Levene reforms' (Moray Stewart, 1988). Their ambition was to increase competition among defence contractors and make them bear more of the risks of failure, notably by reducing the incidence of cost-plus contracts. While no other European country approached the task with such zeal, all grappled with the problem of making the defence industries behave 'as if' they were operating in a commercial environment.

These domestic reforms were soon followed by the Independent European Programme Group's (IEPG's) announcement in November 1988 of its 'Action Plan on a stepwise development of a European Armaments Market'. The IEPG was formed in 1976. Its membership comprises the 13 European countries in NATO (including Turkey but excluding Iceland), although IEPG is not formally a part of NATO. From the outset, its remit has been to promote European cooperation in defence R & D and production, although its role had hitherto been minor. The Action Plan had four main aims: to open domestic markets to competition from other European defence contractors; to encourage technology transfer among European countries; to support a common defence R & D programme; and to provide assistance for 'developing defence industries' in the less advanced European countries (Greece, Portugal and Turkey).

While couched in general terms, the Action Plan implied two important changes in industrial orientation. One was the attempt to overcome the historic separation of the French and British defence sectors which had recently been highlighted by the failure to integrate their respective fighter aircraft development programmes (Rafale and the European Fighter Aircraft). The Action Plan had been preceded in 1987 by an Anglo–French reciprocal purchasing agreement. The other was the attempt to tie the more peripheral European industries (notably those in the third and fourth categories identified above) into European industrial arrangements, thereby weakening their links with the United States. Throughout, the British and French governments have been most enthusiastic supporters of the Action Plan, even if their objectives have been rather different. Thus while Britain has emphasized the importance of 'free trade', France has shown more concern over the condition of European technological capabilities and has insisted that the opening of markets should be accompanied by a strong cooperative R & D programme (EUCLID). Other governments have been concerned lest the Action Plan replace dependence on the United States by dependence on British, French or German suppliers but went along in the broader interest of European integration.

The unified European armaments market existed in only embryonic form by the end of the 1980s. While its intent was similar, the Action Plan was very different from the Single European Act. It had no legal basis and was thus not binding on member countries. With only the small IEPG secretariat to oversee it, responsibility for its execution remained largely in the hands of national defence ministries which retained complete discretion. Certain areas of industrial activity, such as those relating to nuclear weaponry and cryptography, remained outside its purview; it lacked the decisive deadline of 1992; it recognized the need for intervention, via *juste retour*, to ensure a balanced flow of trade; and it had no central funds at its disposal. It was thus less a formal regulatory structure, backed by law and with recourse to sanctions, than a loose set of intentions and obligations. The right of access of the United States and other non-European producers to the enlarged market also remained unclear.

VI Industrial realignments

The changes in industrial structure were more dramatic. Between 1985 and 1989, a wave of industrial transformation occurred within and across European boundaries, involving new alliances, cross-shareholdings and takeovers. Notable among them were the following: the GEC/Siemens takeover of Plessey, and GEC's subsequent absorption of Ferranti; Daimler-Benz's absorption of AEG (electronics), Dornier and MBB (aerospace), making it the dominant German defence contractor – its aerospace and electronics activities were brought together in a new company, Deutsche Aerospace; Thomson-CSF's acquisition of Philips's defence interests and of the Ferranti sonar division, and its formation of a joint company with British Aerospace to develop and manufacture guided weapons; the formation by Daimler-Benz and Aerospatiale of a joint helicopter company (Eurocopter); the acquisition by GEC and Daimler-Benz of shareholdings in the French missile and satellite producer, Matra, and the fusion by GEC-Marconi and Matra of their space technology interests; and the British Aerospace's takeover of the British Royal Ordnance factories.

A tightly knit inner circle of large defence contractors therefore began to take shape in this period. The key players were Daimler-Benz in Germany, Aerospatiale and Thomson-CSF in France (Dassault remained aloof from the realignments) and British Aerospace and GEC in Britain. Each had annual defence sales in 1988 in the region of £2–3.5bn, setting them apart from other European aerospace and electronics companies. It even looked as if two major trans-national and trans-sectoral axes might emerge, the one involving British Aerospace and Thomson-CSF, the other GEC and Aerospatiale, both involving links with Daimler-Benz and possibly Siemens.

Substantial industrial concentration therefore took place across the major European defence markets. The attempt to create an open European armament market helped legitimise these moves, while at the same time threatening to abort the competitive environment that it aimed to establish (Walker and Gummett, 1989). Be this as it may, Table 22.2 shows that the major European defence contractors were by the end of the decade beginning to match the scale and scope of their American counterparts.

VII The end of military confrontation in Europe

What happened to the defence sector in the mid- and late-1980s essentially amounted to an attempt to play the old game but according to modified rules and procedures. Changes occurred in the organisation of supply and in the ways suppliers and consumers (the defence ministries) handled their transactions. Furthermore, the boundaries around defence sectors were shifted to allow a gradual internationalisation of R & D and production. Reform, however, did not extend to the demand side – to the types and quantities of equipment being developed and purchased; it remained much as before, on the assumption that the East–West conflict would continue indefinitely. Indeed, the adjustments to procurement practices and industrial structures were in some degree motivated by the desire to sustain the given composition of demand in the face of higher costs.

Table 22.2 20 largest defence contractors in the United States and Japan, 1988

Company	Country	Industry*	Arms sales ($m)	%Total sales
McDonnell Douglas	USA	AC EL MI	8,500	56
Lockheed	USA	AC	8,400	79
General Dynamics	USA	AC EL MI SH	8,000	84
General Electric	USA	AC ENG	6,250	13
General Motors	USA	MV ENG EL MI	6,000	5
Raytheon	USA	EL MI	5,500	67
Rockwell	USA	AC EL MI	5,000	42
Boeing	USA	AC EL MI	4,500	27
Northrop	USA	AC	4,500	78
United Technologies	USA	AC EL MI	4,500	25
Martin Marietta	USA	MI	4,300	75
Mitsubishi Corporation	Japan	AC MV EL MI	3,100	5
Grumman	USA	AC EL	3,000	82
Litton Industries	USA	EL MI SH	2,920	60
Westinghouse	USA	EL MI SH	2,600	21
Unisys	USA	EL	2,500	25
Kawasaki Heavy Industry	Japan	AC ENG SH MI	2,230	37
Texas Instruments	USA	EL	2,150	34
IBM	USA	EL	2,100	4
Honeywell Marine	USA	EL SA	1,800	25

* Industry Code: A = artillery, AC = aircraft, EL = electronics, ENG = engines, MI = missiles, MV = military vehicles, SH = ships

Source: Extracted from Table 8.2 in Anthony *et al.*,1990a, with adjustments where there are obvious inaccuracies; the figures for arms sales should be treated as estimates

The end of the Cold War means that reform will henceforth have to encompass demand. The amounts and types of equipment deployed in Europe will be partly limited by international agreement through the Conventional Forces in Europe (CFE) negotiations. This and the cuts in defence expenditure expected during the 1990s mean that hard choices can no longer be evaded. At the same time, however, the events of 1989 have left governments and industries without a clear politico–strategic framework within which to make those choices. How is the future 'threat' to be defined and by which political entities (individual nations, NATO, the European Community ...)? How should that threat be translated into equipment requirements? What should be the balance between European and 'out of area' defence commitments, especially in view of lessons learned during the Gulf crisis? Where collective decisions are required, how can agreement be reached when there is no longer a dominant set of presumptions upon which to base the discourse? In short, the 'market' for military technology has lost many of its bearings. It could be many years before a reasonably stable set of guides to project selection emerges again. Recognising these uncertainties, we shall here identify some of the general pressures that are likely to impinge on the

Table 22.2 continued 20 largest defence contractors in Europe, 1988

Company	Country	Industry*	Arms sales ($m)	%Total sales
British Aerospace	UK	AC EL MI	5,470	54
Daimler Benz[a]	Germany	AC EL MI ENG MV	5,410	12
Thomson SA[b]	France	A EL	5,370	47
GEC[c]	UK	EL SH	5,090	41
Aerospatiale	France	AC	2,300	49
DCN[d]	France	SH	2,210	100
IRI	Italy	AC ENG EL SH	2,100	6
Dassault	France	AC EL	2,080	70
EFIM	Italy	AC MV EL	1,520	43
Fiat	Italy	ENG MV A	1,500	4
Rolls-Royce	UK	ENG	1,410	40
Siemens[c]	Germany	EL	1,240	4
Thorn-EMI	UK	EL	1,200	20
GIAT	France	A MV	1,150	100
CASA	Spain	AC	980	65
Oerlikon	Switzerland	AC A EL	930	32
Vickers SEL	UK	SH MV	830	100
Nobel	Sweden	A EL MI	810	23
SNECMA	France	AC	770	45
Hawker Siddeley	UK	AL ENG	680	20

Notes
[a]Includes MBB, acquired in 1989
[b]Includes Philips defence electronics, acquired in 1989
[c]Plessey, acquired by GEC and Siemens in 1988, divided equally between them
[d]Direction des Constructions Navales

defence sector and the kinds of responses that may result. Four aspects will be considered: the technology base, the composition of demand, industrial structures and the regulatory framework.

The civil technology base: the main strategic resource

Before the Gulf crisis erupted, predictions of a substantial decline in defence spending – a 'peace dividend' – were commonplace. Defence procurement budgets might fall, across the NATO countries, to as little as one-half their present levels in real terms by 2000. Over the same period, the scale of civil high-technology markets could come close to doubling. As a proportion of high-technology activity the defence market could therefore be one-quarter of its present size by the end of the century. Events in the Gulf mean that this should perhaps be treated as an outer limit. A fairer estimate is that defence's share of high-technology markets will have fallen to around one-half its present level by the end of the century.

By common consent, defence R & D today contributes less to the devellopment of the generic technology base than in the 1950s and 1960s. Given the expected rate of contraction, its role will be reduced further in the coming decade. As a result, defence technology will increasingly have to ride on the back of civil R & D, rather than the other way round. Furthermore, the defence industry's ability to set standards applicable elsewhere in the economy will diminish. This can already be seen, for instance, in relation to software, where NATO's efforts to have ADA adopted as the standard software language are hampered by industrial preference for languages developed for civil purposes.

The implication is that the civil technology base will increasingly become the primary strategic resource in the context of military as well as economic security. Countries that today have large commitments to defence R & D (particularly France, the United Kingdom and the United States) are therefore faced with a double vulnerability. A large proportion of their high-technology resources are at present allocated to a shrinking market which has become a relatively inefficient developer and user of technology. Unless the balance of resources is successfully shifted in favour of civil activities, a loss of industrial output and technological capability will result along with a weakening of firms previously propped up by defence spending. Countries with accumulated strength in civil markets could in time challenge the established military industrial powers, especially if they develop more effective institutional mechanisms for transferring technology between civil and military activities.

These concerns have surfaced particularly in the United States in relation to Japan (OTA, 1990). The American defence industry has become increasingly dependent on imports of Japanese high-technology components and materials. Despite the relatively low proportion of Japan's economic resources devoted to defence production, it is beginning to acquire a formidable defence capability based precisely on the strength of its civil industries. It is no longer claimed that Japan will remain weak in the defence area because it lacks the capability to develop complex weapon systems. On the contrary, command over component technologies such as VLSI, which are becoming central to systems design, provide Japan with a potential advantage in systems innovation. Experience in systems building is also being rapidly acquired in civilian markets such as telecommunications and by collaboration in aerospace projects such as the FSX fighter aircraft and the V-2500 jet engine.

These trends in the composition of the technology base have implications in Europe too. Civil R & D programmes such as ESPRIT and EUREKA are in an important sense more strategic than R & D programmes with specific military goals; and the latter will have to be drawn up with increasing reference to what is happening in the civil domain. The centre of gravity in the European defence industry will tend to shift towards those firms, regions and countries with the greatest command of civil technologies. One consequence is that Britain and France may find themselves ceding ground to Germany if they fail to invigorate their civil industries, irrespective of the stance that Germany takes on military security.

The importance of space programmes should also be noted in this context. With much talk of hypersonic vehicles, space stations and planetary missions, the pressures on governments to increase spending on space ventures are increasing. The growth of space markets provides an opportunity for firms with large defence commitments to diversify their activities and keep their design teams together. Space ventures are also promoted as important technology 'drivers', especially in the context of new materials. And space technology remains, particularly in countries aspiring to become Great Powers, a symbol of industrial virility. Overall, increased expenditure on space technology is claimed to bring long-term commercial and strategic opportunities, which are worth the financial outlays. Here again we find Germany and Japan growing in prominence. Both are beginning to define their own independent programmes and, where cooperation is involved, are becoming less willing to accept subordinate roles.

Thus recent trends in the relationship between civil and military technology have paradoxical effects. The 'secularisation' of defence technology coincides with an increasing tendency to regard civil technology as the main strategic resource, while it is also the possible springboard for the development of future military capabilities. This in turn raises questions about the degree to which countries or groups of countries will seek to erect protective barriers around their civil technology bases or will use military security as an excuse for so doing.

The composition of demand for defence products

During the post-war period large weapon and communications systems have come to absorb the major part of defence procurement budgets. Examples in Western Europe are Tornado and its putative successor, the European Fighter Aircraft; the British and French programmes to modernise their submarine-borne nuclear deterrents; and the various tank, warship and communications projects. The trend towards technological giganticism has been even more pronounced in the United States – the B-1 and Stealth Bombers, MX missiles, the Trident programme and Star Wars are but a few examples.

These programmes have a number of features. They have long lead times – the life cycle of a weapon system from initial conception to final obsolescence can exceed 40 years. Their design and project management are extremely complex, making them prone to serious cost overrun. And, once begun, they are hard to stop. With defence budgets on a downward path, the burden will become less and less acceptable. Commitments to large systems projects are therefore bound to come under scrutiny.

One option is to broaden international cooperation, notably by seeking joint programmes with the United States or Japan in order to share costs and enlarge the market. But as past European experience shows, such grand collaborations do not necessarily save resources. Given current American sensitivities, they would also probably run into political difficulties over property rights. The other response is to see whether military objectives can be met at lower cost by following other technological trajectories.

Criticism of large, centralised weapon systems goes back a long way. The advent of missiles with electronics-based detection and guidance technologies and the increase in their accuracy and destructiveness have profoundly challenged traditional weapon platforms such as surface ships and manned aircraft. The damage caused by Exocet and Stinger missiles in the Falkland and Afghan wars, and by guided weapons in the Gulf War, are reminders of how far the balance of advantage has shifted. But the 'creative destruction' of technology has here been much less decisive than in competitive civil markets. Traditional platforms often assimilate new weapons (for example, ships and aircraft becoming launch-pads for the cruise missiles that were meant to displace them), and innovations in offensive capabilities are countered by new defensive technologies (for example, tank hulls that are better able to withstand missile attack). The spiral of innovation and counter-innovation, within an environment predisposed towards conserving established arrays of equipment, has been central to the escalation in costs.

With the loss of a credible threat of invasion from the East and with budgets being squeezed, the armed forces are likely to come under more severe pressure to discriminate among technologies. Ironically, the effect could be to push military technology on to a more 'progressive' track, with cost savings being achieved alongside an improvement in military effectiveness. The possibilities discussed by defence analysts include the adoption of more decentralised force structures armed with 'smart' weapons and smaller platforms and taking advantage of the many advances in sensors, robotics, distributed computing and mobile communications (Barnaby and Ter Borg, 1986). The technologies of electronic warfare and surveillance are also becoming more prominent. The potency of these new technologies has been well demonstrated by the Gulf War.

The consequences for the defence sector of such changes would be profound, if very uncertain. They imply a still more decisive shift towards an electronics-based industry, particularly at the systems level; the displacement of industrial resources from hardware to software, and from technology systems that are lumpy and loosely connected, to those that are more diffuse but strongly interconnected through modern communications techniques. They raise questions about who will command the technological frontier in future – the existing defence contractors, or the predominantly civilian-based IT firms or some new configuration of industrial assets.

European industrial structures

Across Europe, a reduction of defence spending will not have great significance in macroeconomic terms. For NATO Europe it accounted for only 3.2 per cent of GDP in 1989. But a recent SIPRI study estimated that between 350,000 and 500,000 jobs in the defence and related industries could be lost in the European NATO countries over the six-year period 1990–5 (Anthony *et al.*, 1990b). While the adjustment may be painful in some contexts, it should be manageable especially if done against a background of industrial growth. In the aerospace sector, for instance, falling defence spending may be partly mitigated by the demand for new civil aircraft.

The countries facing the greatest difficulty but, by the same token the greatest opportunity, are undoubtedly Britain and France. They have the largest commitments to defence procurement, the structural imprint of defence spending is visible in many areas of their industrial economies and their civil industries are weak by German, Japanese and US standards. Where engineering is concerned, defence production has become their area of greatest international competitive advantage. In the British context, for instance, defence was the only market for engineering goods where the trade balance improved during the 1980s. There is therefore a strong instinct to hang on. This said, Britain and France stand to gain from the extensive resources that will be released and from the diminished status of defence production. While painful for a time, emancipation from being held in thrall by defence markets can only strengthen their industrial economies in the longer term – provided the resources are not simply squandered.

While Britain and France remain for the time being the largest players in the European defence industry, the initiative now seems to be shifting in Germany's direction, for a number of reasons. First, Germany has moved to centre stage in the debate about European security. Moreover its preference, driven by both internal and external factors, is for military commitments to be much reduced, certainly where Central Europe is concerned. It is thus prepared to countenance more radical steps towards de-militarisation than Britain and France. Second, as the country involved in the largest number of collaborative projects, Germany wields considerable influence over their fate and over that of those still on the drawing board or not yet in production. Among these, the Anglo–German–Italian–Spanish European Fighter Aircraft is the most significant. There are strong pressures in Germany to abandon the project. Third, the German high-technology industry, in particular, Daimler-Benz and Siemens, is larger and more robust than its British and French counterparts. Germany also has greater underlying strength in the supply of materials, components and sub-systems (such as marine diesels) so that its share of value added in international defence production is probably greater than measures of final output (that is, defence sales) suggest. And fourth, Germany seems more prepared than Britain and France to consider novel military strategies. During the 1980s in response to the misgivings over NATO war-fighting strategies in Central Europe there was a much more extensive debate in Germany than in other countries about alternative strategies and their technological implications. Germany may thus be politically better prepared for the technological shifts that the decline in defence spending may bring in its wake.

All this suggests that the relationship between Germany on the one hand and Britain and France on the other will be critical in deciding how the European defence sector will develop in the 1990s. The situation is paradoxical. The British and French military establishments are concerned about developments in Germany because they doubt its commitment to maintain defence spending at 'adequate' levels and its willingness to sanction defence exports and thus its reliability as a collaborative partner. At the same time, they are alarmed by the growing strength and assertiveness of German high-technology industries (civil and military) and what this might portend. While Britain and France may therefore try to bind Germany firmly into the European armaments market in order to foster interdependence, they also seem likely to take steps to increase the integration of their own defence sectors and attach other European countries to the Anglo–French axis.

Given pressure on resources and possible strains within the European defence sector the question also arises whether extra-European alliances will become more prominent. For the time being it is not expected that there will be much strengthening in trans-Atlantic cooperation. The American defence industry will be preoccupied with its own internal rationalisation in the face of spending cuts and is becoming anxious about the loss of technological property that collaboration can entail. In this regard, the joint American-Japanese FSX fighter aircraft project, which involves a substantial transfer of technology to Japan (controversial in the United States for that reason) may prove a landmark in American attitudes towards international collaboration (Samuels and Whipple, 1989).

For Europe, the relationship with Japan in the defence field is the unknown quantity. With regard to high-technology industries relevant to defence, the ties with Japan are already substantial and deepening: the heavy Japanese multinational investment in semiconductor production in Europe; the growing presence of Fujitsu in the European computer industry, especially after its takeover of ICL; the collaboration in aeroengines with Rolls-Royce and MTU; the Japanese shareholdings in Rolls-Royce, British Aerospace and other firms; and the grand technological alliance that was mooted by Daimler-Benz and Mitsubishi during 1990. The question is whether these developments will also lead to collaborative projects in the defence field, partly in response to growing strains between the United States and Japan in the area of high technology or whether European governments and firms will follow the United States in becoming nervous about Japanese intentions.

Regulatory structures: the role of the EC

Two broad conclusions can be drawn from the above discussion. First, the defence sector will become a less autonomous entity than in recent history; its technology base will be less distinguishable, and it will have to function more in line with normal practices in the market economy. Second, the process of internationalisation will continue. However, the balance between intra- and extra-European initiatives is hard to predict as is the way in which relations between the European centres of excellence will evolve.

In regulatory terms, the implication is that the authority of national defence ministries over industrial affairs will weaken. They will have to work more closely with, or in some contexts even defer to, those parts of government concerned with broad R & D and industrial policies (ACOST, 1989). Moreover, issues relating to defence procurement will increasingly have to be addressed in an international setting. The question is – which setting?

The IEPG has in the last few years been the main forum in which initiatives on the European defence market and on cooperative R & D have been taken. Arguably, however, the EC will have to accept greater responsibilities in this area and could even absorb the functions of the IEPG in time. Under the Treaty of Rome, the EC was given no jurisdiction over defence matters. The Single European Act, however, offers a pretext for future intervention. Article 30 states that Parties are 'ready to co-ordinate their positions more closely on the political and economic aspects of security' and are 'determined to maintain the technological and industrial conditions necessary for their security'. This helps legitimise the gradual shift in the EC's interpretation of its duties from purely economic and political integration to cooperation in the field of security, broadly defined (Tsakaloyannis, 1989).

The Gulf crisis appears to be hastening the search for new forms of military cooperation in Europe. While it is increasingly regarded as natural for the EC to acquire a military arm, this may still be a long time coming. In the medium term it is likely that the Western European Union (WEU) will take the lead, particularly in view of its recent prominence in coordinating Western Europe's response to Iraq's invasion of Kuwait. Nevertheless, the EC's involvement in defence affairs will still grow. As Rupp has written:

> no matter whether EC governments will make a deliberate decision to bestow on the European Commission security policy responsibilities or not; such responsibilities will accrue to the Commission by the nature of its obligations and day-to-day management of an economically and politically more united Europe [Rupp, 1990].

There are good reasons why this is likely to be the case. Civil and military R & D policies will need to be more closely harmonised than in the past. As the CEC is now the central sponsor of civil R & D in Europe, it is the main forum in which decisions are taken about the strategic technologies for Europe. Thus, it is already involved in defence-related R & D and finds itself increasingly drawn into dialogues with military authorities over the organisation and content of R & D programmes. Regular meetings are now held, for instance, between the CEC and IEPG to discuss R & D policy.

Furthermore, the CEC already has broad regulatory responsibilities that impinge on defence trade and procurement. The implementation of the SEA forces it to consider its stance on issues affecting the defence sector: which goods should be exempt from tariffs on the grounds that they are 'defence-related' (as defined in the out-dated annex to Article 223 in the Treaty of Rome)? How forcibly should the CEC's Competition Directorate intervene where actions taken by firms threaten to limit competition in defence and, by extension, high-technology markets? Should the defence sector be exempt from the EC's procurement rules? Could the right to insist on *juste retour* under the IEPG's Action Plan infect other areas if allowed to stand?

A further reason for anticipating deeper CEC involvement is that the IEPG may prove too weak an organisation to achieve open trading in the European market or to withstand the growing concentration of market power in the hands of large trans-European defence contractors. Already, strains are developing among the Action Plan's main protagonists over the amount of access that will in practice be granted to one another's defence markets. The powers of the CEC may increasingly be called upon to ensure fair play among industrial suppliers, notably to prevent governments staging bogus international competitions for defence contracts and to confront monopolistic practices.

Finally, it is plausible that, despite their concern over sovereignty, Britain, France and Germany will find it in their interest to increase the EC's responsibilities in this context for broader political reasons. France will be concerned to bind the burgeoning German defence industry into a strong European framework. The German government and industry for their parts may support developments at a European level in order to shield their defence industrial interests against domestic pressures to wind down military commitments. And Britain will be anxious to prevent a protectionist, French model of defence procurement from gaining ascendancy. Thus EC-supported market integration could simultaneously serve a range of interests – political interdependence, countering pacifism in the united Germany and maintaining some vestige of competition – even in the absence of external threats to European security.

IX Conclusion

Three questions are worth posing in conclusion.

Is there now a distinct 'European' defence industry? Yes and no. Cross-boundary linkages have increased significantly in recent years, and many firms have tried to acquire a European identity – hence the formation of companies with names like Eurodynamique and Eurocopter. However the instinct in each country to protect its industrial assets remains strong. At present, European industry is a federation more than a fusion of national industries and still a long way from achieving a Europe-wide division of labour. Nevertheless, a Rubicon has been crossed: national autarky is no longer sought, and the presumption is that Europe should in time gain its own common pool of defence technologies.

Is the European defence industry technologically progressive (in both commercial and military terms)? Again, yes and no. It commands most of the technologies relevant to modern warfare and has a significant presence in export markets. However, it is still essentially a follower rather than a leader in military technology. An important issue is thus whether Europe should try to come out from the shadow cast by the United States and define its own technological agenda. Innovation is not just a matter of the level of defence R & D: it depends on the choice of technological trajectory. In the long run, Europe's grasp of military technology will be a function of the dynamism of Europe's civil high-technology industries.

What and whose purpose does the European defence sector serve? This is the key question. The Europeanisation of the defence sector has not so far been accompanied by the emergence of a distinct European military policy nor of a European institution responsible for one. Under the aegis of the IEPG, European Community, the Western European Union and the CSCE the search has begun for new institutional forms, but it is likely to be a gradual evolution. In the meantime, NATO still provides the main but increasingly anachronistic framework for military strategy and consultation.

Behind all this lies the unresolved problem of Europe's future stance in the world order. If Europe decides to construct its own military capability outside the framework of NATO, it will have to approach the task with the greatest circumspection. This century has been scarred by the military excesses of the Great Powers. Despite the Gulf crisis and the instabilities in Eastern Europe and the Soviet Union, there is now an opportunity to reduce substantially the role that military institutions play in Europe's political development. At all costs, we must prevent a fresh eruption of the militaristic policies that have been so disastrous in Europe's history.

23 The policy agenda – challenges for the new Europe

Margaret Sharp and William Walker

I Introduction

In many respects Europe is fortunate. It is moderately well endowed with natural resources, it is not threatened with over-population, and within its boundaries it contains a rich array of political, economic and social cultures. At present, it shows every sign of renaissance. The combined effects of economic integration in the West and of political liberation in the East have given rise to considerable optimism.

The challenges facing Europe during the coming decade nevertheless remain formidable. A new Europe, embracing East and West, has rapidly to be constructed in both political and economic spheres if old divisions are not to reassert themselves. Within the international economy, Europe has to find its place alongside the United States and Japan, and, if it is to hold its own, find ways of matching the creativity and dynamism of these countries. And Europe shares with other regions the task of avoiding environmental destruction. Come what may, meeting these challenges will require great effort and good judgement.

What has technology to do with this? As the earlier chapters in this book have illustrated, technology spreads its wings widely. The renaissance in Western Europe is not, we have argued, unconnected with broad global developments in technology and Europe's response to the competitive challenge from the United States and Japan. But competition is not everything and, as the Introduction made clear, it is facile to suggest that the market mechanism is universally applicable. The problems facing Europe as it approaches the twenty-first century are difficult and complex. If this renaissance – this new confidence – is to be maintained (and indeed extended to the new Europe embracing East and West), four key policy objectives have to be met. These objectives are:

(i) *To maintain a dynamic and competitive industrial base at or near the technological frontier*. This is necessary so that Europe can remain a creative, high-income, resource-efficient region with a strong bargaining position in the international system. It requires addressing questions about weakness of Europe's ICT industries, the degree to which Europe needs to be self-reliant in science and technology and the respective roles of indigenous and foreign multinationals.

(ii) *To help protect and restore the environment.* The economist's practice of treating the environment as an 'externality' – an extra issue to be taken into account

– is no longer sufficient. The externality has to be internalised so that thought and action routinely lead to development paths that are sustainable and in tune with nature. Responsibility for leadership lies particularly with the advanced industrial nations, both in the development of new sustainable technologies and also in their global diffusion.

(iii) *To address concerns over widening disparities in living standards.* New technologies are tending to exacerbate the problems of core and periphery. Already there are disparities between the countries of (richer) North and (poorer) South, but to these will be added the problems of deindustrialisation in some older industrial regions, and particularly now in Eastern Europe. The industrial base of Europe needs to comprise more than Germany, yet, unless the rigour of the Single Market and economic and monetary union is offset by a substantive redistribution of activity via regional and social policy initiatives, there is serious risk of imbalance.

(iv) *To meet Europe's international responsibilities.* Western Europe already wields considerable influence within the international system; a new, united Europe, no longer reliant upon hegemonic ties to the superpowers, will have even greater influence. This needs to be used wisely and responsibly, not least by supporting the open international trading system which has contributed so much to Western Europe's post-war prosperity. Europe needs to find ways therefore to counter the growing tendencies towards protection and technological chauvinism both among its own member states and on the part of the United States and Japan.

The economic, political and cultural strengths of Europe depend on each of these challenges being met. Far from being inconsistent with one another, they should be regarded as mutually sustaining. A prosperous and competitive Europe will be better able to deal with its environment than one that is weak and impoverished, just as an internationalist Europe will be more secure than one that is protective and isolationist.

II Guiding principles

Inevitably, there have been gaps in this book. Important sectors, issues and regions (such as aerospace and textiles, labour relations and transport, agriculture and the problems of the poorer EC countries) have received little mention. We would nevertheless like to draw attention to a number of general conclusions that seem to us to emerge from these chapters and which point to broad general guidelines for policy.

Addressing the democratic deficit

In Chapter 21, Mary Kaldor emphasises the technological stagnation of Eastern Europe in recent decades. Economies can develop rapidly, if temporarily, under autocratic regimes, by exploiting known and well-tried technologies; but lack of accountability to consumers and electorates leads in the longer run to inefficiency, stagnation and cynicism. Democracy is also a significant issue for Western Europe. Most of the sectoral case studies in this book have noted the concentration of industrial power in the hands of a few giant firms and of executive power in the hands of the Commission in Brussels. The Single European Act has, if anything, accentuated these trends. It has encouraged firms to restructure to take advantage of the Single Market. Likewise, it has, at least at present, considerably increased the responsibilities of the Commission. It is now generally recognised that one of the major priorities of the European Community is to contrive mechanisms which ensure greater democratic accountability. Without them, there is danger that these institutions could, like those of Eastern Europe, become self-serving and inefficient.

Within our sphere of interest – science and technology – such concentrations of power could be particularly detrimental if they are not paralleled by increases in the breadth and sophistication of public debate. Too much intrusive regulation is also a danger, but to achieve a balance is a complex political and administrative problem, as Erik Millstone shows so clearly in his chapter (20) on the evolving regime of food regulation. However, the effects of technical change are now so far-reaching that many decisions need to be firmly grounded in the political domain. Parliament itself can frequently act as a surrogate for the consumer. If nothing else, fear of public criticism and litigation can act as a powerful spur to industry and its regulators to consider the consequences of their actions.

Competition, rivalry and regulation

Many chapters have shown that industries do not remain dynamic without competition and rivalry. Policies that support national or regional champions by sheltering them from international competition have not worked well in Europe in recent times. (They have a better track record for newly industrialising countries such as Korea.) Open markets allow the new to challenge the old and create the space within which institutional changes can occur. In this regard we endorse the view of the Cecchini Report that moves towards the creation of the Single Market and the spur to competition arising from it are beneficial for European industry and offer the potential for major improvements in performance.

In a number of sectors, semiconductors and telecommunications switching systems being prime examples, the massive R & D expenditures necessary to launch new products, combined with short product lives and the uncertainties inherent in developing new technologies, have nevertheless increased the importance of economies of scale and scope. It is foolish to pretend that the Single Market is itself sufficient to cope with such situations. The firms that are emerging from the process of restructuring and re-positioning are large and powerful corporations, wielding considerable influence at both national and Community levels. Such firms, like all multinationals, notoriously elude control.

The Single Market therefore needs to be supplemented. Competition and merger controls need to be strong and effective to prevent the abuse of positions of power, and in some cases, such as telecommunications, markets must be regulated to maintain competition and 'level playing fields'. There are other cases where markets have long proved incapable of taking care of essential goals such as environmental protection and consumer welfare. Indeed, they may hinder the attainment of those goals, as Surrey has argued in the context of energy security (Chapter 17). The balancing act between these various demands is not an easy one and we shall return to it when we consider the more specific policy issues in the final section.

Education and training

A number of chapters have emphasised the role of education and training as a vital factor underlying competence and performance in all fields of new technology. Increasingly it is recognised that the economies of Japan, and, within Europe, of countries such as Germany, Sweden and Switzerland, derive strength from their commitment to education and training at all levels.

Education has a dual function: it is a source of creativity, and determines the capacity to absorb and apply new knowledge – to learn; and by widening horizons and broadening understanding, it encourages ready adaptation and flexibility. The new technologies of the present time have greatly increased the importance of education. Industry has become more reliant on scientific knowledge. And far from causing the de-skilling that was once feared, flexible automation and analogous developments are raising skill requirements at all levels of production and distribution as the chapter on mechatronics (15) makes clear. The old distinctions between skilled and unskilled labour are breaking down: there are (or should be) only gradations and varieties of skill.

Education and training therefore become a key factor in the absorption and diffusion of new technologies. Within Europe, the excellence of the German-style training system has led to emulation elsewhere, but such systems often have deep cultural roots and cannot readily be copied and transposed. Britain's poor record in training is now recognised to underlie many of its failures in the engineering and other sectors (see for example the chapter on Construction (14)). Britain may be the worst case, but many other European countries still have some way to go before they match the German or Scandinavian provision for training. Moreover, Europe as a whole is still caught in its traditional, hierarchical mould of higher education. There is urgent need to raise general educational levels, in particular to create a more flexible and responsive system of higher education and to weld this to an extensive and high-quality training system on German or Japanese lines.

Long term investment in the science base and R & D

Many chapters have stressed the extensive investments that now have to be made in R & D and the links necessary between firm-based R & D and academic science.

Governments remain the principal financiers of basic science and, as Table 3.2 of Chapter 3 shows, European governments in general match, or even surpass, their competitors in such funding. Given the importance of the science base to new technologies, there is pressure on governments to increase their support. We support these pressures and endorse the view that the returns on investment in basic science, although long term and often intangible, can be substantial, especially where there are good mechanisms of knowledge transfer.

Margaret Sharp stresses in Chapter 4 the importance of the European Community's R & D programmes such as ESPRIT, EUREKA and RACE in stimulating R & D linkages between firms and between firms and academic science both within and across countries. The success of these programmes has led the Community to emphasise the need for good university/industry links and to increase its own contributions towards academic research. Indeed, the 1989 review of the Framework Programme urged the Community to do more to enrich its 'scientific humus', suggesting a switch of emphasis from firm-based R & D to academic research. While welcoming the extra funding such initiatives bring and the stimulus collaborative research provides for cross-fertilisation of ideas, there are grounds for discouraging too major a role by the Commission in funding basic research. Good science, like industry, benefits from competition. There are relatively few areas where equipment is so expensive that, as with CERN, it only makes sense to fund it collaboratively at a European level. In general, therefore, responsibility for basic science should properly remain at the national or regional level, limiting the Community's role primarily to that of promoting the interchange of personnel, ideas and knowledge.

National governments, however, are playing a lesser role in direct support for firm-based R & D, a role which may further diminish if commitments to defence fall away. Increasingly emphasis is focusing on the EC's own Framework Programme supporting pre-competitive research and the other major intergovernmental European programme – EUREKA. Besides creating new alliances and providing channels for diffusing knowledge, these programmes have also brought to European firms a common awareness of the task that lies ahead in meeting competition from the United States and Japan. They also increasingly pick up areas such as environmental R & D, where the corporate sector, left to itself, does not automatically provide.

Important though these programmes are, the key actor is the firm. A dynamic industrial base relies on commitments made by firms of all sizes – large, medium and small – towards technological innovation. While much has been written about the 'globalisation' of production, Pavitt and Patel have shown in Chapter 3 that the performance of multinational companies still rests heavily on the quality of their innovative activities and the supporting institutions within their home territories. One of the reasons for this is that success most frequently comes not from one-off innovations but from the deliberate, patient, cumulative development of technological resources. As a result countries such as Britain and the United States, which have leaned towards gaining quick returns from investment and shunned those with protracted and uncertain pay-offs, have over the long term suffered a loss of competitive advantage compared with countries such as Japan and Germany, whose financial institutions tend to be more supportive of long-term investments. As in education and training it is not easy to transpose systems. Nor should we be blind to the corporatist nature of such systems which, in Germany in particular, sometimes make for conservatism and slow response to new opportunities. Nevertheless, there is cause for concern that liberalisation of capital markets across Europe will lead to the progressive dominance of 'short termism' in European industry and we urge consideration of policies and mechanisms which support the longer view of corporate affairs.

Diversity, decentralisation and regional initiatives

We have already noted the degree to which the Single European Act has concentrated power in Brussels and on the Community institutions and how this will be enhanced by the moves towards economic and monetary union and a single currency. Likewise the concentration of industrial assets in pan-European firms during the 1980s in its turn increased pressures on the Community to act as the main industrial regulator, further concentrating power at the centre. The emergence of certain regions in Europe as dynamic instigators of economic activity has been an equally important, if less noticed, feature of the past two decades. In areas such as Baden-Württemberg, Grenoble and the Scottish Lowlands, actions to attract foreign investment, to forge linkages between higher education, industry and finance, and to encourage the formation of new firms, have brought renewed vitality. Some of the most innovative thinking about economic development, including the networking of small firms to address large markets, is now to be found in these regions. The essence of their success is not to be found in massive subsidies for investment but in more subtle mechanisms of institutional innovation, coordination and support.

Actions at both the centre and the periphery are also clearly needed in addressing environmental problems. Much of the progress in dealing with the environment has resulted from what might be called the 'social periphery' exerting pressure on the industrial and political centre. In Chapter 19, Boehmer-Christiansen and Skea note the existence of environmental policy cycles. The public and non-government organisations usually set the agenda in the early stages; thereafter it falls to governments, industries and regulators to devise and implement remedial measures. Boehmer-Christiansen and Skea also stress the importance of transparency and political pluralism in achieving a rapid response to environmental threats. In general, action within the decentralised and relatively open political structures found in Germany and Scandinavia has been swifter and more effective than in Britain and France where public administration is more centralised and closed.

III The policy agenda

In this final section we return to the policy agenda set out at the beginning of the chapter and focus on the policies which seem to us to be appropriate to begin to tackle the issues identified there. In each case there is one easy solution – to let the market decide. Such a solution would let firms or regions sink or swim in markets that are open to all comers; would trust in prices and consumer response to shift resources from activities which endanger the environment; would leave the redevelopment of Eastern Europe to local entrepreneurship and foreign investment. While market disciplines are essential, in none of these cases do they appear as a wholly viable response. But move away from this simple response and the answers become far more complex.

Remaining a strong player in global high-technology sectors

During much of the post-war period the United States has provided the main competitive challenge to Europe in high-technology markets. In the 1950s and 1960s its high productivity, large internal market and technological dynamism led the major European countries to rationalise their industries in the pursuit for scale economies. As a result monopoly or duopoly – national champions supported by public procurement and other non-tariff barriers – became the characteristic market structure in most high-technology sectors.

By the 1980s, these policies came to be regarded as unsustainable. As Margaret Sharp explains in Chapter 4, the force of Japanese and American competition, with firms now acting on a global scale, led European industrialists to realise that their firms would not survive unless they too operated globally, with Europe their base market. The result was the 'twin-track' approach of the Single European Act; on the one hand, the Framework Programme, encouraging technological cooperation across the EC countries; on the other, the Single Market with its emphasis on liberalisation, deregulation and an end to preferential treatment.

The response has been a rapid concentration of industrial assets in a few large companies in many industrial sectors. As the chapters in Part II have shown, there are now just a few significant European players in each of the electronics sectors – semiconductors, computers, telecommunications and consumer electronics. Yet in spite of both overt and covert protection, and extensive subsidies from both national governments and the Framework Programme, this trans-European structure, though outwardly more viable, still seems vulnerable as has recently been demonstrated by Philips' problems and ICL's takeover by Fujitsu.

The dilemma facing the policy-makers is that a pluralist, European-owned electronics industry may not survive in an open European market. Indeed the main beneficiaries of the Single Market in this sector may end up by being the American and Japanese producers whom the 'twin-track' policy was supposed to hold at bay. The presence of American multinationals in the European market is well established; that of the Japanese is increasing. Does increasing dependency on foreign-based firms for supplies in this vital sector matter? Does a policy of protection and support really enhance the competitiveness of European-based firms?

There is no simple or easy answer. The case for support and protection derives from three strands of thought – dependency, strategic trade theory and the traditional infant industry arguments. The dependency argument rests its case on the systemic nature of the ICT industries and the synergies between upstream and downstream activities; competence and a manufacturing presence upstream is required to operate effectively downstream. Strategic trade theory suggests that given the size and integrated operations of the multinational players in this sector, it might be possible for one firm deliberately to withhold supplies from another to gain competitive advantage and that the threat of potential protection or support is needed to modify this behaviour. Infant industry arguments emphasize the importance of dynamic economies of scale which make it difficult to re-establish, without support, a presence in state-of-the art technology once this has been lost. All these arguments can be used to justify a 'special case' for support for the European electronics industry and have been deployed in various forms by the different authors in Part II.

By contrast, the free-trade case rests on there being 'a level playing field' among international players. This is clearly not the case; all the major players operate in imperfect markets with varying levels of support and protection from their own governments. Moreover, as large multinationals they are in many senses a law unto themselves. Equally, as Cawson and Holmes argue forcibly in their chapter on consumer electronics, Philips and Thomson, though European based, are also large, integrated multinationals, some of whose main manufacturing activities are located in South-East Asia. What is the case for favouring them against Japanese producers who are establishing manufacturing bases and employment in Europe?

There is no 'right' solution to this problem. Circumstances will clearly vary from sector to sector and case to case. The best we can do is to lay down some general principles which we think should inform policy decisions.

(i) Europe's basic commitment should be against protection and in favour of maintaining an open international trading system. There are many arguments in favour of this stance, not least the need to maintain competition and rivalry among large, monopolistic firms.
(ii) Europe should welcome foreign investment – it helps to transfer technology, helps disperse industrial activity, provides new sources of finance, trains a new generation of management in fresh ideas and, through its demonstration effect, stimulates improved operating performance.
(iii) Equally, Europe needs to ensure that it captures these potentially beneficial spin-offs from multinational investment. Hence both American and Japanese multinationals establishing manufacturing facilities should be required to make commitments also to establish research, development and design laboratories, to source components locally and to train Europeans in both management and technical grades.
(iv) We should not rule out protection or support to European firms in specific cases, but where support is offered it must be qualified and provided on condition:

    (a) that the recipients demonstrate that they are using it to upgrade technology and manufacturing processes;
    (b) that it is progressively withdrawn according to a pre-arranged and monitored timetable.

(v) Persistent efforts should be made to achieve 'level playing fields' at the international as well as the European level. At present there is a real danger that the trade friction in high-technology areas between the United States and Japan could spill over to have an adverse effect on Europe's ability to compete. We shall return to this in discussing later Europe's international responsibilities.

Within Europe, emphasis should be put on developing 'diffusion oriented' policies which concentrate on developing a strong infrastructure for innovation. Of particular importance are policies which support education and training; which promote investment in the science base and schemes to promote technology transfer linking academic science to industry; which encourage new, small high-technology firms and more generally industrial investment in research, development and design.

Countering the 'natural' economic tendency to consume the environment

Historically, technical change has resulted in fewer resources being used per unit of economic output. This is well illustrated by comparisons between today's Eastern and Western European energy economies, the latter being several times more efficient than the former. However, pressures on the environment have still increased greatly along with the growth of incomes and output.

If these pressures are not to become even more acute, 'environment-saving' will have to become more deeply embedded in the personal behaviour and commercial routines which determine the allocation of income and investment and the directions in which products and processes are developed. In the face of rising real incomes and falling real costs that accompany economic progress, this requires that the 'natural' economic propensity to consume, or otherwise distort, the environment is overridden. The fundamental problem is that the economic actor is often unaware of, or cannot be held accountable for, the environmental harm caused by his or her decisions.

This is not a new problem, but one which presents itself with new intensity as the possibilities of global warming and other environmental dislocations become apparent. The chapters by Surrey, Berkhout, Boehmer-Christiansen and Skea, and Millstone suggest that the following are necessary but not sufficient conditions for combating environmental deterioriation:

(i) *Measurement and assessment* Improvements will depend heavily on understanding the causes and effects of environmental damage and the measures proposed to redress or forestall it. Techniques for measuring and analysing environmental impacts are improving greatly, although understanding will always be incomplete where long-term and highly complex mechanisms are involved as in the cases of nuclear-waste management and global warming. As Berkhout stresses in Chapter 18, solutions have to be negotiated and will only gain lasting legitimacy if submitted to broad debate. It also needs stressing that the 'information deficit' can operate at all levels of society: it is as important for the householder to understand how he or she uses energy as for the chemical manufacturer.

(ii) *Standards* Stricter and more progressive standards are required to regulate performance and practice in areas ranging from building design to motorway speed limits; from waste disposal to the use of pesticides. They are also useful in stimulating the development of new, clean, resource-efficient technologies if properly set and monitored. Increasingly, these standards are being devised and implemented at Community level. The next stage will be to seek global standards that ensure that competitive advantage cannot be gained through their evasion (not least by firms locating harmful processes outside the regions where the standards apply). While essential tools, regulatory authorities nevertheless have to guard against standards becoming too rigid and conservative and too encumbered with bureaucracy. Again, legitimacy depends upon public acceptance and public credibility.

(iii) *Incentives* Incentives of various kinds can encourage investment in environment-saving technologies, hastening the replacement of resource-inefficient parts of the capital stock and retrofitting where that is not practicable (as in housing). These investments need to target the most energy- and material-intensive activities and those that give rise to the most damaging pollutants. New ideas about taxation, involving for instance raising the prices of harmful products and processes, and pricing the use of roads in towns deserve a thorough airing. Existing R & D priorities also need to be reassessed. It seems wholly inappropriate, for instance, that spending by the EC on nuclear fusion R & D should dwarf that on renewable energy sources and conservation.

In the end, however, the most important objective must be to strengthen Europe's resource-saving culture, ensuring that conservation is practised routinely at every level of the economy and society. By comparison with North America and most other regions, this culture is already strong in parts of Europe. However, propensity to consume still too readily takes precedence over propensity to conserve.

While resource saving can involve economic sacrifices, particularly in the short term, we are confident that progress can be made and that it will bring clear commercial advantage to industries and countries that manage it successfully. The 'greening' of Germany has already given German firms an advantage in developing export markets and seems to have few serious cost disadvantages. Japan's current dominance in electronics, which stems in part from its strong commitment to energy-saving development paths after the crises in the 1970s, also inspires confidence in how much can be achieved. In general, the discipline of resource saving is an important source of engineering creativity since it focuses attention on the fundamentals of how energy and materials are consumed and transformed.

Coping with disparities between countries and regions

The major disparities in income and wealth that already exist within the EC reflect not only variations in relative economic development and/or endowments of natural resources but also the quirks of history and politics. Geography plays its part – hence the centre tends to be more prosperous than outlying countries and regions – but within the EC it is no longer necessarily true that town will be richer than country, industry than services, or for that matter North than South. Europe has always been a continent of diversity and gains much in economic and cultural strength from that diversity. But when the diversity is reflected in major and growing variations in prosperity, it can feed, not creativity, but alienation, dissidence and discontent. Within the present EC, the tensions and 'fault lines' are already there. The next decade will bring, on the one hand, the extra strains of the Single Market, on the other the need somehow to accommodate within the umbrella of the Community, the ex-communist countries of Eastern Europe.

The Single Market will if anything exacerbate the problems of core and periphery. Some chapters of this book have described how in the last decade emphasis has shifted from *national* programmes of support and intervention to 'the level playing field' issues of standards, regulations and control. Technology is at the centre of many of the issues, and the process of negotiation is often complex, lengthy and difficult because there are both winners and losers and a balance has to be found between their interests. The relative industrial weakness of the periphery means that these regions are often among the losers, a tendency which will be added to by economic and monetary union.

The obvious policy response, which we support, is to put emphasis on the redistributive programmes of the Community – regional policy, the European

Investment Bank and the Social Charter. It is important to push forward the 'level playing field' issues for they are an essential part of establishing the framework of competition which we consider so important. However, Europe also gains greatly from its diversity of peoples, life styles and cultures. Measures which offset some of the harsher effects of these policies on individuals and regions are a necessary part of the package.

The issues go wider than disparities in income and wealth. The concentration of decision-making power in Brussels also needs to be countered if the decisions are to gain public acceptance and credibility. In discussing the democratic deficit we urged the need for public debate on the complex issues of technology lest they too lose public credibility. As is often the case, the German language has a convenient shorthand expression for what we seek – *kritik-fähige Öffentlichkeit* – a well-informed public which is capable of giving serious and critical consideration to complex issues. To promote such an open and informed public debate, and to maintain the diversity and pluralism of approach, it is important both to decentralise decision-taking, wherever possible, to lower tiers of government which can show greater sensitivity to local conditions and local preferences and to make the decision-taking processes open and explicit.

One way of promoting this *kritik-fähige Öffentlichkeit* is to strengthen the small technology assessment unit that assists the European Parliament and to give it the status of the Office of Technology Assessment attached to Congress in the United States. Such an institution can help identify the important issues; can mobilise resources so that issues are properly analysed and can provide independent advice to parliamentary bodies. Offices of Technology Assessment already exist in Denmark and the Netherlands and in embryonic form in a number of other countries. The European OTA should augment rather than displace the national bodies, whose role should be developed to ensure pluralism and where necessary to focus attention on specific national and regional issues.

The problem of Eastern Europe – a special case?

The preparation of this book began in 1989 before the Soviet withdrawal from Eastern Europe and the effective collapse of the Soviet-dominated trading system. Although Mary Kaldor's Chapter 21 alone focuses on Eastern Europe, the problems facing these countries are so enormous and their resolution so vital for the whole of Europe that it seems appropriate to provide some comment. Developments are currently rapid and what we say may well be overtaken by events. We therefore limit ourselves to a few fairly general comments and policy suggestions.

From the technological point of view, the situation in Eastern Europe might well be seen as a special case of the core–periphery problem. The major failure of the Soviet system was the failure to generate internally the innovative creativity which has been such a marked feature of Western and Japanese capitalism. (In relation to our earlier remarks about the importance of consumer views, it is interesting to note that the key feature in this failure was breaking the link between consumers and producers.) As a consequence their industries have been caught in a technological 'time warp'. The result today is an industrial system that is profoundly uncompetitive. There is little alternative than to rebuild industrial capabilities more or less from scratch. The need for foreign currency to finance this reconstruction is immense. Many of these countries are already saddled with heavy foreign debt and with the task of recovering from decades of environmental mismanagement. Without reconstruction, their industries cannot survive in international competition. At present the semi-protected mixed economy formula by which Western Europe pulled itself up after World War II has been discredited in their eyes by their experience of socialism. Yet it is difficult to see any other solution. Change cannot be achieved overnight. The approach has to be gradualist and people have to be offered hope that a new society can be created from the chaos of the present.

How can this be achieved? First, it must be emphasised that the economic aid currently being offered to countries other than East Germany is quite insufficient. As Mary Kaldor shows, financial aid currently proposed is meagre compared to American Marshall Aid in the 1950s. Just as Marshall Aid benefited the United States because that country supplied the resultant orders for materials, equipment and services, so Western Europe stands to be the main beneficiary from the reconstruction of Eastern Europe. It is in its own interests to be much more generous.

Second, financial aid on its own will not be enough. Resources are best concentrated on rebuilding the infrastructure (especially telecommunications and transport), on specific industrial projects and on meeting the costs of environmental improvement. The basic education systems are strong but emphasis will need to be given to upgrading technical skills and capabilities and particularly to training a new cadre of managers for industry. It is also necessary to increase the efficiency of the innovation process, in particular ending the segregation of R & D from production and marketing.

It is clear also that economic revival will not be achieved without substantial foreign investment. Already a number of collaborations and joint ventures have been announced, but many Western companies are wary of such involvement unless they see rapid returns on their investments and the chance of repatriating profits. This tends to lead to demands for monopoly sales rights and equivalent restrictions; there needs also to be limitations on the sale of industrial property if a complete takeover is to be avoided. It is important that some indigenous companies survive and eventually prosper in alliance with foreign firms to provide the entrepreneurial core for the future.

Finally, thought needs to be given to the relationship of these countries to the Community. Their immediate membership of the EC is not feasible; rather, as with Greece, it may be appropriate to bring them initially into an 'associate' status, allowing for a long transition period as they reorientate their economic systems. It has to be recognised that economic recovery will not be rapid. It is necesssary to think in terms of years, or even decades to bring these economies up to Western European levels. This task should now be approached with the spirit of 1992 – for both East and West there needs to be a 'vision', a clear set of strategies and a timetable for its achievement. Without such a focus and its promise of better things to come, the dangers of political liberation slipping into economic anarchy are all too apparent.

The international agenda

The new Europe – West and East – will be one of the most powerful trading blocs in the world. The influence of the EC within international negotiations is already considerable and will become more so. As such, the EC should not ignore the importance of sustaining competition itself on a world scale. The benefits from trade and specialisation depend upon a sufficient degree of competition existing somewhere in the world. The number of companies capable of serious technological competition in industries such as aeroengines, very large-scale integrated circuits (VLSI), supercomputers, or large electronic switches is already very small. It is only within the 'Triad' of the United States, Japan and the EC (and to some degree in the Soviet Union) that such companies exist today.

There are, however, tensions and strains within the Triad, many associated with the rise of Japan's capabilities in new technologies and the relative decline of the United States. It is, of course, tempting for Americans (and Europeans) to attribute Japanese success primarily to copying (or stealing) American (or European) designs and ideas and to sharp practice in international trade, especially through dumping and hindering the penetration of the Japanese domestic market. These types of explanation have certainly influenced legislation in Congress and the stance of the United States' government in trade negotiations. Books such as those by Prestowitz (1988) and Balassa and Noland (1989) provide some evidence that Japanese firms and the Japanese government have followed a well-coordinated long-term strategy in their penetration of the United States and other markets while maintaining non-tariff barriers to imports in their home market.

The EC cannot afford to be starry-eyed or lax in trying to uphold the rules of the game in international trade. However, the bending (or violation) of the GATT and other rules is by no means confined to any one country, and the enormous success of Japanese firms in international trade has more fundamental causes. The more serious American and European analyses recognise this fact and resist nationalistic and xenophobic interpretations. In the long run European firms will have to compete primarily on the basis of their own technological and industrial competence.

There are, however, some fundamental issues underlying these international frictions. Global oligopoly has both advantages and disadvantages. Its advantages stem from the technological rivalry that it engenders and which has proved such a fertile source of creativity and economic progress. Its disadvantages arise from its own nature – large multinational companies, as we said earlier, often elude control. Inevitably, there are fears that such companies will not 'play by the rules', but will cross subsidise activities in order to pre-empt and capture market share.

Attempts to counter such fears through technological protectionism, whether of trade or of intellectual property, are self-defeating. All countries would be losers if there were a serious shift towards protection. The post-war world has benefited immensely from the progressive liberalisation of trade and open access to American (and European) science and technology. All parts of the Triad stand to gain from the continuation of this system – from the pluralism of approach, the technological rivalry and the presence of high value-added activities within their borders.

Fears of 'unfair competition' from multinationals are, nevertheless, legitimate. One way to counter them would be to extend the concept of the 'level playing field' from the European to the international arena. In particular, Europe could take the lead in seeking to establish an international organisation whose responsibilities would be the monitoring and supervision of a global competition policy. Such an organisation was originally envisaged as part of the 1945 Bretton Woods programme. Part of its remit might be the establishment of a mutually recognised and enforceable code of conduct for multinational companies which would have substantial benefit for Third World as well as First World countries.

Indeed, in a number of areas where Europe is currently struggling to establish Community standards and regulations – food and drugs, electrical goods, ICT – it might make sense to shift the focus from the European to the global arena. This is not to underestimate the difficulties involved, or the dangers of ending up (see Chapter 20 on food standards) with the lowest common factor, but to emphasise the long-run advantage for many countries, including the Third World, in extending the level playing field of the Single Market to the world as a whole.

IV In conclusion

Technological capabilities are likely to remain, as they are now, one of the fundamental determinants of wealth and prosperity. This book has surveyed Europe's current capabilities across a range of sectors, and this chapter has sought to highlight those issues which, in our view, come top of the policy agenda for the next decade.

There are, as we said earlier, both challenges and opportunities. Europe still has to resolve for itself how it balances the conflicting pulls between dynamism and conservatism, between equity and efficiency, between materialism and sustainability.

We have made our position plain. We do not believe these conflicts are insoluble; nor do we delude ourselves in thinking they are easy to resolve. Our preferences lean, where there is choice, towards pluralism which we see not as the enemy of dynamism but rather as a source of creativity. Hence our emphasis on meeting the democratic deficit and devolving decision-taking down to lower tiers of government and, where it makes sense, to the market. And hence our general presumption against protection and in favour of maintaining an open trading environment. Equally, we are conscious also of the limitations of the market system and the need to intervene and to regulate both to make markets work better and to achieve the complex trade-offs which society demands.

The current conjuncture of events offers we believe an opportunity for Europe to take the lead in three important areas of policy – in gearing technology towards sustainability; in organising a 'Marshall Plan' for Eastern Europe; and the setting up of a new international competition authority. All three are areas where developments in technology are creating tensions and conflicts which will not necessarily be resolved by themselves or by the market. We believe that European governments, working through the Community, have a unique opportunity to show leadership which will eventually be of universal benefit.

References

Abernathy, W. (1977), *The Productivity Dilemma: Roadblock to Innovation in the Automobile Industry*. Johns Hopkins University Press, Baltimore.

Abraham, J. and Millstone, E. (1989), 'Food Additive Controls: Some International Comparisons', *Food Policy*, 14, (Feb), pp. 43–57.

ACARD (1980), (Advisory Council for Applied Research and Development), *Biotechnology: Report of Joint Working Party of ACARD, the Advisory Board for the Research Councils and The Royal Society* (The Spinks Report) HMSO, London.

ACARD (1986), Advisory Council for Applied Research and Development, *Software: A Vital Key to UK Competitiveness*, HMSO, London.

ACDA (1989), 'World Military Expenditures and Arms Transfers 1988', US Government Printing Office, Washington DC.

ACOST (1989), (Advisory Council on Science and Technology), 'Defence R & D: a National Resource', HMSO, London.

Aigrain (1989), *Report of the Framework Programme Review Board*, (commonly known as Aigrain Report), Brussels.

Aksoy, A., Jagger, N. and Miles, I. (1990), 'VANS Suppliers and Users: results of the CICT Data Communications Survey' SPRU, Brighton, February.

Allen, D. (1988), 'New telecommunications services: Network externalities and critical mass', *Telecommunications Policy*, September, pp. 257–71.

Allen, G. (1990), 'Our Chemical Industry in 2001', *Chemistry & Industry*, 4 June, pp. 349–53.

Alton, T. (1985), 'East European GNP Origins of Product, Final Uses, Rates of Growth and International Comparisons', in *East European Economies: Slow Growth in the 1980s, Vol. Economic Policy and Performance*, Joint Economic Committee, US Congress, Washington, DC.

Altshuler, A., Anderson, M., Jones, D.T., Roos, D. and Womack, J. (1985), '*The Future of the Automobile: The Report of MIT's International Automobile Programme*, George Allen and Unwin, London.

Alvey (1987), *Alvey Programme Annual Report 1987*, Department of Trade and Industry, London.

Amery, C. (1985), Die Okologische Chance, Suddeutscher Verlag, Munich.

Anderson, A. and Hersleb, A. (1980), *Computer Manpower in the 1980s*, NEDO, London.

Anderson, P. (1974), *Lineages of the Absolutist State*, New Left Books, London.

Anizon, D. (1990), 'Chimie: l'avance des groupes européens', in *Marche Unique: Marche Multiple*, Guy Crespy (ed), Economica, Paris.

Anthony, I., Allebeck, A., Gullikstad, E., Hagmeyer, G. and Wulf, H. (1990a), 'Arms Production', in *World Armaments and Disarmament: SIPRI Yearbook 1990*, Oxford University Press, Oxford.

Anthony, I., Allebeck, A. and Wulf, H. (1990b), 'Western European Arms Production: Structural Changes in the New Political Environment', SIPRI Research Report, Stockholm.

Arnold, D.L., Krewski, D. and Munro, I.C. (1983), 'Saccharin: a Toxicological and Historical Perspective', *Toxicology*, 27, pp. 179–256.

Arnold, E. and Guy, K. (1986), *Parallel Convergence: National Strategies in Information Technology*, Frances Pinter, London.

Arthur, W.B. (1988), 'Competing Technologies, Increasing Returns and Lock-in by Historical Events', *Center for Economic Policy Research*, Stanford University, Stanford May.

Arthur, W.B. (1988), 'Competitive Technologies–An Overview', in Dosi *et al.* (eds) *Technical Change and Economic Theory*, Pinter Publishers, London, pp. 590–607..

Baba, Y. and Takai, S. (1990), 'Information Technology Introduction in the Big Banks: The Case of Japan', in C. Freeman and L. Soete (eds), *New Explorations in the Economics of Technical Change*, Pinter Publishers, London.

Bahro, R. (1974), *The Alternatives in Eastern Europe*, New Left Review, London.

Balassa, B. and Noland, M. (1988), *Japan in the World Economy*, Institute for International Economics, Washington, DC.

Baldwin, R. (1990), 'Promoting New Industrial Activities, a Survey of the Evidence', OECD Economic Papers, 14 (Spring), pp. 87–126.

Ball, M. (1988), *Rebuilding Construction*, Routledge, London.

Bangermann, M. (1989/90), Comment to the 1989 Annual Meeting of the International Sweeteners Association reported in *Sweetener Update*, Winter, p.1.

Bangermann, M. (1990), Answer given by Mr Bangermann to European Parliament Questions QXW1152/89EN and QXW1154/89EN, 21 February 1990, *Official Journal*, C 303, 3 December.

Baranson, J. (1990), 'National Differences in the Business Environment for CIM', in *Proceedings of UN/ECE Seminar on Computer-integrated Manufacturing, Botevgrad, Bulgaria, 25–29 September, 1989.* UN/ECE, Geneva.

Barbour, J. (1989), cited in 'Integrated Pollution Control to put Britain Ahead in Europe?', *Haznews*, no. 13, January, p. 11.

Barclays de Zoete Wedd (1990), *Pharmaceutical Industry Perspectives*, BZW, Ebbgate House, 2 Swan Lane, London EC4 3TS.

Barnaby, F. and Ter Borg, M. (1986), *Emerging Technologies and Military Doctrine: A Political Assessment*, Macmillan Press, London.

Bauer, T. (1978), 'Investment Cycles in Planned Economies', *Acta Oeconomica*, 21, No. 3.

Bell, M., Scott-Kemmis, D., Satyarakwit, W. (1982), 'Limited Learning in Infant Industry: A Case Study' in F Stewart and J. James (eds.), *The Economics of New Technology in Developing Countries*, Frances Pinter, London.

Bennett, E. (1989), Director of Nuclear Safety, Chemical Control and Wastes Management, Commission of the European Communities, speaking at Future Trends in Waste Management, Conference, 21 September.

Bentley, R. (1984), *Technological Change in the GDR*, Westview Press, London and Boulder.

Berend, I. and Ranki, G. (1982), *The European Periphery and Industrialization, 1780–1914,* Cambridge University Press, Cambridge.

Bernard, K.E. (1989), 'The EC Green Paper: What the Future Holds *Telecommunications Policy,* December 13, no. 4, pp. 313–17.

Bessant, J. (1984), 'Making IT Fit', in G. Winch (ed.), *Information Technology i Manufacturing Processes*, Rossendale, London.

Bessant, J. (1988), 'Pushing Boxes or Solving Problems? Some Marketing Issue in the Adoption of Computer-integrated Manufacturing Technology', *Journal o Marketing Management*, 3, no.3.

Bessant, J. and Grunt, M. (1985), *Management and Manufacturing Innovation in th UK and West Germany*, Gower Press, Aldershot.

Bessant, J. and Haywood, W. (1988), 'Flexible manufacturing in Europe', *Europea Management Journal*, 6, no. 2.

Biofutur (1990), *Dix Ans de Biotechnologis* (Special Edition on Ten Years c Biotechnology), numero 90, May.

Bobrow, D.E. (1989), 'The Object of Desire', *Datamation*, 1-5, pp. 37-41.

Boddy, D. and Buchanan, D. (1984), *Organisations in the Computer Age*, Gowe Press, Aldershot.

Boehmer-Christiansen, S. (1990), 'Vehicle Emission Regulation in Europe – The Demise of Lean-burn Engines, the Polluter Pays Principle ... and the Small Car?', *Energy and Environment*, 1, no. 2.

Boyer, R. (1989), Paper given to the OECD Conference on Technical Change in Helsinki, December.

BP Statistical Review of World Energy (1990), June.

Brady, T. (1990), 'In-House Development and Growth of the Software Industry', CICT Working Paper no. 12, SPRU, Brighton.

Brech, M. and Sharp, M. (1984), *Inward Investment: Policy Options for the UK*, Chatham House Paper no. 21, Routledge and Kegan Paul for the Royal Institute of International Affairs, London.

British Nutrition Foundation (1988), *Food Fit to Eat*, Sphere Books, Harmondsworth.

Brooks H. (1990), 'CIM: Implications for Governments' in W. Haywood (ed.), *CIM: Revolution in Progress*, Conference Proceedings, IIASA, Vienna.

Buckroyd, B. and Cornford, D. (1988), *The IT Skills Crisis: The Way Ahead*, NCC Publications, London.

Building (1988), 'Principal British Construction Interests in Spain', 30 September.

Building (1990), 'The Top 500 European Companies', 16 November, pp. 53–62.

Bundesministerium fur Umwelt, Strahlenschutz und Realtorsicherheit (BMU), (1989), Die Technische Anleitung Abfall, *Umwelt*, nr 2, pp. 80–82.

Bureau of European Consumer Unions (1989), *BEUC News* March/April, p. 2.

Carroll, J.E. (ed.) (1988), *International Environmental Diplomacy*, Cambridge University Press, Cambridge.

Cawson, A., Morgan, K., Webber, D., Holmes, P. and Stevens, A. (1990), *Hostile Brothers: Competition and Closure in the European Electronics Industry*, Clarendon Press, Oxford.

CEC (1970), La Politique Industrielle de la Communauté COM (70) 100 Final (Colonna Report).

CEC (1980), Council directive on the protection of groundwater against pollution caused by certain dangerous substances, Directive 80/68/EEC, Brussels.

CEC (1984), Council directive on the supervision and control within the European Community of the transfrontier shipment of hazardous waste, Directive 84/631/EEC, Brussels.

CEC (1985), Council directive on the assessment of the effects of certain public and private projects on the environment, Directive 85/337/EEC, Brussels.

CEC (1986a), Council directive amending Directive 84/631/EEC on the supervision and control within the European Community of the transfrontier shipment of hazardous waste, Directive 86/279/EEC, Brussels.

CEC (1986b), EUREKA and the European Technology Community, Communication from the Commission to the Council, Commission of the European Communities, COM (86), Final, Brussels.

CEC (1987a), Council directive amending Directive 75/439/EEC on the disposal of waste oils, Directive 87/101/EEC, Brussels.

CEC (1987b), The ESPRIT Programme: Project Synopses, Sub-programme 1, Advanced Microelectronics, Directorate General X111, Telecommunications, Information Industries and Innovation, Commission of European Communities, Brussels.

CEC (1987c), 'Towards a Dynamic European Economy', Green Paper on the Development of the Common Market for Telecommunications Services and Equipment, Brussels, DG XIII/195.

CEC (1988a), Proposal for a Council directive on hazardous waste, draft directive 88/C 295/04, Brussels.

CEC (1988b), *17th Report on Competition policy*. European Community Publications Office, Luxemburg.

CEC (1988c), 'Standardization in Information Technology', Fact Sheet 3, May.

CEC (1988d), 'Towards a Competitive Community-Wide Telecommunications Market in 1992. Implementing the Green Paper on the Development of the Common Market for Telecommunications Services and Equipment.' COM (88) 48 final, Brussels, 9 February.

CEC (1988e), 'Telecommunications: the New Highways for the Single European Market', *European File*, 15/88.

CEC (1988f), *Research and Technological Development Policy*, European Documentation, Luxemburg.

CEC (1988g), *The Internal Energy Market*, COM (88) 238, Brussels, May.

CEC (1989a), *ESSI: The Case for a Major European Systems and Software Initiative*, report by the Industrial Working Group to DG XIII, Brussels.

CEC (1989b), Guidelines for improving the synergy between the public and private sectors in the information market DG XIII, Luxemburg.

CEC (1989c), EUR 11883: EDI in Perspective, Luxemburg 1989.

CEC (1989d), The Review of ESPRIT 1984 to 1988: the Report of the ESPRIT Review Board, Brussels.

CEC (1989e), *Employment in Europe*, Brussels, July.

CEC (1989f), *Report of the Framework Programme Review Board*, Brussels, June.

CEC (1990a), Economic and Social Committee, *EDI – 1992 and Beyond*, proceedings of a conference, Brussels September 1989, Brussels.

CEC (1990b), *Panorama of Industry*, EC, Luxemburg.

CSP (1986), *Extent and Nature of International Trade in Electronic Specialised Information Services*, London.

Cecchini, P. (1988), 'The European Challenge – 1992', in P. Cecchini *et al.*, *The Benefits of the Single Market*, The European Commission, Brussels.

Clark, K.B. and Fujimoto, T. (1988), 'The European Model of Product Development: Challenge and Opportunity', International Motor Vehicle Program, MIT, Cambridge, Mass.

Cockbill, C. (1990), 'The Impact of 1992 on Food Law', *BNF Nutrition Bulletin*, 15, p. 33.

Cohendet, P., Ledoux, M.J., Zuscovitch, E. (1988), *New Advanced Materials*, Springer-Verlag, Berlin.

Cole, H.S.D., Freeman, C., Jahoda, M. and Pavitt, K. (eds) (1973),*Thinking about the Future; a Critique of Limits to Growth*, Chatto and Windus, London.

Collingridge, D. (1980), *The Social Control of Technology*, Frances Pinter., London.

Collins, P. (1989), 'Use of formal methods in CICS', paper presented at the 'Research into Practice' Conference, Centre for Software Reliability, Bournemouth, 25–26 May.

Cooper, P. (1988), 'British Contractors Fear New EC Public Works Directive', and 'Public Works Controversy', *Building*, 14 October.

Cooper, P. (1989a), 'West German Bankruptcies Leave Room for UK Builders', *Building*, 3 February.

Cooper, P. (1989b), 'Laing Set to Join the UK's European Housebuilders', *Building*, 3 February.

Coopers and Lybrand (1989), *Computing Services Industry 1986–1996: A Decade of Opportunity – 1989 Update*, Report to the Department of Trade and Industry, Coopers and Lybrand, London.

Cosumano, M.A. (1985), *The Japanese Automobile Industry: Technology and Management at Nissan and Toyota*, Harvard University Press, Cambridge, Mass.

Council of Ministers (1990), *Common Position With a View to the Adoption of a Directive on Nutrition Labelling for Foodstuffs*, Document Number 4281/90, 22 February.

Crampton, S. (1989), 'The Participation of Consumers in National and European Community Policy–Making in the Run Up to 1992 and Thereafter', Submission to the Committee in the Environment, Public Health & Consumer Protection of the European Parliament, Consumers in the European Community Group, London, February, p. 10.

Currie, A. (1983), 'Viewdata at the London Stock Exchange' *Aslib Proceeeding*, 35, no. 8, August.

Dallago, B. (1988), 'The Role of the Non-socialized Sector in Hungary and Poland', NATO Colloquium, The Economies of Eastern Europe Under Gorbachev's Influence, NATO, Brussels.

Dang Nguyen, G. (1985), 'Telecommunications: a challenge to the old order', in M. Sharp, (ed.): *Europe and the New Technologies*, Frances Pinter, London.

Dankbaar, B. (1990), 'International Competition and National Institutions. The Case of the Automobile Industry' in C. Freeman, and L. Soete, (eds), *New Explorations in the Economics of Technical Change*, Pinter Publishers, London.

Datamation (1989) 1 June.

Datamation (1990) 15 June.

David, P.A. and Bunn, J.A. (1988), 'The Economics of Gateways' Technologies and Network Evolution, Lessons from the Electricity Supply Industry', in *Information Economics and Policy 3*, North Holland, April, pp. 165–202.

Davidson, W.D. (1987), 'Japanese Telecommunications Policy – New Direction and Old Dilemmas', *Telecommunications Policy*, June, pp. 147–160.

Davies, A.C. (1989), Ford OECD/BRIE Telecom User Group Case Studies, mimeo, SPRU, Brighton.

Department of the Environment (DoE.) (1985), Management Information System for Ministers (MINIS) Report, reported in ENDS Report 135, April, Environmental Data Services Ltd., London, p. 3.

Department of the Environment (1985), *Report of the Review of the Control of Pollution (Special Wastes) Regulations 1980*, Annex 2.2, DoE, London.

Department of Trade and Industry, (DTI) (1990a), 'Competition and Choice: Telecommunications Policy for the 1990s', HMSO, London.

Department of Trade and Industry, (DTI) (1990b), *A Guide to European Community Industrial Research and Development Programmes*, London.

Dertouzos, M., Lester, R. and Solow, R. (eds) (1989), *Made in America; Regaining the Productive Edge*, MIT Press, Cambridge, Mass.

DiMasi, J.A., Hansen, R.W., Grabowski, H.G. and Lasagna, L. (1991), 'The Cost of Innovation in the Pharmaceutical Industry: New Drug R & D Cost Estimates', *Journal of Health Economics*, June.

Dodgson, M. (1990), *Celltech: The first 10 years of a Biotech Company*, February, SPRU Special Report, Brighton pp. 95.

Dorfman, N.S. (1987), *Innovation and Market Structure: Lessons from the Computer and Semiconductor Industries*, Ballinger, Cambridge, Mass.

Dosi, G. (1981), *Technical Change and Survival: Europe's Semiconductor Industry*, Sussex European Papers No. 9, SPRU, Brighton.

Dosi, G., Pavitt, K. and Soete, L. (1990), *The Economics of Technical Change and International Trade*, Harvester Wheatsheaf, Hemel Hempstead.

Dowling, M. and Linerooth, J. (1987), 'The Listing and Classifying of Hazardous Waste' in B. Wynne (ed.), *Risk Management and Hazardous Waste*, Springer-Verlag, Berlin, pp. 114–149.

Drewer, S. (1990), *Habitat* Special Issue on UK/Japan Construction, 14, no. 2/3.

East, H. and Forrest, V. (1988), Statistics of Online Expenditure', CCIS, Polytechnic of Central London, London, mimeo

Eatwell, J. (1982), 'Whatever happened to Britain?: the economics of decline', BBC, London (Published to accompany a Television series with the same title).

Ebel, K. (1985), 'Social and Labour Implications of FMS', *International Labour Review*, 124, no. 2.

ECE, (1990), Annual Review of Engineering Industries and Automation, Geneva.

Economist, The (1990), 14 July.

ENDS Report 110, (1984), March, p. 2.

Ergas, H. (1984), 'Why do Some Countries Innovate More than Others?', *Centre for European Policy Studies*, Paper no. 5, Brussels.

Ergas, H. (1986), 'Information Technology Standards: The Issues', *Telecommunications*, 20, no. 8, pp. 38–42.

Euromatica (1986), 'The Businessman's Guide to Electronic Data Interchange', *Euromatica Publications*, Brussels, in M. T. Cunningham & K. L. Culligan, *Relationships, Networks and Competitiveness: The Implications of Electronic Data Interchange Systems*, UMIST, mimeo.

European Commission (1985), *Completion of the Internal Market: Community Legislation on Foodstuffs*, Commission Communication COM(85)603 Final, Brussels, November.

European Commission (1989), *Discussion Document on The Consolidation of Measures on Food Additives and their Purity Criteria: Part 1 Council Directives*, III/3671/88-Rev.2., Brussels.

European Commission (1990), 'Proposal for a Council Regulation laying down the health rules for the production and placing on the market of minced meat, meat preparations and comminuted meat for industrial uses', COM(89)671 final, Brussels, 1 February.

European Study Service (1985), *EEC Waste Management Policy*, Rixensant, Belgium.

Eurostat (1989), Energy Statistical Yearbook, EC, Brussels.

Eurostat (1990), Energy Statistical Yearbook, EC, Brussels.

Eurostrategies (1989), *Employment structures and trends in telecommunications Volume 1: Main Report*, a study carried out for the Commission of the European Community, Brussels.

Fagerberg, J. (1987), 'A Technology Gap Approach to Why Growth Rates Differ', in C. Freeman (ed.), *Output Measurement in Science and Technology: Essays in Honor of Y. Fabian*, North-Holland, Amsterdam.

Fagerberg, J. (1988), 'International Competitiveness', *Economic Journal*, 98, pp. 355–374.

Feeny, D. and Brownlee, C. (1987), *Competition in the Era of Interactive Network Service*, The Oxford Institute of Information Management, Templeton College, Oxford.

Ferguson, C.H. (1988), 'From the People who Bought You Voodoo Economics: Beyond Entrepreneuralism to U.S. Competitiveness', *Harvard Business Review*, May–June.

Ferné, G. (1989), 'R & D Programmes for Information Technology' *OECD Observer*, Aug/Sept., pp. 11–13.

Fertig, R.T. (1985), *The Software Revolution: trends, players and market dynamics in Personal Computer Software*, Elsevier, Amsterdam.

Financial Times (1989), 'AT&T Chief Calls for Truce in Unix Wars', March 3.

Financial Times (1989): May.

Financial Times (1990): September, Section III.

Fisher, A. (1989), 'Measuring up to high standards', *Financial Times*, 23 January.

Fishlock, D. (1990), 'Commitment to go back to the basics: R & D in BASP', *Financial Times*, 3 October.

Flamm, K. (1987), *Targeting the Computer: Government Support and International Competition*, The Brookings Institute, Washington DC.

Fleck, J. (1988), 'The Development of Information Integration: Beyond CIM?', PICT Working Paper, Edinburgh University, p. 9.

Ford, J.G. (1990), European Parliament Written Question Number QXW1154/89EN, *Official Journal*, C303, 3 December.

Foster, A. (1990), '1992 and the Consumer', *BNF Nutrition Bulletin*, January, p. 24.

Freeman, C. (ed.) (1987), *Output Measurement in Science and Technology: Essays in Honor of Y. Fabian*, North-Holland, Amsterdam.

Freeman, C. and Jahoda, M. (eds) (1978), *World Futures*, Martin Robertson, London.

Gann, D.M. (1989), *New Technology, Employment and Operative Skills in Building Services*, SPRU, Brighton.

Gann, D.M. (1990), 'Intelligent Buildings and Smart Homes' in G. Locksley, (ed.), *The Single European Market and the Information and Communication Technologies*, Belhaven Press, London.

Gann, D.M. (1991), 'New Management Strategies in the British Construction Industry' in H. Rainbird and G. Syben,(eds), *Restructuring a Traditional Industry: Construction Employment and Skills in Europe*, Berg Publishers, Oxford.

Garnett, N. (1988), 'A Bumpy Ride over Europe's Traditions', *Financial Times*, 30 October.

Gartner Group (1987), *Top 100 Data Processing Almanac*, Stanford, California.

Gerwin, D. (1986), 'An Agenda for Research on the Flexibility of Manufacturing Processes', *International Journal of Production and Operations Management*, 7, no. l.

Gilhooly, D (1990) 'ETI Mobile Breakthrough Comes After Infighting' *Communications Week International*, 26 March, p.2.

Glaziev, S. (1990), Flexible automation in the concept of structural changes in a modern economy, *Proceedings of a Seminar on Computer-Integrated Manufacturing, Botevgrad, Bulgaria 25–29 September 1989*, UN/ECE, Geneva.

Goldsmith, E. (1988), *The Great U-Turn*, Green Books, Bideford.

Goodhart, D. (1988), 'An Attack of the 1992 Jitters', *Financial Times*, 7 December.

Graham, N.A. and Mody, B. (1990), 'Export Policy Liberalization, Telecommunications Infrastructure, and International Competitiveness: The Computer Software Industry in India', Michigan State University, East Lansing (mimeo).

Graves, A.P. (1987), 'Comparative Trends in Automotive Research and Development,' International Motor Vehicle Program, MIT, Cambridge, Mass.

Graves, A.P. (1988), 'European Design and Engineering Capabilities – A Continuing Strength', International Motor Vehicle Program, MIT, Cambridge, Mass.

Gray, P. (1984), 'A Commission View of the Food and Drink Industries', in A. Swinbank and J. Burns (eds), *The EC and the Food Industries*, Food Economics Study No. 1, Department of Agricultural Economics and Management, University of Reading; cited in Swinbank 1990a:45.

Gray, P. (1988), *Report on a meeting of the Parliamentary Food and Health Forum*, Parliamentary Food and Health Forum, Westminster, 28 June, p. 6.

Green, B. (1988), 'Britain leads the way in the boom of the decade', *Construction News*, 26 January.

Greenaway, D. and Hindley, B. (1985), 'What Britain Pays for VERs', Trade Policy Research Centre, London.

Grindley, P. (1988), *The UK Software Industry: A Survey of the Industry and Evaluation of Policy*, London Business School, London.

Guest, D. (1990), 'EDI: It's Not that Simple' *Datacom*, February, pp. 38–40.

Haefele, W. (1974), 'Hypotheticality and the New Challenges: the Pathfinder Role of Nuclear Energy', *Minerva*, 12, pp. 301–325.

Hagelin, B. (1989), 'Politics, Technology and Small Power Armaments in Future Europe', paper presented to the Conference on Restructuring Arms Industries in Western Europe, Haus Rissen, Hamburg, June.

Haigh R. (1978), 'Harmonization of Legislation on Foodstuffs, Food Additives and Contaminants: II Achievements and programme', *Journal of Food Technology*, 13.

Hamel, G., Doz, Y. and Prahalad, C.K. (1986), *Strategic Partnerships: Success or Surrender? The Challenge of Competitive Collaboration*, Working Paper Series, number 24, Centre for Business Strategy, London Business School.

Hansen, S.C. (1979), 'Conditions for Use of Food Additives Based on a Budget for an Acceptable Daily Intake', *Journal of Food Protection*, 42, no. 5, pp. 429–34.

Hare, P., Lauchlan, J. and Thompson, M. (1988), 'An Assessment of ESPRIT in the UK', Technological Change Research Centre, Heriot-Watt University, Edinburgh.

Harris, R.G. (1990), 'Divestiture and Regulatory Policies: Implications for Research, Development and Innovation in the US Telecommunications Industry' *Telecommunications Policy*, April, 14, no. 2, pp. 105–24.

Hauser, H. (ed.) (1988), *Deutsches Lebemsmittelbuch: Leitsaetze 1988*, Bundesanzeiger, Cologne.

Hayward, K. (1986), *International Collaboration in Civil Aerospace*, Frances Pinter, London.

Haznews, (1988a), no. 1, April.

Haznews, (1988b), no. 4, June.

Haznews, (1989a), no. 10, January.

Haznews, (1989b), no. 11, February.

Her Majesty's Stationery Office, (1980), *Convention on Long Range Transboundary Air Pollution, Geneva, 13–16 November 1979*, HMSO, London.

Her Majesty's Stationery Office (HMSO) (1989), *Toxic Waste: The Government's Response to the Second Report of the Environment Committee*, Cm 679, London, p 6.

Hills, J. (1989), 'Universal Service – Liberalization and Privatization of Telecommunications', *Telecommunications Policy*, June, pp. 129–44.

Hills, J. (1986), '*Deregulating Telecoms – Competition and Control in the United States, Japan and Britain', Frances Pinter*, London.

Hoare, C.A.R. (1982), *Programming is an Engineering Profession*, Technical Monograph PRG-27, Oxford University Computing Laboratory, Oxford.

Hobday, M. (1990), 'Semiconductors: Creative Destruction or US Industrial Decline?' *Futures*, July/August, pp. 571–85.

Hobday, M. (1990), 'Dynamic Networks, Technology Diffusion and Complementary Assets: Explaining US Industrial Decline in Semiconductors', SPRU, Brighton, (mimeo).

House of Commons (HoC) (1989), *Toxic Waste*, Second Report of the Environment Committee, vol 1, 22–I, HMSO, London, p. lxvi.

House of Commons Select Committee on Energy, (1989), *Energy Policy Implications of the Greenhouse Effect*, HC 192–4, House of Commons, HMSO, London, July.

House of Lords Select Committee on the European Communities (1985), ESPRIT (European Research and Development in Information Technologies), session 1984–85, 8th Report, HMSO, London, p. 169.

House of Lords Select Committee on Science and Technology (1981), *Hazardous Waste Disposal*, vol 1, HMSO, London, p. 6.

House of Lords Select Committee on the European Communities, (1989), *Efficiency of Electricity Use*, House of Lords, HMSO, London, p. 18.

IDC Eurocast (1988), *Software and Services Market Place Western Europe 1987-1993*, IDC European Technology Services Group, London.

IDC-Italia (1989), 'Il mercato dell'information technology dalla vendita di hardware alla fornitura di servizi' Milan, (mimeo).

IEPG (1986), 'Towards a Stronger Europe', report by an Independent Study Team established by Defence Ministers of Nations of the IEPG, Brussels.

IMO (1990), *Databases in Japan Information Market*, Observatory Working Paper 90/2, May, CEC DG XIII, Luxemburg.

Ince, D. (1989), 'Five Years On', *Datalink*, 15 February, p. 12.

International Atomic Energy Agency (IAEE) (1970), *Standardisation of Radioactive Waste Categories*, TRS no. 101, STI/DOC/10/101, Vienna.

International Coal Report (1989), *Coal Year 1989, Financial Times*, Business Information, May.

International Commission on Radiological Protection (ICRP) (1977), *Recommendations of the ICRP*, ICRP Publication 26, *Annals of the ICRP*, 1, no. 3, Pergamon Press, Oxford.

International Commission on Radiological Protection (ICRP) (1985), *Radiological Protection Principles for the disposal of solid radioactive wastes*, ICRP Publication 46, Pergamon Press, Oxford.

International Environmental Affairs, (1989) 1, no. 1.

International Institute of Strategic Studies (IISS) (1990), *The Military Balance 1989–90*', IISS, London.

Irvine, J., Martin B. and Isard, P., (1990), *Investing in the Future: An International Comparison of Government Findings of Academic and Related Research*, Edward Elgar, Aldershot.

Jacobbsson, S. and Edquist, C. (1986), *Flexible Automation*, Basil Blackwell.

Jagger, N. (1989), 'Prestel's Alter Egos: the Diffusion of Private Videotex in the UK, CICT Working Paper no.6, SPRU, Brighton.

Jagger, N. (1990), 'Videotex Strategies', *Datacoms '90 Yearbook*, VNU, London.

Jaikumar, R. (1986), 'Post-industrial Manufacturing', *Harvard Business Review*, November.

JIPDEC (Japan Information Processing Development Center) (1990), 'EDI in Japan', *Japan Computer Quarterly* no. 80, Tokyo.

Jones, B. and Scott, P. (1987), 'Working the System' in C. Voss (ed.) *Managing Advanced Manufacturing Technology*, IFS Publications, Kempston.

Jones, C.D. (1987), 'Tariff and Non-tariff Barriers to Trade' Government Economic Service Working Paper, no. 97, Department of Trade and Industry, London.

Jones, D.T. (1988), 'Measuring Technological Advantage in the Motor Vehicle Industry', International Motor Vehicle Program, MIT, Cambridge, Mass.

Jones, D.T. (1990), 'The Competitive Outlook for the European Auto Industry', *International Journal of Vehicle Design*, II, no. 3, pp. 222–33.

Jones, K. (1988), 'IPSEs Start to Come Into Their Own' *Computer* Weekly, 4 February, pp. 24–5.

Jowett, P. and Rothwell M. (1986), *The Economics of Information Technology*, Macmillan, Basingstoke & London.

Julius de, Anne (1990), *Global Companies and Public Policy. The Growing Challenge of Direct Foreign Investment*, Pinter Publishers for Royal Institute of International Affairs.

Kalantzopoulos, O. (1986), *The Costs of Voluntary Export Restraints*, The World Bank, Washington, DC.

Kaldor, M.H. (1982), *The Baroque Arsenal*, Andre Deutsch, London.

Katz, M. and Shapiro, C. (1985), 'Network Externalities, Competition and Compatibility', *American Economic Review*, *75*, pp. 424–40.

Keepin, W. and Katz, G. (1988), 'Greenhouse Warming : Comparative Analysis of Nuclear and Efficiency Abatement Strategies', *Energy Policy*, 16, no. 6.

Kitchenham, B. (1989), 'Software Metrics for Industrial Applications', paper presented at the Research into Practice Conference, Centre for Software Reliability, Bournemouth, 25–6 May.

Kornai, J. (1982), *The Economics of Shortage*, North-Holland, Amsterdam.

Kornai, J. (1985), *Contradictions and Dilemmas*, Corvina, New York.

Kornai, J. (1986), 'The Hungarian Reform Process: Visions, Hopes and Realities', *Journal of Economic Literature*, XXIV.

Kornai, J. (1990), *The Road to a Free Economy*, W.W. Norton, New York.

Köves, A. (1989) 'Problems and Prospects of East-West Economic Cooperation: An Eastern European View' in: M. Kaldor, G. Holden and R. Falke, (eds), *The New Detente*. Verso and United Nations University, Tokyo, pp. 253–72.

Krafcik, J. (1987), 'The Triumph of Lean Manufacturing, *Sloan Management Review*, 30, no. 1.

Krafcik, J. (1988), 'European Manufacturing Practice in a World Perspective' International Motor Vehicle Program, MIT, Cambridge, Mass.

Krafcik, J. (1989), 'Comparative Analysis of Performance Indicators on World Assembly Plants.' International Motor Vehicle Program, MIT, Cambridge, Mass.

Krafcik, J.F. and MacDuffie, J.P. (1989), 'Explaining High Performance Manufacturing: The International Automotive Assembly Plant Study' *IMVP Forum Paper*, May.

Krugman, P. (ed.) (1986), *Strategic Trade Policy and the New International Economics*, MIT Press, Cambridge, Mass.

Laurence, D. and Wynne, B. (1989), 'Transporting Waste in the European Community: A Free Market?', *Environment*, 31, no 6, July–August, p. 15.

Lay G., Fix-Sterz, J., Schulz-Wild, R. and Wengel, J. (1990), 'Flexible Manufacturing Systems and Cells in the Federal Republic of Germany' in M. Warner, W. Wobbe and P. Brodner (eds), *New Technology and Manufacturing Management*, John Wiley, Chichester.

Layton, C. (1969), *European Advanced Technology: A Programme for Integration*, Allen and Unwin, London.

Leadbeater, C. (1990), 'Scrapped Factories "Will feed" East Europe's Steel Industry', *Financial Times*, 4 September, London.

Lehman, J.P. (1981), 'Hazardous Waste Definition and Recommended Procedures', in J.P. Lehman (ed.), *Hazardous Waste Disposal*, Plenum Press, New York, p. 52.

Leibowitz, M.R. (1987), 'Chip Takeover Targets', H*igh Technology Business*, September.

Link (1986), *The European Electronic Information Industry 1986–1990*, Link, London.

Lowe, K. (1989), 'IPSEs Take a Step Backwards as Genos Stumbles', *Computing*, 13 July, p. 16.

Lundvall, B-A, (1988), 'Innovation as an Inter-active Process, Firm User–Producer Interaction to the National System of Innovation', in G. Dosi, *et al.*, *Technical Change and Economic Theory*, Pinter Publishers, London.

MacDuffie, J.P. and Krafcik, J. (1988), 'The Team Concept: Models for Change', *Jama Forum*, 7, no. 3.

Mackenzie, F.W. (1985), 'Microelectronic Components: the Issues for Europe', Centre for Business Strategy, London Business School, (mimeo).

MacKerron G. (1989), 'The International Steam Coal Market and UK Coal', *Energy Policy*, 17, no. 2.

McKinsey Inc. (1983), *A Call to Action: The European Information Technology Industry*, report to the Commission of the European Community, January.

McLoughlin, I. and Clark, J. (1988), *Technological change and work*, Oxford University Press, London.

MAFF (1990), '*Government Fights for British Mince'*, press release 131/90, 28 March.

Malerba, F. (1985), *The Semiconductor Business: the Economics of Rapid Growth and Decline*, Frances Pinter, London.

Malerba, F. and Torrisi S. (1989), 'Innovazione e competitività internazionale dell'industria informatica europea', *Quaderni di Economia e Politica Industriale*, December.

Malerba F. and Torrisi, S. (1990), 'Complessità tecnologica organizzazione industriale e diversità tra imprese: il caso delle grandi imprese informatiche', *Economia e Politica Industriale*, no. 65.

Marcel, T. (1990), 'Pharmacie: la Montée en puissance' (Pharmaceuticals: gaining strength), interview with Christian Vincent, *Biofutur*, May, pp. 49–53.

Marer, P. (1984), 'Intra-Bloc Economic Relations and Prospects' in D. Holloway, and J. Sharpe, (eds) *The Warsaw Pact Alliance in Transition*, Macmillan, London.

Margirier, G. and Hollard, M. (1986), L'autonomisation avancée de la production dans les activités d'usinage, FOP No. 124, FAST programme, EC, Brussels.

Marsh, P. (1989), 'Chemical Groups Make Slow Progress on Cutting Waste', *Financial Times*, 22 September.

Martin, B., Irvine, J., Narin, F., Sterritt C. and Stevens, K. (1990), 'Recent Trends in the Output and Impact of British Science', *Science and Public Policy*, 17, pp. 14-26.

Mattera, A. (1980), 'The 'Cassis de Dijon' Judgement: A New Step Towards Completing and Protecting the Common Market', *Revue Du Marché Commune*, no. 241, November, pp. 3–12.

Meadows, D.H., Meadows, D.L., Randers, J. and Behrens, W. W. (1972), *The Limits to Growth*, Universe Books, New York.

Meat and Livestock Commission (1990), Letter to MAFF on 'EC Proposal COM(89)671', Milton Keynes.

Mettelet, C. (1989), 'Production, traitement, recyclage et transferts transfrontaliers de dechets dangereux dans la communauté Européene, paper at STOA Workshop, *Hazardous Waste Management Beyond 1992*, Luxemburg 25–26 April.

Metz, G. (1984), 'Solving Structural Problems of the European Chemical Industry', *Chemistry and Industry*, 17 December, pp. 871–77.

Metz, G. (1989), Presidential Address to CIFIC. Quoted in *Chemistry and Industry*, 19 June, p. 361.

Michnik, A. (1990), 'The Two Faces of Europe', *New York Review of Books*, XXXVII, 12 November.

Miles, I. (1989), 'Teleshopping: just around the corner?', *Journal of the Royal Society of Arts*, CXXXVIII, no. 5403, February, pp. 180–203.

Miles, I. and contributors (1990), *Measuring and Mapping the Information Economy*, British Library Research and Development Department, Boston Spa.

Millstone, E. (1988), *A Comparative Study of Industry-Government Relations: The Case of Food Additive Regulations*, a report to the ESRC, SPRU, Brighton.

Millstone, E. (1989), 'The Regulatory Environment: Science and Politics in the Control of a Technology' in *Strategies for New Technology: Case Studies from Britain and France*, Philip Allan, Hemel Hempstead.

Millstone, E. (1990a), *Study of the Demand for Differentiated Food Products in Holland*, report to the Consumer Policy Service of the European Commission, SPRU, Brighton.

Millstone, E. (1990b), *Study of the Demand for Differentiated Food Products in West Germany*, report to the Consumer Policy Service of the European Commission, SPRU, Brighton.

Millstone, E. (1990c), *Study of the Demand for Differentiated Food Products in Belgium and Luxembourg*, report to the Consumer Policy Service of the European Commission, SPRU, Brighton.

Millstone, E. and Abraham, J. (1988), *Additives: A Guide for Everyone*, Penguin Books, Harmondsworth, pp. 186–190.

Molle, W. and Wever, E. (1984), *Oil Refineries and Petrochemical Industries in W. Europe: Buoyant Past, Uncertain Future*, Gower, Aldershot.

Moray Stewart, J. (1988), 'Defence Procurement in Britain', *RUSI Journal*, Winter, pp. 43–47.

Morgan, K. and Davies, A. (1990), 'Seeking Advantage from Telecommunications: Regulatory Innovation and Corporate Information Networks in the UK', CICT Working Paper no. 10, July, SPRU, Brighton.

Morgan, K., Harbor, B., Hobday, M., von Tunzleman, N. and Walker, W. (1989), 'The GEC-Siemens Bid for Plessey: the Wider European Issues', CICT Working Paper, no. 2, January, SPRU, Brighton.

Morgan, K. and Sayer, A. (1988), *Microcircuits of Capital: Sunrise Industry and Uneven Development*, Polity Press, Cambridge.

Murphy, K.J. (1983), *Macroproject Development in the Third World*, cited in Strassmann and Wells (1988).

National Council of Building Material Producers and Building Magazine (1988), 'The UK Construction Industry and the European Community', June, 15.

NATO's Sixteen Nations (1990), 'Defence Expenditures of NATO Countries 1970–1989' January, 34, 8, pp. 50–51.

Naur, P. and Randell, B. (1969), 'Software Engineering' in *Proceeds of Nato Conference*, Scientific Affairs Division, NATO, Brussels.

NCC (1990), National Consumer Council, 'Consumer Electronics and the EC's Anti-Dumping Policy', Working Paper, National Consumer Council, London.

Nelson, R. (1990), 'US Technological Leadership: Where Did it Come from, and Where Did it Go?, *Research Policy*, 19, pp. 117–132.

Noam, E.M. (1987), 'The Public Telecommunications Network: A Concept in Transition', *Journal of Communications*, Winter, 38, no. l, pp. 30–48.

Noble, D. (1982), *Forces of Production*, Knopf, New York.

Nora, S. and Minc, A. (1981), *The Computerization of Society*, MIT Press, Cambridge, Mass.

Northern Business Information (1990), European Telecom Market, 1990 Edition, New York.

O'Brien, R. (1989), 'Notes on the European Computer Industry', CEC, DG XIII, Brussels.

OECD (R & D statistics).

OECD (1979), *The Petrochemical Industry: Trends in Production and Investment*, OECD, Paris.

OECD (1983), *The Economics Aspects of the International Control of Chemicals*, OECD, Paris.

OECD (1985a), *Environmental Policy and Technical Change*, Paris.

OECD (1985b), *Software: 'An Emerging Industry'*, OECD, Paris.

OECD (1987a), *The Internationalisation of Software and Computer Services*, OECD, Paris.

OECD (1987b), *Environmental Policy and Technical Change*, OECD, Paris.

OECD (1988), *The Economic Dimension in Information Technology Standards*, ICCP, Paris.

OECD (1989a), *Trade in Telecommunication Network-based Services: International Charging Practices and Procedures*, Paris, June.

OECD (1989b), *Telecommunication Network-based Services: Policy Implications*, ICCP Report No. 18 by R. Mansell, Paris.

OECD (1989c), *Energy Balances of OECD Countries*, Main Economic Indicators, OECD, Paris.

Office of Technology Assessment (OTA) (1984), *Commercial Biotechnology–An International Analysis*, Government Printing Office, Washington DC.

Office of Technology Assessment (OTA) (1986), *Serious Reduction of Hazardous Waste*, Government Printing Office, Washington.

Office of Technology Assessment (OTA) (1988), *New Developments in Biotechnology 4: US Investment in Biotechnology*, Government Printing Office, Washington DC.

Office of Technology Assessment (OTA) (1990), *Arming our Allies: Cooperation and Competition in Defense Technology*, US Congress, Government Printing Office, Washington DC.

Ohmae, K. (1990), *The Borderless World*, Collins, London.

Oliver, N. and Wilkinson, B. (1988), 'The Japanisation of British Industry', Blackwell, Oxford.

OSI (1988), Observatoire des Strategies Industrielles, *Strategies et competitivités dans l'industrie mondiale*, Economica, Paris.

Ovum (1989), *Object-Oriented Systems: The Commercial Benefits*, Ovum Ltd., London.

Ovum (1990), *Software Europe*, July–August, Ovum Ltd., London.

Patel, P. and Pavitt, K. (1988), 'The International Distribution and Determinants of Technological Activities', *Oxford Review of Economic Policy*, 4, pp. 35–55.

Patel, P. and Pavitt, K. (1990), 'The Innovative Performance of the World's Largest Firms: Some New Evidence', SPRU, Brighton.

Patel, P. and Pavitt, K. (forthcoming), 'Large Firms in the Production of the World's Technology: an Important Case of "Non-Globalisation"', *Journal of International Business Studies*.

Pavitt, K. (1984), 'Sectoral Patterns of Technical Change: Towards a Taxonomy and a Theory, *Research Policy*, 13, no.6, pp. 343–73.

Pavitt, K. (1986), 'Chips and Trajectories', in R. MacLeod (ed.), *Technology and the Human Prospect*, Frances Pinter, London.

Pavitt, K. (1988a), 'Uses and Abuses of Patent Statistics' in A. Van Raan (ed.), *Handbook of Quantitative Studies of Science and Technology*, North-Holland, Amsterdam.

Pavitt, K. (1988b), 'International Patterns of Technological Accumulation' in N. Hood and J-E Vahlne (eds), *Strategies in Global Competition*, Croom Helm, Beckenham.

Pavitt, K. and Patel, P. (1990), 'Large Firms in the Production of the World's Technology: An Important Case of 'Non-globalisation; SPRU, Brighton, (mimeo).

Pearce, D., Markandy, A. and Barbier, E.B. (1989), *Blueprint for a Green Economy*, Earthscan, London.

Pelaez, E. (1988), 'A Gift from Pandora's Box: The Software Crisis' PhD. Thesis, University of Edinburgh.

Perez, C. (1983), 'Structural Change and the Assimilation of New Technologies in the Social and Economic System', *Futures*, 15, no. 5, pp. 357–75.

Perez, C. (1990), *Technical Change, Competitive Restructuring and Institutional Reform in Developing Countries*, World Bank Strategic Planning and Review discussion paper, no. 4, World Bank, Washington, DC.

Persson, M. (1988), *New Patterns of Work*, Swedish Work Environment Fund, Stockholm.

Policard, C. (1990), 'Diagnostic in vitro: premier marche d'application des biotechnologies', interview with Christian Vincent, *Biofutur*, May, pp. 57–60.

Porter, M. (1990), *The Competitive Advantage of Nations*, Macmillan, London.

Posner, M. (1961), 'International Trade and Technical Change', *Oxford Economic Papers*, 13, pp. 323–41.

Prais, S. (1981), 'Vocational Qualifications of the Labour Force in Britain and Germany', *National Institute Economic Review*, 98, pp. 47–59.

Prestowitz, C.V.(1988), *Trading Places: How We Allowed Japan to Take the Lead*, Basic Books, New York.

Prior M. (1989), 'Power Privatisation and the UK Coal Industry', *Energy Policy*, 17, no. 3.

Programme of the Federal Republic of Germany for Specialized Information (1985–1988). The Federal Minister for Research and Technology, Bonn.

Quintas, P. (1990), *Software Engineering – The Experiences of Member Countries and the International Policy Challenge*, Committee for Information, Computer and Communications Policy, OECD, Paris.

Quintas, P., Hobday, M., Guy, K. (1989), *Evaluation of the Alvey Software Engineering Programme*, report to the Department of Trade and Industry, SPRU, Brighton.

Ranta, J. and Tchijov, I. (1988), 'Economics and success factors of FMSs' *IIASA discussion paper*, Vienna.

Ray, G. (1989), 'Full Circle: The Diffusion of Technology', *Research Policy*, 18, pp. 1–18.

Rev, I. (1984), 'Local Autonomy or Centralism – When was the Origional Sin Committed?', *International Journal of Urban and Regional Research*, 8, no.1, March.

Reis-Arndt, E. (1987), 'A Quarter of a Century of Pharmaceutical Research: New Drug Entities 1961–85, *Drugs Made in Germany*, 30, no.3.

Rosario, M.C. (1989), 'Standards in ICT: the EDI Case', unpublished MSc Thesis, SPRU, Brighton.

Rostowski, J. (1988), 'Economic Structure and Materials and Energy Intensity', NATO Colloquium, *The Economies of Eastern Europe Under Gorbachev's Influence*, NATO, Brussels.

Rothwell, R. and Gardiner, P. (1983), 'The Role of Design in Product and Process Change', *Design Studies*, 4, no. 3, pp. 161–9.

Roussel, A-M. (1989), 'France Telecom Rebuts Teletel Drain', *Communication Week International*, 14 August.

Royal Institute of International Affairs (RIIA) (1987), *Europe's Future in Space: A Joint Policy Report*, Chatham House Special Paper, Routledge and Kegan Paul for RIIA.

Rupp, R. (1990), 'Europe 1992: Potential Implications for the North Atlantic Alliance', *NATO's Sixteen Nations*, 34, no. 8, January, pp. 22–26.

Rush, H., Bessant, J. and Hoffmann, K. (1990), *Evaluation of the DTI's Flexible Manufacturing Systems Scheme*, report to the Department of Trade and Industry, Centre for Business Research, Brighton Business School, Brighton.

Saloner, G. (1989), 'Economic Issues in Computer Interface Standardization', Rand Corporation Paper, Massachusetts Institute of Technology, Cambridge, Mass.

Samuels, R.J. and Whipple, B.J. (1989), 'The FSX and Japan's Strategy for Aerospace', *Technology Review*, 92, no. 7, pp. 42–51.

Sarich, A. (1990), 'An Update on EDI in 1989', in '*EDI–1992 and Beyond*', CEC, Brussels, pp. 17–21.

Sauvant, K.P. (1986), *International Transactions in Services: The Politics of Transborder Data Flows*, Westview Press, Boulder, Colorado.

Saxenian, A. (1990), 'Regional Networks and the Resurgence of Silicon Valley', *Californian Management Review*, Fall, pp. 89–112.

Schmidt, S.K. (1989), 'Liberalisation of the Telecommunications Services Market in the European Community and the Federal Republic of Germany', unpublished MSc Thesis, SPRU, Brighton.

Schneider, V. and Werle, R. (1988), 'Regime oder korporativer Akteur? Die EG in der Telekommunikationspolitik', Max-Planck-Institut fur Gesellschaft-forschung, discussion paper 88/4, Cologne.

Schofield, J. (1988), 'A Taxonomy of Computer Evolution'. *The Guardian*, 1 December.

Schonberger, R.J. (1982), 'Japanese Manufacturing Techniques',*The Free Press*, New York/London.

Scibberras, E. and Payne, B. (1986), *International competitiveness in the UK Machine Tool Industry*, Technical Change Centre, London.

Scrip (1989), Review Issue, January.

Senker, P. (1985), *Towards the Automatic Factory*, IFS Publications, Kempston.

Servan-Schreiber, J.J. (1967), *Le Défi americain*, De Noel, Paris.

Sharp, M. (1985), *Europe and the New Technologies*, Frances Pinter, London.

Sharp, M. (1988), 'Inward Investment and Industrial Competitiveness'. *Inter Economics*, 23 no.5, Sept–Oct.

Sharp, M. (1989a), *Corporate Strategies and Collaboration – the Case of ESPRIT and European Electronics*, SPRU, Brighton, (mimeo).

Sharp, M. (1989b), *European Technology: Does 1992 Matter*? SPRU/Imperial College, Papers in Science, Technology and Public Policy, no. 19.

Sharp, M. (1989c), *The Community and New Technologies*, in Juliet Lodge (ed.), *The European Community and the Challenge of the Future*, Pinter Publishers, London.

Sharp, M. (1990), 'The Single Market and European Policies for Advanced Technologies', *Political Quarterly*, special issue, October.

Sharp, M. (1991), *Technological Trajectories and Corporate Strategy in the Diffusion of Biotechnology*, in E. Deiaco, E. Hornell and G. Vickeny, G. (eds), *Technology and Investment: Crucial Issues for the 1990s*, Pinter Publishers, London.

Sharp, M. and Shearman, C. (1987), *European Technological Collaboration*, Chatham House Paper no. 36, Routledge and Kegan Paul for the Royal Institute of International Affairs.

Sharp, M. and Shepherd, G. (1985), *Managing Change in British Industry*, Employment, Adjustment and Industrialisation Series, no. 5, International Labour Office, Geneva.

Shaw, B. (1987), 'Strategies for User-Producer Interaction' in R. Rothwell and J. Bessant (eds), *Innovation, Adaptation and Growth*, Elsevier, Amsterdam.

Shearson, Lehman, Hutton (1990), *SLH Universe League Tables*. Shearman, Lehman, Hutton Securities, New York.

Shell, (1989), *Energy in Profile*, Shell Briefing Service, no. 3, London.

Sheriff, A. (1988), 'The Competitive Producer Position of Automobile Manufacturers: Performance and Strategies,' International Motor Vehicle Program, MIT, Cambridge, Mass.

Shortland, M. (1987), *Custom Integrated Circuits: the User Company Perspective*, Michael Shortland Associates, Guildford, Surrey.

Sigurdson, J. (1986), *Industry and State Partnership in Japan: the Very Large Scale Integrated Circuit (VLSI) Project*, discussion paper no. 168, Research Policy Institute, Lund, Sweden.

Slack, N. (1987), 'Flexibility of Manufacturing Systems', *International Journal of Operations and Production Management*, 7, no. 4, pp. 35–45.

Slack, N. (1989), 'Focus on Flexibility', in R. Wild (ed.), '*International Handbook on Production and Operations Management*, Cassell, London.

Soete, L. (1981), 'A General Test of Technological Gap Trade Theory', *Weltwirtschaftliches Archiv.*, 117, pp. 638–66.

Soete L. and von Tunzelmann, N. (1988), 'Convergence in Firms and Technologies in the European Telecommunications Equipment Industry', Limburg & SPRU, Brighton, (mimeo).

Sorge, A., Warner, M., Nicholas, I., Hartman, G. (1982), *Microelectronics and Manpower in Manufacturing*, Gower Press, Aldershot.

Sousa, L.J. (1988), 'Problems and Opportunities in Metal and Materials: An Integrated Perspective'. United States Department of the Interior and United States Bureau of Mines, Washington, DC.

Stamps, D. (1987), 'CASE: Cranking Out the Productivity', *Datamation*, 33, no. 13, pp. 55–8.

Statistical Overview of all EUREKA Projects (1989), UK EUREKA Office, Ashdown House, 123 Victoria Street, London SW1.

Steedman, H. (1988), 'Vocational Training in France and Britain: Mechanical and Electrical Craftsmen', *National Institute Economic Review*, 126, pp. 57–70.

Steedman, H. (1990), 'Improvements in Workforce Qualifications: Britain and France, 1979–88', *National Institute Economic Review*, 133, pp. 50–61.

Stewart, A. (1989), 'The cost of ISDN for Linking Workstations', *Communications International*, 16, no. 8, August, p.10.

Strassmann, W.P. and Wells, J. (eds) (1988), *The Global Construction Industry*, Hyman, London.

Swinbank A. (1990a), 'The Economics of 1992: Food', *BNF Nutrition Bulletin*, January, p. 40ff.

Swinbank, A. (1990b), 'Brave Attempt at an Impossible Task', *Food Policy*, April, pp. 176–7.

Szücs, J. (1988), 'Three Historical Regions of Europe', in J. Keane, (ed.),*Civil Society and the State*, Verso, London.

Taylor, A. (1989), 'Builders Lay Foundations for the Link-ups of 1992', *Financial Times*, 25 January.

Tchijov, I. (1989), 'FMS World Data Bank', *IIASA WP*, Vienna, pp. 89–133.

Tchijov, I. (1990), 'The World-wide Diffusion of FMS' in W. Haywood (ed.), *CIM: Revolution in Progress*, Conference Proceedings, IIASA, Vienna.

Temmerman, G. (ed) (1989), 'Composition et denomination des dentrées alimentaire', *Legislation Alimentaire*, section 2, La Charte, Brugges.

Thomas, G. and Miles, I. (1989), *Telematics in Transition: the Development of New Interactive Services in the United Kingdom*, Longman, Harlow.

Thomas, S. (1988), *The Realities of Nuclear Power*, Cambridge University Press, Cambridge.

Tidd, J. (1990), *Flexible Manufacturing Technologies and International Competitiveness*, Pinter Publishers, London.

Toepfer, K. (1989), 'Marktchancen durch moderne Umwelttechnik nutzen', *Umwelt*, no. 3, pp. 119–120.

Tsakaloyannis, P. (1989), 'The EC: from Civilian Power to Military Integration', in J. Lodge (ed.), *The European Community and the Challenge of the Future*, Pinter Publishers, London, pp. 241–55.

Tudler, R. and Empel, E. (1984), 'European Multinationals in the Semiconductor Industry: Their Position in Microprocessors, A Pilot Study', University of Amsterdam, Amsterdam, (mimeo).

Tully, C. (1989), 'Proposals for ESSI: A European Systems and Software Initiative', *Proceedings*, Centre For Software Reliability 'Research Into Practice' Conference, Bournemouth, 25–6 May..

UN (1990), *Proceedings of Seminar on Computer-Integrated Manufacturing (Botevgrad, Bulgaria, 25–9 September, 1989)*, UN/ECE, Geneva.

United Nations Economic Commission for Europe (UN/ECE) (1988), *Overall Economic Perspective to the Year 2000*, New York.

UN/ECE (1989), *Annual Review of Engineering Industries and Automation*, Geneva.

UN/ECE (1990), *Economic Survey of Europe, 1989-90*, Geneva and New York.

UNESCO (1987), *Statistical Yearbook 1987*, Paris.

Ungerer, H. and Costello, N. (1988), *Telecommunications in Europe*, Commission of the European Communities, Brussels.

United States Environmental Protection Agency (USEPA) (1987), *Waste Minimization: Environmental Quality with Economic Benefits*, Washington, DC.

USA (1989), *Statistical Abstract of the United States*, Washington, DC.

Vanguard (1987), *The Economic Effects of Value-Added and Data Services*, HMSO, London.

Vanguard (1988), *EDI and X.400 Study*, HMSO, London.

Vanguard (1989), *EDI Standards: a Guide for Existing and Prospective Users*, HMSO, London.

van Miert, K. (1989), *Consumer Policy*, EC Commission, Brussels, 11 April, p. 5.

van Tulden, R. and Junne, G. (1987), *EC European Multinationals in Core Technologies*, Wiley, Chichester

Vernon, R. (1966), 'International Investment and International Trade in the Product Cycle', *Quarterly Journal of Economics*, 80, pp. 190–207.

Viesti, G. and Zanzottera, L. (1989), 'Japanese Multinationals and EEC: The case of car industry', CESPRI Working Paper, Università Commerciale 'Luigi Bocconi', Milan.

von Hippel, E. (1988), *The Sources of Innovation*, MIT Press, Cambridge, Mass.

von Tunzelmann, N. (1988), 'Convergence of Firms in Information and Communication: a Test Using Patents Data', SPRU, Brighton, (mimeo).

von Tunzelmann, N. (1989), 'Patterns in Patenting: the European Electronics Industry, 1969–86', SPRU, Brighton, (mimeo).

Vredeling Report (1986), 'Towards a Stronger Europe', report by an Independent Study Team established by Defence Ministers of Nations of the IEPG, Brussels.

Wakeling, R. (1989), 'The Tedis Program – A Progress Report, April 1989', *IES News*, CEC DG XIII, April.

Walker, C. and Cannon, G. (1985), *The Food Scandal*, Century Publishing, London, esp. Part 1, Ch. 5.

Walker, W., Graham, M. and Harbor, B. (1988), 'From Components to Integrated Systems: Technological Diversity and Interactions between Military and Civilian Sectors', in P. Gummett and J. Reppy (eds), *The Relations between Military and Civil Technologies*, Kluwer Academic Publishers, Dordrecht, pp. 17–37.

Walker, W. and Gummett, P. (1989), 'Britain and the European Armaments Market', *International Affairs*, 65, no. 3, pp. 419–42.

Walko, J. (1989), 'Network Failures', *Communications Systems Worldwide*, July, pp. 5–6.

Weatherill, J. (1989), 'Food Scares and the Treaty of Rome', *Law Society Gazette*, no. 14, 12 April 1989, p. 14ff.

Wetstone, G.S. and Rosencranz, A. (1983), *Acid Rain in Europe and North America: National Response to an International Problem*. International Law Institute, Washington, DC.

White, M. (1990), 'The European Information Industry' in C. Oppenheim, C.C. Citroen and J.M. Griffiths (eds) '*Perspectives in Information Management*', 2, Bonker-Saus, London.

'*White Papers of Japan*', Annual Abstract of Official Report and Statistics of the Japanese Government (1987), Japan Institute of International Affairs, Editorial Section of English Annual, Tokyo.

Wilkinson, B. (1983), *The Shopfloor Politics of New Technology*, Heinemann, London.

Wilkinson, C. (1983), *Proceedings 10th International Congress of Plant Protection*, 1, p. 46, quoted in *Chemistry and Industry*, 17 December 1984, p. 864.

Williams, R. (1973), *European Technology: the Politics of Collaboration*, Croom Helm, London.

Witcoff, H.A. and Reuben, B.G. (1980), *Industrial Organic Chemicals in Perspective*, John Wiley, London.

Womack, J.P. (1988), 'Multi-national Joint Ventures in Motor Vehicles' in D.C. Mowery (ed.) *International Collaborative Ventures in US Manufacturing*, Ballinger, Cambridge, Mass.

Womack, J.P., Jones, D.T. and Roos, D. (1990), *The Machine that Changed the World*, Rawson Associates, New York.

World Coal Study Report (1980) *Coal – Bridge to the Future*, Ballinger, Cambridge, Mass.

World Development Report (1987), World Bank, Washington.

World Military Expenditures and Arms Transfer (1988), US Arms Control and Disarmament Agency, Washington, DC.

World Nuclear Industry Handbook 1990 (1989), Nuclear Engineering International, Reed International Publishers, London.

Wynne, B. and Hortensius, D. (1987), 'The Rationalities of Problem Definition: The Netherlands and Hazardous Waste Management' in B. Wynne (ed.), *Risk Management and Hazardous Waste*, Springer-Verlag, Berlin, p. 87.

Index